Selection of Writings of
MIGUEL IRADIER

Volume II

PHILOSOPHY, SCIENCE and CULTURE

(2020 - 2022)

Edited by *Hurqualya*

ISBN: 9798309921782

www.hurqualya.net

Cover: Adelaida Rondán

Índex

Selection of Writtings of
MIGUEL IRADIER
Volume II
PHILOSOPHY, SCIENCE and CULTURE
(2020 - 2022)

Selection of Writings of
MIGUEL IRADIER

Volume II

PHILOSOPHY, SCIENCE and CULTURE

(2020 - 2022)

Edited by *Hurqualya*

THE LITTLE FINGER STRATEGY

8 June, 2020

TechTsunami

There is a technological war, but whoever thinks it is only technological has already lost it. Now China seems to have taken the lead over the United States in the fight for control of communication channels, and many would celebrate it were it not for the fact that this 5G is only intensifying what already was a huge wave overflow.

Not only do we oppose the indiscriminate deployment of technologies but, in a now distant article, we even suggested another line of biophysical research to evaluate the impact of electromagnetic radiation on humans and other living beings [1]. It is clear that the big corporations promoting this deployment are only concerned about market shares, but who says that in a few years they will not be sued for damages?

We already see the American attempts to blame China for the coronavirus; if they have not already waged a campaign to demonstrate the harmfulness of this coming generation and file trillion-dollar lawsuits, it is simply because 1) the deployment has barely begun and it is too early to accumulate evidence against it, 2) it would clip the wings of their own future developments, and 3) it could equally be applied retrospectively to previous generations and American companies.

And yet one can be sure that there are law firms discussing how a full-blown legal war might be orchestrated. But, leaving aside the vileness of this machinations, I think we would all be more grateful to anyone who showed us that this kind of radiation is or is not safe than to anyone trying bombarding us no matter the consequences.

Maybe the leadership of 5G is good for China's strategic interests, but it would be much better if those interests coincided with those of most human beings. And most people do not want more and more technology, but rather some kind of protection against it; anyone should understand that.

China, for example, now hires some 50,000 foreign employees, specialists and engineers to conduct a large part of the research and development that takes place throughout the country. The mastery of 5G technology, a great display of muscle, has drawn on a good deal of that workforce. But surely only a small fraction of that number is needed to make a much more important conquest than that of a fleeting technological advantage.

Today the technological tsunami is synonymous with the digital revolution and with a category, information, which seems to pervade everything. All previous developments in science and technology converge in it and pass through it. Even physicists now think of the universe as a gigantic computer; there was much discussion about whether the world was made of atoms or stories but in the end it was decided that it was made of bits and case closed.

Actually the universe seems to matter very little now. Rockets are almost the same as they were fifty years ago, but in that time the capacity of computers has increased by a factor of billions. Hence the feeling for the user that the world wants to get into the computer. The developer, on the other hand, thinks very differently.

Information is not a powerful concept, but an extremely gaseous one that fits all. Today it does not seem we can put limits on it, but limits it has: material, mental and operational, to mention only the most obvious. If it did not, we would not be talking about the information economy.

What is the purpose of information technologies? To reprogram the humans who use them. Science and technology have always done so, but now we have the confirmation as never before.

What is important about information theory is not so much its definitions as the direction it imposes on everything; changing that direction is equivalent to changing the direction of technology as a whole. The unequivocal direction, which obviously inherits from statistical mechanics, is that of decomposing everything into minimal elements that can then be recomposed at will.

For statistical mechanics there is no direction in time: if we do not see a shattered vase recompose itself and return to the table, it is only because we do not live long enough; if the pieces of an eviscerated corpse do not get back together and walk again as if nothing had happened, it is only because we are not in a position to wait $10^{10.000.000.000}$ years or something similar.

Information theory is not concerned at all with the reality of the physical world, but with the probability in its constituent elements, or rather the probability *within its accounting of the constituent elements*. The physical world itself is on the contrary a pool of resources for the sphere of computation, aimed to be independent of the former.

Is this statistical view a neutral position or is it simply a pretext to manipulate everything unencumbered? There is no single answer to this. Statistical mechanics and information theory are successfully applied in countless cases, and in countless cases they could not be more irrelevant. What is worrying is the overall trend they create.

For my part, I am more than convinced that a shattered corpse is not going to recompose itself, no matter how fabulous the period of time we choose. And this is not rhetoric; rather, large numbers are the rhetoric of probability; an inflated and poor rhetoric, as it ignores the interdependence of things.

Although in principle we should not confuse levels of scientific discourse with others openly ideological, in practice we observe a sort of pre-established harmony between liberalism and, for example, neoclassical economics, the present neo-darwinian synthesis of evolution, or statistical mechanics and information theory. It is not only that power tends to takeover the sense of any possible tool; theories like these were creatures hatched in the same nest and under the same watchful eye.

The link is even more explicit and exaggerated in today's neoliberal environment. In the now booming field that tries to bridge quantum mechanics and information theory, "a resource theory of quantum thermodynamics without a background temperature, *so that no states at all come for free*" has already been proposed. The goal is to design a quantum system in which information is used as currency to trade between different material resources quantified as states. The initiative comes from the University College London but it could equally have come from the London School of Economics [2].

It is clear that there is more than affinity; there is consanguinity, there is a same logic, there is a same destiny. The real world is only an instrument for computation, and computation for trade. And trade, on this totalitarian scale, exists for the concentration of capital, and capital for the concentration of power. This is a despicable and insane logic, but we have allowed it to be installed by default as our operating system, thanks, among other things, to the fact that we have accepted a more than dubious "neutrality of science".

The problem is much more difficult to grasp in depth since it does not only affect the soft, descriptive sciences or even the statistical ones, but is already inscribed since the beginning of the modern world in the utilitarian bias of the harder sciences such as calculus or mechanics, which create a disposition of the whole from which we cannot find the exit anymore. In fact the disposition *is* the whole for which any number of elements exist.

There are open, natural totalities, and there are artificial and closed totalities gravitating towards death. Today's globalized marketplace is almost the diametrically opposite of "an open system". The fiction of an "open society" is maintained by a horizontal narrative in terms of competition that is deemed convenient for the "atomic states" of the social body, the individuals. But there is a much more implacable and concretely structured vertical logic that thinks in terms of marketing, social engineering and population ecology —in the exploitation of niches and ecosystems.

The horizontal vision aims to break everything down into parts while the vertical one is conceived by principle in terms of the whole. "Big fish eats little fish" means very different things depending on the coordinate axis we choose.

The narrative in the horizontal axis of an unstructured collection is for the masses; the vertical logic of a structured whole is for the rampant techno-feudalism and financial techno-fascism. That is why we continually hear about "the markets" but we never hear about the power laws that governs the distribution of wealth or the size of companies, and that shows inescapably that 80 percent of all pie is owned by 20 percent, 80 percent of that 80 percent is owned by a fifth of a fifth, and so on. *Ideally*, this law tends towards a singularity [3].

Clearly, the logic of globalization has so far been an all-or-nothing totalitarian logic even if now is at a standstill. Adding to the haze, mainstream discourses try to confuse the analysis of totality with the totalitarian discourse, and promote social atomism as if it were more egalitarian, when the social atom is been shaped from above. Double standards have always existed for this.

A similar duality works for the two extremes of consumer technology development, data mining or those plantations of the mind that are the social networks: developers arrange and shape the whole, consumers interact with the parts. This resonates with Simondon's modes of minority and modes of majority at the ends on the technological circuit.

In other articles we have also talked about the double circuit in the scientific worldview, consisting of both hard sciences such as physics, based on mathematics and prediction, but without the ability to describe reality, and descriptive and narrative sciences, such as cosmology or the theory of evolution, which try to fill the great gap between abstract laws, nature and the real world. If the former do the hard work, the latter have a preponderant role in our imaginary.

In current lingo we would say that the former are like hardware and the latter like software, with the statistical aspects, like those of information theory, mediating between both.

These planes of duality overlap and reproduce themselves at various levels and make orientation and judgment about the whole techno-scientific phenomenon extremely difficult. But probably the greatest of all oppositions, between man and nature, is the one that has receded the most in our conscience.

Neither science nor technology alone has created this opposition; on the contrary, science and technology are the channelled development of certain passions. However, this economy of impulses has been transmitted to the interior of the principles and has been perpetuated in "the spirit of the laws" and their values.

What could seem more natural today than Newton's law of motion? But this laws of mechanics, as Poincaré noticed well, are neither true nor false, but only a disposition of the whole. We could dispense with the principle of inertia completely, replace it with a principle of dynamic equilibrium, and still have the same knowledge of observable phenomena and their corresponding laws. Then, why change them?

To have another vision of the whole, of that whole from which the parts and individuals receive their form. Apart from the fact that technoscience is not only the study and control of the external nature, but also a symptom of our inner nature and that of power. As in all surface phenomena, there is always a struggle between what wants to be expressed through it and what wants to gain access.

Newton's laws of mechanics can be summed up as follows: nothing moves unless something else moves it. Which is just an educated way of saying that we are all dead shit, at the expense of being moved by a God, an initial explosion, an invisible hand or whatever. The consequences of this go on until today. It seems incredible that we have been able to calmly subscribe to this for a third of a millennium, but it is what we have.

Everything that is said today about the environmental crisis or climate change pales in comparison with the mechanical disposition of nature that we have accepted. If we do not rebel against this in our innermost selves, we will hardly do anything about everything else, which is only a by-product. Our *external* relationship with nature always begins with how we understand it. And not only with nature.

What has been applied to nature, with the complicit feeling of the wise to increase his sense of power, then gradually but uncontrollably spreads to the interior of the social apparatus through the increasingly dense technological network. And it is utterly superficial to pretend that the most recent developments in science change the panorama because they are nothing but a refinement of the same presuppositions, seeking a greater performance.

Science has not changed in essence, which could be great news if one thinks there is still something left for a surprise. Science cannot stop being a surface phenomenon, but neither can it stop echoing deeper impulses. In it, the relationship between subject and object, as well as our conception of totality, can still be decisively transformed.

To put it bluntly, science has a noble side and a side that is at the service of power. What side prevails, depends largely on the scientists themselves, but not only on them. They are no less divided than the masses, for serving two masters at the same time is always more than difficult, and all kind of currents are interwoven in their midst.

The twin questions are what we want to do with what we know, and what we want to know with what we can do. Science and technology have always formed a continuum, even though today it seems to turn in on itself much more quickly. This has much of an illusion because the most important concepts have changed very little; and precisely because of this, a grand reversal could be in order. Or not.

Our liberal-materialism, or material liberalism, pretends to "give life" to a passive nature in order to fulfill its virtuous circle of "improving the world" and realizing the self-transcendence of the species. This process that oscillates between a false ideal pole and another false material pole is fulfilled through a recurrent cycle of idealizations and rationalizations thought to be exhaustive.

The theory and philosophy of information cannot hide their idealistic background. No wonder, since even classical mechanics has it, and the only way to "overcome" this heritage is through the compulsive practical application to material objects. The same is true for the designer-user circuit in the development of technologies.

The feedback that theory and practice create in technoscience does not have to be a virtuous circle, though; it can also be a circle of growing restriction and shrinking horizons. In this respect we still have naive notions, which will not allow us to go much further if we do not learn certain basic lessons.

Idealization and rationalization are like the mythical Symplegades rocks that destroyed ships and navigators; only the one who understands their dangers and avoids them will be able to pass to the other side.

No doubt modern neo-Babylonian science is the crudest inversion of Plato and Pythagoras and the "all is number" program, but even inverted it remains their heir. Today numbers no longer seem to exist to understand the world, but to crush and squeeze it, and all of us with it.

In this indescribable situation, in which mathematics is already the last whore for the lowest tasks —for what is so dirty, that only through numbers can be ignored-, we need a radically different vision of number and math, of theory and practice, of the whole and the parts, of quantity and quality, of knowledge and rationality.

Since everything is hype in the sphere of information, let us confront it with something even more hyperbolic, which nevertheless starts from a surprisingly simple formula.

An analytical totality: the Riemann zeta function

It has been said that if the Riemann hypothesis were solved, all the keys to cryptography and cybersecurity could be broken; no one has specified how this could happen, but at least it reminds us of the close relationship between a hitherto intractable problem, cryptography and information theory.

Such speculations are only based on the fact that the Riemann zeta function establishes a connection between prime numbers and the zeros of an infinitely differentiable function that provides the most powerful method for exploring this field, prime numbers being the basis of classical cryptography. But the zeta function is way, way more than any kind of cryptographic application; it seems to be a code itself, or even a code of codes [5].

There is even a theorem, due to Voronin, that proves that any kind of information of any size can be encoded with any precision needed within this function —and not only once, but an infinite number of times. Another thing is the extent to which that information can be made effective, about which several studies have been done. This is called the *universality* of the zeta function [6]. Besides, the Riemann zeta function is just the main case of an infinite number of similar functions.

One could also find any kind of information in white noise; but white noise lacks any structure, while the zeta function has an infinite structure. Had Hegel known both cases he would have spoken of false and true infinity; we will see that these Hegelian apperceptions are not out of place here.

It seems that there is a "zeta code" that could be the object of the algorithmic theory of complexity; but this function, intimately linked to the elementary act of counting, resists being an object. Even being analytical, it cannot be broken down in pieces, which is why all the usual methods have turned out to be pathetically inadequate. The zeta remains irreducible. It is the Great White Hope for all those who do not identify with that or any other color. The white hope of those who do not want to be reduced to white noise.

So far all the non-trivial zeros of the function computed, amounting to many trillions, lie exactly on the critical line, with a real part equal to 1/2; but no certain reason is known why this should be so, and there could still be an infinite number of zeros off the line.

Were we to put all the particles in the universe of the "cosmic computer" that many dream of computing zeros, having all the time in the world for the task, we could still be waiting infinitely for an answer. The story has it that Turing designed an archaic analog computer to falsify the hypothesis; possibly it came with a crank.

It is inevitable that the word "infinite" appears repeatedly when talking about this problem. The zeta function has a singular pole at unity and a critical line with a real value of 1/2 in which all the known and unknown non-trivial zeros appear, with a duality between the zeros and the pole: we either approach the problem from the point of view of unity, or from the point of view of infinity; but it is only at unity that the function ceases to have finite values. So far mathematicians, who in a very definite sense descend from the tradition of infinitesimal calculus inaugurated by Leibniz, have naturally opted for the second point of view, although it can be assumed that both extremes are equivalent.

A well-known mathematician said that, without knowing the truth of Riemann's hypothesis, it is as using a screwdriver; but having it, it would be more like a bulldozer. Some will never stop thinking about the next real estate development.

But surely the problem has a significance that goes beyond our "powerful methods" and our usual practices. The zeta function questions from top to bottom the relationship we assume between mathematics and reality. As is known, this function presents an enigmatic and totally unexpected similarity to the random matrices that describe subatomic energy levels and other collective signatures of quantum mechanics.

Until now, the mathematical physics from which the entire scientific revolution emerged has been applying mathematical structures to physical problems in order to obtain a partial solution by means of an artful reverse engineering. Calculus itself emerged in this process. On the contrary, what we have here are physical processes that spontaneously reflect a mathematical reality for which there is no known solution.

Since no physical theory explicitly justifies the reproduction of this type of function, the enigma of the underlying dynamics is as transcendental as the very resolution of the hypothesis, although the relationship between both aspects is also mere conjecture. Hence, this problem appeals to physicists and mathematicians alike.

We said that science advances through an alternate process of idealization and rationalization, issuing hypotheses for an idealized case and going later with this case from coast to coast in imperialistic generalizations. Calculus is a good example: infinitesimals are idealizations, while the concept of limit allows to rationalize and "give a foundation" to what in the end are still heuristic procedures.

Statistical mechanics is another of the most flagrant cases of rationalization, since, after extreme idealizations, it has been given explanatory value for everything; and a theory that explains everything cannot say more clearly that it actually explains nothing.

In philosophy, Hegel gives us the best example of the large-scale sway between idealization and rationalization. In an interesting article, Ian Wright attempts to explain the nature of the Riemann zeta function, with Hegel's *Science of Logic*, as the mediation between being and non-being through becoming within numbers. Old-fashioned as this can sound to many, three things can be said:

First, that from the point of view of pure arithmetic, this is admissible since if there is a part of mathematics that can be considered a priori, that is arithmetic; while, on the other hand, the very definition of this function affects the totality of integers, and its extension to real and complex numbers.

Second, that calculus is not pure mathematics like arithmetic, although many specialists seem to think so, but mathematics applied to change or becoming. Were it not, there would be no need to compute zeros. The zeta function is a relationship between arithmetic and calculus as close as it is uncertain.

Third, for experimental sciences such as physics the opposition between being and not being does not seem to have systemic implications, and for formal sciences such as information theory this is reduced at most to fluctuations between ones and zeros; however the distinction between reversible and irreversible, open and closed systems, which is at the very heart of the idea of becoming, is decisive and lies at the crossroads of the issue.

Famous physicists like Michael Berry have long noted that the dynamics of the function should be irreversible, bound and unstable. There are even elementary reasons for the former, which we have already pointed out elsewhere; and in spite of everything, and the fact that no Hamiltonian fits exactly into this dynamic, physicists still insist on working with the conservative assumptions of fundamental physics. Why? Because for physicists fundamental physics can only be conservative.

The *seriousness* of a problem such as the zeta demands that the problem of irreversibility be considered in depth. And considering it in depth means precisely including it at the most fundamental level. Conservative systems are still toy models for this.

This breaks all the patterns of the establishment in physics but is necessary to move forward. No doubt in this way both the idea of reversibility and that of exact mathematical representation lose much of their status, but on the other hand, seeing that the appearance of reversible behavior emerge from a background of irreversibility is the closest thing to finding the magic ring at the heart of nature. We all know that Physics today is not about nature, but about certain laws that affect it.

It is said, for example, that the Riemann zeta function could play the same role for chaotic quantum systems as the harmonic oscillator does for integrable quantum systems, although it also remains unclear what lies at the bottom of the harmonic oscillator model. To get closer to the subject, quantum thermodynamics can be considered in the sense that it has been done, for example, by the Keenan school at MIT, especially with Hatsopoulos, Gyftopoulos and Gian Paolo Beretta.

The Brussels school, led by Prigogine, is much better known to the public almost everywhere, including the United States, than the MIT school, even though the latter is much more sound when dealing with fundamental physics. That such a development, having emerged from MIT, is not more widespread is not a sign of its lack of relevance, but rather the opposite: it affects too much the position of large branches of physics to be easily admitted, and goes against the grain of everything that is so actively promoted.

The quantum thermodynamics of this school is totally opposed to the rationalization of entropy by statistical mechanics. The dynamics is irreversible at the most fundamental level. The number of states is incomparably greater than in the standard model, and only a few of them are selected. The selection principle is very similar to that of maximum entropy, although somewhat less restrictive: it is the attraction in the direction with the steepest-entropy-ascent. No great changes are needed in the usual formalisms, what is transfigured is the sense of the whole [8].

The equilibrium formalisms are also retained, but their meaning changes completely. The uniqueness of the stable equilibrium states amounts to one of the deepest conceptual transformations of science in the last decades. The approach is contrary to the idealism of mechanics and much more in line with the daily practice of engineers, for whom entropy is a physical property as real as energy. The theory has many more advantages than sacrifices, and it can be applied to the entire domain of non-equilibrium and to all temporal and spatial scales.

But there are other much more basic ways of including irreversibility and entropy in both quantum mechanics and classical mechanics through calculus, as we will see soon.

The connection between entropy and the zeta function is quite recent and only in the 21st century is beginning to take shape. At a very basic level, one can study the entropy of the sequence of zeros: a high entropy would lead to a low structure, and vice versa. It is observed that structure is high and entropy low, and even low level neural networks were already successful in their prediction [10].

There are other non-extensive types of entropy independent of the amount of material, such as Tsallis entropy, which can be applied to the zeta; these types of entropy are generally associated with power laws, more than exponential laws. But as in thermodynamics in general, the definition of entropy can vary from author to author and from application to application.

In order to unify and generalize so many definitions of entropy, Piergiulio Tempesta has proposed a group entropy [11]. Whether in physical contexts or in complex systems, in economics, biology or social sciences, the relations between different subsystems depends crucially on how we define the *correlation*. This is also the central issue in data mining, networks and artificial intelligence.

Each correlation law may have its own entropy and statistics, rather than the other way around. Thus, it is not necessary to postulate entropy, but its functional emerges from the kind of interactions that one wants to consider. Each suitable universal class of group entropy is associated with a type of zeta function. The Riemann zeta function would be associated with Tsallis entropy, which contains classical entropy as a particular case. But these correlation laws are based on independent elements, while irreversibility at the fundamental level seems to reject that assumption. Fundamental irreversibility assumes universal interdependence.

The zeta function itself, under other criteria, can be seen as an infinite encyclopedia of correlations and correlation laws.

Riemann's hypothesis is said to involve the most basic connection between addition and multiplication. So basic that it is not even apprehensible — at least not as yet. So, if there is a possibility to get to the core of the problem, it would have to be through the simplest arguments, rather than the more sophisticated and complex ones. Additionally the overall impact of an idea is all the greater the simpler its nature is.

Calculus or analysis has developed between the idealization of the infinitesimals and the rationalization of the limit, but it has rejected the foundation stone, the constant differential, which by definition must be a unit interval, 1. Let us look at an example that puts into play both the idea of calculation and that of natural and artificial intelligence.

Think about the problem of knowing where to run to catch fly balls—evaluating a three-dimensional parabola in real time. It is an ordinary skill that even recreational baseball players perform without knowing how they do it, but its imitation by machines triggers the whole usual arsenal of calculus, representations and algorithms.

However, McBeath et al. more than convincingly demonstrated in 1995 that what outfielders do is to move in such a way that the ball remains in a constant visual relation —at a constant relative angle of motion- instead of making complicated time estimates of acceleration as the heuristic model based on calculus intended [65]. Can there be any doubt about this? If the runner makes the correct move, it is precisely because he does not even consider anything like the graph of a parabola.

Although he has not taken this example as a starting point, the only one who has put this "non-method" or direct method into practice is Miles Williams Mathis, simplifying as much as possible ideas already latent in the methods of *finite differences* [13]. Mathis is the first to admit that he has not been able to apply the principle even in many cases of real analysis, let alone complex analysis, which he does not even deal with. Needless to say, the mathematical community does not even consider such limited proposals.

However, the constant differential is the naked truth of calculus, without idealization or rationalization, and we should not dismiss something so precious even if we do not succeed in seeing how it can be applied. If a constant differential is not found in the tables of values of the function, we can still estimate the dispersion, and whoever says dispersion, can also say dissipation or entropy.

In this way it is possible to obtain an intrinsic entropy of the function, from the process of calculus itself, rather than from correlations between different aspects or parts. This would be, at the most strict functional level, the mother of all entropies. It should be possible to apply this criterion to complex analysis, to which Riemann made such fundamental contributions, and from which the theory of the zeta function emerged.

The criterion of constant differential may even remind us of the mean-value methods of elementary calculus or of analytical number theory. However the core of this method is not averages, it is on the contrary standard calculus that relies on averages —which without this reference lack a common measure. Finite-difference methods have also been applied to the study of the zeta function by leading specialists. But to make the matter even more interesting, we should remember that both methods do *not* always yield the same values.

This would simultaneously fulfill several major objectives. It would be possible to connect the simplest and most irreducible element of calculus with the most complex aspects. Something less appreciated, but no less important, is that it puts the idea of function in direct contact with that which is not functional —with that which does not change. Analysis is the study of rates of change, but change with respect to what? It should be with respect to what does not change. This turn is imperceptible but transcendental.

The direct method laughs in the face of the "computational paradigm" and its fervent operationalism. The now prevailing idea of intelligence as "predictive power" is completely reactive and servile; in fact, it is no longer known whether it is the one we want to export to machines or the one humans want to import from them. Perceiving what does not change, even if it does not change anything, gives depth to our field. To perceive only what changes is not intelligence but confusion.

I don't know if this will help the understanding of the function, but it is certainly the most basic level for the problem I can conceive of. At least to me, only the simplest can be relevant here. Researchers now work from the other side, that of complexity, where of course there is always so much to be done. The question is to what extent both ends can be connected.

There are also good reasons to think that the zeta function is linked to classical mechanics: the frequent recourse in this area to "semi-classical" physics is just one of them. Quantum mechanics is adamant in affirming its independence from classical mechanics, but it is incapable of telling us where one begins and the other ends.

Mario J. Pinheiro has proposed an irreversible reformulation of classical mechanics, as an equilibrium between energy and entropy that can replace the Lagrangian. Apart from the fact that the Lagrangian is also fundamental in quantum mechanics, it would be of great interest to evaluate the intrinsic entropy of classical trajectories with respect to the various formulations [14].

In problems such as that posed by the zeta function in relation to fundamental physics, crucial experiments capable of unraveling the relationship between the reversible and irreversible in systems could soon be conducted — provided that they are expressly sought. For example, a Japanese team very recently showed that two distinct probes of quantum chaos, noncommutativity and temporal irreversibility, are equivalent for initially localized states. As is well known, this function has been studied a great deal from the point of view of non-commutative operators, in which the product is not always independent of order, so this would bring together two areas as vast as distant up to now [15].

Irreversibility and Reversibility

With the rise of quantum computing and quantum thermodynamics, along with all the associated disciplines, there will be no shortage of opportunities to design key experiments. Until now, the dominant trend has always been to favor reversible aspects, since they are the most manipulable and exploitable; only by changing our logic and duly valuing the role of irreversibility can we reach higher levels of understanding.

The reversible and irreversible, concepts referring to time, may be intimately related —non-commutative aspects in between- with that so basic of the additive and multiplicative properties of the sequence of numbers that evades the mathematicians confronted with the zeta function. It is as if there were two ways of considering time, according to its sequential ordering, and according to the simultaneous coexistence of all its elements.

The relationship between the reversible and the irreversible is critical in the evolution of complex systems: think only of aging, or the difference between reversible or irreversible damage to the health of an organism or any evolution of events. If we believe that this distinction is irrelevant in fundamental physics, it is only because physicists have carefully segregated thermodynamics so as not to be tainted by problems having something to do with real time. And because the projection of mathematics from above led directly to a fatal misunderstanding.

It is not that hard either. It is said that electromagnetism is a reversible process, but we have never seen the same light rays returning intact to the bulb. The irreversible is primary, the reversible, important as it is, is just a settling of scores.

But physicists have always been fascinated by reversible artifacts, independent of time, which some have seen as the last legacy of metaphysics. And what else? There are no closed, reversible systems in the universe, which are just fictions. Metaphysics is an art of fiction and physics has continued metaphysics with other means with the perfect excuse that it now descends into reality. And no doubt it has, but at what cost?

Curiously, it is the discovery and application of reversible physical laws that has exponentially increased the possibility of accumulation and with it of progress understood as an irreversible process. This reversibility is as fully in line with the logic of exchange and equivalence as the unrepeatable is alien to it.

No less curiously, only the understanding of the irreversible would allow us to partially reverse certain dynamics —partially because it goes without saying that no one ever steps in the same river twice. But power is always the least interested in the possibility of rectification for anything. It is decidedly committed to accumulation and historical irreversibility, which also means committed to the business of a reversible and interchangeable nature.

Curious, very curious, and even more curious then. Although it was always seen that "there is no way back" is the way to make us jump through the hoops. But in all this of the reversible and irreversible we must distinguish between an external and internal perspective. What distinguishes both perspectives is the definition of equilibrium as a sum or a product.

A zero-sum equilibrium, although derived and not necessarily the most important, is the one Newtonian mechanics presents between action and reaction. It defines a system from the most external point of view possible, which usually coincides with our idea of reversible processes.

An equilibrium of densities in which the product equals unity is, for example, the right perspective for continuous media mechanics; an internal perspective to a primitively homogeneous medium that is always available even if physics has obscured it with convenience tools like vector calculus and successive and increasingly opaque formalisms —tensors, connections, operators, etc. This way reversible processes can also be described, as in fluid mechanics, the theory of elasticity or electromagnetism —the latter clearly exhibiting both reversible and irreversible behaviors.

Suffice it to say here that the definition of equilibrium allows for the establishment of different types of relationships between the part and the whole, between local and global adjustment. Classical entropy, for example, is an extensive quantity, and additive for independent systems without interaction. The basic quantity is the logarithm because additive relationships are much more manageable than multiplicative ones.

Ilya Prigogine showed that any type of energy can be broken down into an intensive and an extensive variable whose product gives us a quantity; an expansion, for example, is given by the PxV product of pressure (intensive) by volume (extensive). The same can be done for relationships such as changes in mass/density with the relation between velocity and volume, and so on.

Without entering now into the complexity of the subject, this double definition of equilibrium has implications for mechanics, arithmetic, calculus and entropy itself. It is related with the external or internal perspective of the same process or problem.

The definition of the equilibrium also depends on whether we are talking about open or closed systems. For the most elemental intuition, which is not mistaken in this, something "alive" is what has a permanent exchange with the environment, while something "dead" is what does not. It goes without saying that the general case of statistical mechanics is that of a closed system with independent elements, that is, something "really dead".

As we have already noted with Pinheiro's proposal, even classical mechanics can be reformulated as an irreversible mechanics with a term for the free energy of the system —that background energy that some would like to proscribe for the improvement of trade and resource management. One could ask what is the need for this, with the standard equations working fine just as they are. But, apart from everything we have said, is it not worth realizing that nothing is dead and that everything has its own intrinsic form of regulation?

But this intrinsic regulation can be shown even without any appeal to thermodynamics, even with the conventional equations of celestial mechanics, as we have already shown sufficiently elsewhere [16]. But the introduction of an irreversible element, unnecessary as it seems, allows other connections and another conception of nature which is not cut off by our utilitarian interests.

Entropy and self-information

In fact all the gauge fields of the standard theory, based on the invariance of the Lagrangian, clearly exhibit a feedback. They don't even hide it, and the theorists have been struggling with terms like self-energy and self-interaction for generations now, shooing them away like flies of their faces because they thought it was a "silly" idea. Now, the renormalization group at the base of the standard model and statistical physics, containing this self-interaction, is more or less the same that now is used in the neural networks of deep learning and artificial intelligence —so something that the electromagnetic field already has built-in is being used externally. Let us call it a missed natural intelligence.

Nobody made the integers, and all else is the work of man. The zeta function is an ideal candidate for creating a collective network of emerging artificial intelligence around it. Why? Because in it the internal and external aspects of the act of counting, what some would call the object and the subject, coincide. In this type of networks will be key to what extent the potential of the physical substrate is tapped, the local and global equilibrium conditions, the correlation spectrum, and a good number of aspects we cannot deal here.

Thus it may be possible to create a collective emerging artificial intelligence alien to human intelligence defined by purpose, and yet with the possibility of communication with humans at different levels through a common act, counting, that, as purpose limits our scope, also for humans use to exist at very superficial levels. Intellection is a simple act and intelligence is a connection or contact of the complex with the simple, rather than the opposite.

Renormalization in physics emerges at the same time as information theory, around 1948. Entropy and information become almost synonymous, and if in thermodynamics entropy was seen as work loss, it will now be seen as information loss. Shannon's information was initially called *surprisal* or *self-information*.

The confusion of entropy with disorder is due to Boltzmann's rationalization in attempting to derive macroscopic irreversibility from mechanical reversibility, quite a tour de force by itself. It was Boltzmann's use of the concept of "order" what introduced a subjective element. Clausius original energetic conception of entropy, stating that the entropy of the world tends to the maximum, was, if nothing else, far more natural.

Order is a subjective concept. But to say world is to say order too, then also the idea of world is inevitably subjective. However subjective they may be, "order" and "world" are not just concepts, they are vital aspects of every organized entity or "open system". Yet even our very intuition of order and entropy are entangled and confused.

Not that entropy increases disorder. On the contrary, the tendency towards maximum entropy is conducive to order, or seen from the other side, order has more potential for dissipation. As Rod Swenson put it: "*The world is in the order production business, including the business of producing living things and their perception and action capacitities, because order produces entropy faster than disorder*".

"Open systems", like the outfielder trying to catch the fly ball, have perception-action cycles, with their constant differentials, and probably even their own pole and critical line. But is there any entity or process that is not an open system? Surely not, except for the peculiar conceptions of science.

Even so, we see that information theory itself, no matter how formal and objective it may be, cannot avoid the reflective turn in the interpretation and use of data. At the same time, this interpretation is part of the system's behavior, although in black box mode.

We could go on indefinitely, but in short, let us say that modern information theory is too broad and void, too unrestrictive, and therefore with very poor capacity to shape anything by itself —by itself it is only a general framework. But understood as *disposition*, it is the biggest weapon of mass destruction. It is a will of nothing pure and simple, a will of nothing willing to sweep the world away.

If this is not obvious enough to us, it is because we have been trained for it for a few centuries and the training has worked. In reality science, and modernity in general, is a dual movement between knowledge and control, but from the beginning it has had to purge half of the potential of the former to better exercise the latter.

There is therefore a permanent process of immunization against the inconvenient right from the start to the present day. The development of the suppressed part, even if it had a higher order rationality, would cause a serious short circuit to the system, which by the same selective logic of exclusion cannot assimilate this inconveniences.

If both the thinker and the layman are intimidated by the complexity of scientific problems, the experts are even more intimidated by the idea of revising the foundations of their own discipline, which is tantamount to undermining their condition of possibility. The logic of specialization is clearly another of those irreversible and increasingly restrictive processes, which has only the alternative of creating new specialities.

However, the devil can only exit by the same way he has come in, and the depth of the problems remains where they were first found, much more so than in the supposedly advantageous, but not selfless, position of hindsight readings. The impossibility of reconsider anything important betrays a definite structural weakness. Technoscience is an elitist project that has to pay the price of isolating itself from reality, even though, so far, both the project and the isolation are paid by all.

The zeta function is just one great example of this kind of indigestible analytical totality for modern methods. Someone, in an Information Age typical turn, might say that, if any possible information is already in that function, we might as well look for the answers there instead of in the universe. But the opposite is true: to better understand the function we need to change the ideas of how the universe "works" and the range of application of our more general methods. That is the beauty of the subject.

There is of course a whole constellation of much simpler, identifiable and treatable problems that can and should be studied independently of that problem, even having much in common: in the foundation of calculus, in the irreversibility in classical and quantum mechanics, in the meanings of entropy and information, in the connections of all this with the theory of complexity and the algorithmic theory of measurement, and so on.

Besides, there are relatively simple open totalities that do not require a great deal of theoretical development but rather a selective change of attention. Think for example of good old Ehret's principle, that says that the vitality of an organism is equal to the power minus the obstruction (V = P-O). In reality, this is the only functional definition of health that has been given, and with its help, a proper measurement of the pulse, and a series of derived and associated formulas, a whole global theory of health and aging can be developed that we totally lack today.

But why develop a consistent and comprehensive theory of health, full of common sense and understandable to all, and which makes primary health care an easy target to meet, when we can have an indecipherable medical industry taking 5 or 10 times more profit out of its customers? And with the immense advantage that they will never understand anything. Those other things have no right to enter our world.

The complexity of nature, so often fabulous, has little to do with the fabulous business of complexity. Anyone can see that high complexity is extremely profitable and inherent to the increase of economic activity. The fact that nature can be disconcerting in detail does not prevent it from having in many aspects an elemental logic that is very easy to extract, but systematically obscured by the accomplices of complexity, who in order to solve problems have to create them first.

The omnipresent praise of intelligence and creativity is entirely at the service of this context of "finding new problems" and veins, where simplicity will always be suspect, and where those concerned will always seek to vilify and degrade it. But no degree of intelligence can replace rectitude, even in the most intellectual of problems.

It is absolutely true that technological solutionism is the great excuse of the present for not looking directly at things, at the real problems, those with not need to be invented. Of course, not only that, it is a way to channel the amorphous frustration and in the meantime re-educate the frustrated. It is clear that our most serious problems are not solved by technology; but technology and its perverse use are already one of the biggest of them all.

This does not mean that we want to flee into the past or react or reduce our horizons, either on a theoretical or practical level. The main reactive forces are those operating from above, and we have to show that they have no superiority whatsoever; not moral superiority, which is obvious, but not even technical superiority, which is decisive for them. They are the ones who are closing horizons, and that encirclement must be broken.

There are some 25 million programmers scattered around the world, a great intellectual force composed of people of all kinds who play an important mediating role between science, technology and the majority. Let us hope that they and the multi-specialists will help to overcome this darkness.

Seeing the bigger picture

Bernhard Riemann's eight-page paper *On the number of primes less than a given magnitude* was published in November 1859, the same month as *The origin of species*. That same year saw the beginning of statistical mechanics with a pioneer work by Maxwell, and the starting point of quantum mechanics with spectroscopy and the definition of the black body by Kirchhoff.

I have always felt that in Riemann's conscientious study, so free of further purpose, there is more potential than in all of quantum mechanics, statistical mechanics and its daughter information theory, and the evolution theory combined; or that at least, in a secret balance sheet, it constitutes a fair counterweight to all of them. Although if it helps to redirect the doomed trend of infonegation of reality it will have already been enough.

These three or four developments mentioned are after all children of their time, circumstantial and more or less opportunistic theories. It is true that Boltzmann fought and despaired for the assumptions of atomic theory to be admitted, but he had his whole battle with physicists, because chemists had already been working with molecules and atoms for a while. In mathematics, however, the lags operate on another scale.

A wordy bestseller, *The origin of species* was discussed in cafés and taverns on the very day of its presentation; but Riemann's hypothesis, although being only the strongest version of the prime number theorem, nobody knows how to go about it after 160 years. Clearly we are talking about different wavelengths.

And what does one text have to do with the other? Nothing at all, but they still present an elusive point of contact and extreme contrast. The usual reading of the theory of evolution says that what drives the appreciable order is chance. Riemann's hypothesis, that prime numbers are distributed as randomly as possible, but that this randomness hides in itself a structure of infinite richness.

Mathematicians, even living on their own planet, are gradually realizing that the consequences of proving or disproving Riemann's hypothesis could be enormous, inconceivable. It is not something that happens every Sunday, or every millennium. But we do not need to ask so much: it would be enough to properly understand the problem for the consequences to be enormous, inconceivable.

And the inconceivable becomes more conceivable precisely as we enter "the Information Age". The inconceivable could be something that simultaneously affects our idea of information, its calculation and its physical support. Both software and hardware: a whole damn short-circuit.

Neither economics nor information are destiny, but the zeta function can be the destiny of the Information Age, its funnel and event horizon. There is another singularity and it is very different from the one they had in mind.

You can bet that the root of the problem is not a technical issue, or not that much in any case. But many of its consequences are, and not just technical. They will affect the dynamics of the system as a whole, and one can foresee in advance the ideological warfare, the distraction, the attempts of cognitive infiltration and the struggle for appropriation.

Like the prime numbers themselves, in the chronology and the sequential order the events cannot seem more fortuitous, but from another point of view they seem destined by the whole, pushed by the ebb from all shores to fit in the very moment they took place. Riemann lived in Germany in the middle of Hegel's century, but in its second half the opposition to idealism could not have been greater in all areas, and in the sciences in particular. The pendulum had turned with all its force and the sixth decade of the century marked the peak of materialism.

Also a son of his time, Riemann could not help but feel acutely the contradictions of 19th-century liberal materialism; but the German mathematician, additionally a theoretical physicist and a profound natural philosopher, was an heir of Leibniz and Euler continuing their legacy by other means. His basic conviction was that we humans do not have access to the infinitely large, but at least we can approach it through the study of its counterpart, the infinitely small.

These were prodigiously fruitful years for physics and mathematics. Riemann died before the age of forty and did not have time to articulate a synthesis in tune with his conceptions and permanent search for unity; a synthesis understood not as an arbitrary construction, but as the unveiling of the indivisibility of truth. But he achieved something of that kind where least expected: in the analytical theory of numbers. A provisional synthesis though, that has given rise to the most famous conditional of mathematics: "If the Riemann's hypothesis is true, then…"

This unexpected and conditional synthesis came about through complex analysis, just in the same years that complex numbers were beginning to emerge in physics, seeking, like atoms and molecules, more degrees of freedom, more room to move. The role of complex numbers in physics is a topic always postponed since it is assumed that their only reason for being is convenience —however, when it comes to dealing with the zeta function and its relationship with physics, there is no mathematician who is not forced to interpret this question one way or another, which physicists generally associate with rotations and amplitudes.

But complex analysis is only the extension of real analysis, and to get to the heart of the matter we must look further back. Riemann's conditional synthesis speaks to us of something indivisible, but it still relies on the logic of the infinitely divisible; not to solve the famous problem, but simply to be in tune with it, it would have to be understood in terms of the indivisible itself, whose touchstone is the constant differential.

Is there anything beyond information and the computer? Of course there is, the same thing that is waiting for us beyond the Symplegades. A much more vast and undivided reality.

Referencias

(www.hurqualya.net)

[1] Miguel Iradier, *Stop 5G* (2019). En ese artículo apuntábamos a la medición de la fase geométrica, descubierta originalmente en potenciales de campos electromagnéticos, pero que también puede detectarse en macromoléculas orgánicas como el ADN, en la movilidad celular, e incluso probablemente en ritmos de orden superior como el ciclo respiratorio y el ciclo nasal bilateral. El estudio de esta memoria de fase a niveles tan diversos podría aportar un nivel de correlación mucho más consistente, y tal vez, concluyente.

[2] Carlo Sparaciari, Jonathan Oppenheim, Tobias Fritz. *A Resource Theory for Work and Heat* (2017).

[3] Miguel Iradier, *Caos y transfiguración* (2019)
Bruce M. Boghosian, *Kinetics of wealth and the Pareto law* (2014)

[4] Miguel Iradier, *La tecnociencia y el laboratorio del yo*(2019), capítulo I, Disposición de la mecánica

[5] Para una buena y breve introducción a la hipótesis de Riemann se puede consultar, por ejemplo, *The Riemann Hypothesis, explained*, de Jørgen Veisdal.

[6] Matthew Watkins, *Voronin's Universality Theorem*

[7] Ian Wright, (*Notes on a Hegelian interpretation of Riemann's Zeta function* 2019)

[8] Gian Paolo Beretta, *What is Quantum Thermodynamics* (2007) También puede visitarse la página http://www.quantumthermodynamics.org/

[9] Matthew Watkins, *Number Theory and Entropy*

[10] O. Shanker, *Entropy of Riemann zeta zero sequence* (2013)
Alec Misra, *Entropy and Prime Number Distribution*; (a Non-heuristic Approach)

[11] Piergiulio Tempesta, *Group entropies, correlation laws and zeta functions* (2011)

[12] McBeath, M. K., Shaffer, D. M., & Kaiser, M. K. (1995) *How baseball outfielders determine where to run to catch fly balls*, Science, 268(5210), 569-573.

[13] Miles Mathis, *A Re-definition of the Derivative (why the calculus works—and why it doesn't) (2003) Calculus simplified*
My Calculus applied to the Derivative for Exponents
The Derivatives of ln(x) and 1/x are Wrong

[14] Mario J. Pinheiro, *A reformulation of mechanics and electrodynamics* (2017)

[15] Ryusuke Hamazaki, Kazuya Fujimoto, and Masahito Ueda, *Operator Noncommutativity and Irreversibility in Quantum Chaos* (2018)

[16] Miguel Iradier, *Pole of inspiration —Math, Science and Tradition.* La versión original en español se encuentra en hurqualya.net.
Miguel Iradier, *¿Hacia una ciencia de la salud? Biofísica y biomecánica.* Miguel Iradier, *El multiespecialista y la torre de Babel*

[17] Rod Swenson, M. T. Turvey, T*hermodynamic Reasons for Perception-Action Cycles*

ART AND THEORY OF REVERSIBILITY

21 October, 2020

According to the World Economic Forum that annually meets in Davos, «nothing ever will be the same» after the coronavirus. The Great Reset awaits us this impending 2021 and we only have to get on board.

How can we know when a change is definitive or merely occasional? How can we foresee if its consequences will be reversible or irreversible?

It is assumed that the miscalled «Spanish flu» of 1918-1919 was incomparably deadlier than this virus, and yet it vanished like a ghost in the memory of an entire generation which however could not forget the First World War and the Treaty of Versailles, or the crash of 29.

The great war and the crisis of the 1930s did have irreversible effects, which would lead to the world of 1945; but it is clear that the flu did not, and as soon as it stopped filling the news, it was little less than obliterated. Some say that the figures were inflated shamelessly to frighten the population and make them forget the terrible question of the responsibility of the conflict, which had met with the general opposition of the unions, and of any person able to evade the much more lethal war propaganda of the press.

*

Is globalization irreversible? Massive immigration from poor to rich countries? Emigration to the cities in most of the planet? The urban exodus in the United States? The split of this nation in two confronted societies? The flight from work? The «conquest» of rights that are often promoted and granted from above? The expropriation of techniques by technology that we call digitalization? Climate change? The hyper-concentration of capital? The corruption and decomposition of the social body? Civilization? Progress? Human domestication?

In principle there is no process that cannot be reversed, either voluntarily or spontaneously, in a smooth or catastrophic way; everything depends, of course, on the time scale that we use.

Climate change itself, so insistently attributed to human intervention, may well be a minor fluctuation on the way to the next glaciation. And even if our species were to alter the climate decisively, a rapid collapse would perhaps be the quickest way to reverse the process.

Emigration from the countryside to the city seems unstoppable in most parts of the world, but in some economically developed countries there is already a trend towards urban exodus in search of more space, more time and fewer restrictions.

Of course, the inhabitants of the city are not going to live in the countryside to grow potatoes as three hundred years ago; on the contrary, they do so because the rural areas are also largely assimilated and in any case well connected to the web.

And so we could say with everything. The relationship between the reversible and irreversible would be largely dialectical, since when we perceive something as irreversible, we immediately explore other directions and even the opposite way if it promises new degrees of freedom.

It is obvious to anyone that the direction of many processes can be reversed, while that does not at all mean that things simply return to their previous state. Strictly speaking, there are no reversible processes because in practice the conditions are never the same again —yet there is always the possibility of partially reversing a situation, and what we take as an absolute trend is nothing but a highly conditional arrange of changing circumstances.

Progress always aspires to irreversible mutations because, for lack of anything else, the destruction of the past as a fait accompli would be its only form of legitimacy. But beyond its destructive rage and trivial secrets, modernity has its own mystery, which so far has resisted all efforts to see through it. It may have much to do with the mutual relationship between the irreversible and the reversible, not just at the superficial level of pairs of opposites, but on the kind of highly selective commerce it has established with Nature, what it wants from her and what it neglects.

*

Let's go first with one of those trivial secrets; so basic, it seems, that we don't need to include it in our maps of reality. Incidentally, it also has to do with irreversibility.

According to Hegel's master and slave dialectics, the master dominates the slave by his contempt for death, while the slave accepts submission to avoid it. But today the opposite is true: the unrecognized masters of the world have no choice but to hide, and it is to compensate for this infamy that they try to sow fear in the majority of the population.

Needless to say that those masters are not the familiar faces of Forbes' list, the big corporations or the Davos Forum, who are there to stand as a front and give some appearance of normality. Even the chief executive of BlackRock would be nothing but an errand boy for some subjects in unfriendly terms with the sun.

There is a simple mathematical reason for this: it is called the Pareto law of wealth distribution. Since economists prefer not to dig in it, it never hurts to remember how it works: 20 percent of the population has 80 percent of the wealth, but this 80/20 law is repeated indefinitely: Eighty percent of that 80 percent is held by 20 percent of that 20 percent (64 percent of the total is owned by a 4 percent), over 51 percent by a 0.8 percent, and so on up to the top of the wealth pyramid.

Thus we arrive at the well known estimate that the 62 richest people have more wealth than half the world's population, the poorest 3.7 billion. Perhaps not so well known is that, following the same elementary logic, most of the wealth of those 62 would be owned by 12, and most of the wealth of those 12 would be owned by just 3 fortunes.

These 3 fortunes would possess, not exactly most of the world's wealth, but most of the surplus buying power and subordination of the smaller fortunes all the way down.

Physicists like B. Boghosian have made dynamic simulations of the Pareto distribution establishing a more or less exact analogy between collisions of molecules and monetary transactions. The asymptotic dynamics tends to a singularity in which all but a vanishingly small fraction of the agents have zero wealth. No doubt it is a very idealized model, but that is the trend.

The Pareto law is only a particular case of the power laws that appears in all kinds of distributions in nature and society, from the size of cities and corporations to the intensity of earthquakes and the caliber of blood vessels in the circulatory system.

Although the appearance of power laws in all kinds of phenomena, and its self-similar, fractal character may make us think that the inequality in the distribution of wealth is a «natural fact», in truth we still don't know for sure which are the factors that determine its existence and the number of powers.

As for its relevance to the economy, it is hard to believe that there are practically no studies or even a remotely consistent theory on the subject, but in its absence it seems quite reasonable to think that the number of iterations —the degree of concentration of wealth- depends critically on credit mechanisms and the continuity or cancellation of accumulated debt. As Michael Hudson points out, since the first great civilizations in Sumer and Babylon, sovereigns periodically cancelled debts, but in modern states the creditor always ends up collecting, if only for the privatization of public goods and effective control of governments.

Thus the present democracies, with central banks that are effectively cartels of private banks from which the state has to borrow its own money, have become the ideal instrument to create irreversible processes of debt accumulation from top to bottom that invariably ends up with a power grab.

Needless to say, private banking not only controls the issue of the monetary base —mainly cash-, which is just about 5 percent of the money supply, but also credit as a whole, from which all the debt-money comes by virtue of the fractional reserve mechanics.

Fractional reserve itself presents an elementary resemblance with power laws and the leverage through credit, since the greater the credit, the greater the leverage and the smaller the base. Given that the state is already a hostage of credit, this power law with its inherent mechanics is the very lever of power both generally and in detail, which through the distribution of money manages to encompass all social relations.

The non-reversible relationship of modern states with debt and the extension of credit to all spheres explains why the present pyramid of inequality is more pronounced than in any previous era.

Only the unprecedented increase in productivity due to technology, the lower cost of some consumer goods and the expansion of credit made possible to somewhat conceal the rampant increase in inequality in recent decades; however, today we also see that digitalization, like the flow of money, conforms to the structure of this huge extractive apparatus, further accelerating the process —since the very meaning of digitalization is to eliminate resistance or minimize friction.

Models like the one presented by Bogoshian would confirm the increasingly widespread suspicion that our system truly behaves like a black hole, whose suction is ultimately indiscriminate although it possesses a rich selective structure of recurrent mediations along the way, in which the big fish eats the small one all down the line —»ideally» at least, and ignoring the undesirable resistances to overcome.

To use a mechanical analogy, the control parameters of this dynamics would be directly linked to the interest rate that governs the accumulation of the debt and generates the load, and therefore the tension, vacuum and suction. We always live between stress and pressure, and the change of sign from one to the other, between the full and the empty, can be extremely fluid, but the institutions that dominate the banking industry attempt to regulate it through the interest rate and the percentage of deposits of the private banks that determine their capacity to create money.

*

The parasite is «closer to you than your jugular vein», and it is clear that although it depends on a liquid medium, it does not respond only to a horizontal logic.

Liberals try to sell the illusion that in a world governed by money and its freedom of movement we are definitely moving in a «liquid» medium and horizontality reigns supreme, and most of the left also buys this argument motivated by a textbook materialism and its desire to eliminate any «verticality» that sounds like a residue of a further past. Thus the new verticality escapes due attention. Today both speak of «inequality» and create all kinds of centers and institutes for its study, dealing with it as if the Great Syphon did not even exist.

The power law of wealth distribution is an absolutely basic fact of the economy and society, in the sense that it is both structure and function, form and dynamics, capillary suction and dripping system, hierarchy with a continuous exchange of favors offered and services rendered. To pretend that capitalism is a purely «liquid» phenomenon or we have neutral markets would be utter folly. If capitalism has come this far, it is because it involves a plutocratic hierarchy that is nevertheless highly functional. And the power law is the x-ray of this hierarchy and this structure, without which the existing system would not stand.

If they are looking for a quantitative model in sociology and economics, there is no need to go any further: here we have a whole lot of quantitative correlations, qualitative gradations, nuances and appreciations waiting to be extracted. And in these Big Data times, how not to think that the engineers and guardians of the Great Syphon are dedicating a good deal of time to it? They would be truly incompetent if they weren't.

«Progress» is the growing differentiation of the productive and social fabric —or the endorsement of this process. More differentiation requires more organization. Oligarchy is organization, and to survive in a context of growing complexity the organization must be diversified and more refined, becoming more fragile and irreversible too.

An organism, as a system, ages through a unique process that presents several different but ultimately equivalent aspects: increasing inability to eliminate, increasing inability to renew because the available space is filled with debris, and *INCREASING RESTRICTION OF THE DEGREES OF FREEDOM*. Modern science ignores widely the concept of evolution as a growing restriction, essential to the life cycle of organisms.

While a horizontal Darwinian and Hobbesian narrative of competition and war of all against all is heavily promoted for the masses, the organized minorities take over the vertical, much more concretely structured logic, governed by the ecological terms of marketing and the exploitation of niches and ecosystems, now coated with the splendors of digital neo-feudalism.

Noblesse oblige. «Big fish eats small fish» has very different meanings depending on the coordinate axis in which we move. The new progressive, horizontal ecology ignores the vertical ecology of elitism that came first demarcating territories. In fact, a good deal of the «horizontal ecology» and the «ecology of the mind» were designed by such dubious characters as Gregory Bateson, a reputed specialist in psychological warfare.

*

Elementary statistics say that approximately half of the great fortunes in the United States, Europe or Russia belong to individuals of Jewish origin. This is a convenient proportion, if only to keep up appearances and to have at least as many human shields in the rare event that someone is to be blame.

One could think that whether this plutocracy is Jew or not can hardly be relevant for the general dynamics and the flow of money. In fact it is also fundamental at this level, and as we have already pointed out the inverted pyramid is first and foremost a great hierarchy, a highly selective filter in both directions, up and down, that determines, positions, priorities, favors, and obediences.

Nor should this 50/50 share be important when, Jews or not, in the end, whether they vote democrat or republican, nearly all of them are fervent Zionists or sympathizers.

If the law of 80/20 and its successive powers offers us the most revealing and eluded x-ray of the state of things in society, this tacit «law of 50/50 percent» is the x-ray of the x-ray, an additional penetration into the complexion of what we use to call «the capital».

This 50/50 share is not only relevant to the effective weight of money and its projection of power but also to the historic and symbolic order, besides insinuating the virtual axis of a dynamic. On the other hand, it should be obvious that the fact that such a numerically small minority holds half of the great fortunes has something to do with the vicissitudes and mysteries of accumulation.

Already the Hebrew Genesis tell us the history of Joseph in Egypt, winning Pharaoh's favor and then speculating with money and grain to achieve the enslavement of the population through debt —a feat crowned by making the slaves say: «You have saved our lives!»

Hegel's master gets recognition by not killing; the financial administrator, by saving a life he has previously starved and strangled. Of course, the former exposes himself, while the latter plays it safe. Violence and deception are opposing ways of using force and spirit, but both have maintained a special relationship from far back in history.

In Christendom, we have documented evidence of this iniquitous alliance since at least the time of Charlemagne's son, when Agobard, Bishop of Lyon, sent five letters to Louis the Pious in 826 denouncing the ludicrous privilege of the Jews at the expense of the christian majority. Coincidentally, this privileges had been on the rise since the Emperor's marriage to a certain Judith of Bavaria. But from the Pharaoh to Trump and from Joseph and Esther to the present day, this pattern sounds too familiar.

As with Faust and Mephisto, a certain eternal archetype, incarnation of the spirit in exile, wastes no time and goes directly to the tap of power promising more wealth by taxing and financing wars, flattering the fatuity and whims of the powerful in charge, carefully studying human weakness and relentlessly exploiting stupidity.

Only later, when the sovereigns gave way to democratic governments, and the creation of false needs that is the salt of the last phase of every civilization spread, like credit, to every layer of the population, all and sundry could take part in the same lie.

Nowhere in modern times has been this alliance so close as in Britain, the origin of liberalism and the industrial revolution, and the United States later; since the conmixtion of Puritanism with the interests of Judah, Anglo-Zionism is the real name of the Empire. It responds both to overt behavior and to a much less visible physiology.

To point this out is not only useful, but also necessary, given that the knot in the heart of Modernity, the effective contraction of its current reality among the whole of possibilities, has the contour and logic of Anglo-Saxon culture and its intimate host; and this is essential on both a functional and historical level.

On the so-called Jewish question, Marx, who was Jewish and bourgeois, came to say that the Jew, as a worshipper of money, was the bourgeois par excellence; but it would have been somewhat more credible if he had said that it is the bourgeois who wants to be halfway between the Jew and the sovereign, the sovereign and the Jew.

The characterization of capitalism as a brainless gluttonous monster with no other logic than greed always diverts us from its indispensable qualities of farsighted planning and «creative destruction». No doubt both components coexist and correspond well with the 50/50 complexion, although in the long run we all know which part will take command. When Lyotard spoke of the libidinal economy of the centaur of capital with its two uses of the power of wealth, one of them reproductive and plundering the other, one «circular, global, organic,» and the other «partial, deadly, jealous… that feeds on the plunder of over-excited energies,» it is as if he had been thinking about the very same thing.

*

The overwhelming reality of this grotesque inverted pyramid must force all kinds of abnormal consequences. One of them is that the minuscule apex of plutarchy is forced to cower in the depths to minimize its exposure. And in this way it is deprived of that recognition which, according to Hegel, makes a master a master.

And here we would enter fully into the greatest comedy of our time, which we are not allowed to appreciate properly. And the comedy resides, naturally, in the fact that if someone points to the «Jewish power» he has all the numbers to be labeled «anti-Semitic», while on the other hand those who hold that power the only thing they lack in order to fulfill their aspirations is general recognition, because until that happens, the pretended master does not cease to be a simple aspirant.

But the comedy of recognition can have a very sharp and dangerous edge for both halves of this strange androgynous Siamese.

It is a well known fact that the Balfour Declaration of 1917, at the origin of the modern state of Israel, was addressed to a member of a famous lineage of bankers who, at the end of the war, preferred to merge discreetly into the background.

That both things overlap in time does not seem to be a coincidence. While some individuals were fleeing from a notoriety that could only be inconvenient for them, a new entity with state pretensions was emerging in fierce struggle for recognition among nations, which nevertheless gloats over placing itself above the compliance of international law.

Whether for Jehovah's ancient chosen people or for the present Zionist entity, the ultimate goal so explicitly stated in the scriptures, «recognition by all nations» coincides with their debasement, degradation and subordination; that today it all comes down to the recognition of the power of money in the world does not deserve further comment. But since even today we realize that, apart from a ruthless use of any advantage, there is no other superiority in this prevalence than that of deceit, usurpation and imposture, the question of recognition can be redirected only to a theological instance able to present what is simple baseness as a voluntary and beneficent descent of the spirit.

In this way Israel as a nation and function among nations would be a pledge and symbol of that recognition that is impossible at a personal level, while at the same time longing for a horizon of transvaluation for a cursed accumulation.

That most of the surplus purchasing power is located at the top of the inverted pyramid takes its potential for corruption to the extreme, since only buying wills can it be mobilized and made profitable, and in a certain sense, «fruitful». Not only this situation produces all sorts of ill consequences; nothing that gives voice to that other underground summit can be nothing but an aberration.

*

The Oligarchy hangs from self-similarity, and gravitates towards that singularity whose horizon of events is the unknown Plutarch.

In many ways, nothing could be less new than the recipes that the Davos boys have in store for us. The transition from classical liberalism to the present cybernetic regime has been taking place since 1945, but the current glitch would be perfect to conclude it and start a new era.

The fact that the spokesman for the Magic Mountain is a German engineer already tells us that the reigning technocracy may still play the role of the villain in the next and surprising turn of the script —something from which we could perhaps redeem ourselves at a given moment thanks to divine, and more specifically Israeli, intervention, although no one yet manages to figure out how that could happen. But to get there we would still have to descend way down.

In Davos they like to talk a lot about governance, which is nothing but the socio-political translation of the cybernetic meme. And of course, an essential part of the global governance on the agenda is that of digital currencies, to which they have been devoting exhaustive studies for a good number of years but that have now found, coincidentally, the ideal scenario for their implementation.

And the global regulation of these currencies is a matter of concern, first of all, lest some fair currency emerge, no doubt by mistake, which is in demand and sets a bad example. «Governance is the central pillar of any form of digital currency,» said Mark Carney, former Governor of the Bank of England and ex of Goldman Sachs. But why, one could ask, was it not a matter of finding alternatives? Judging by his insistence on the «need for governance» of something that is barely at the beginning of this road, there seems to be a very real risk here that too many possible outflows will open up for a suction system that needs to function without leaks.

The family to which the Balfour Declaration was addressed had already taken control of the Bank of England before 1830, and their money has not been sleeping since. The name of the game in capitalism, more than the publicized spirit of enterprise, is the implacable exploitation of every advantage gained; and certainly these people are not about to be lectured on how money is created and distributed.

It may be true that large accumulations work to a large extent without the need for a brain; but not thanks to the «horizontal logic of the markets», generally rigged, but to the interaction of their open component with that vertical logic and dynamics of which we have spoken.

As we have already seen in other articles, the implementation of digital money as they understand it and the sharp reduction of cash would be the consolidation of the current banking impunity, a shielding and closure of the system that would install us fully in a financial prison. With the danger of banks runs precluded, users are losing the last resource they have to call to account while the banks are being totally freed from any responsibility.

Although everything already pointed to this situation, its consummation would effectively mean the closure of the system with all this implies. To get it done, not even concessions are needed, since it seems the monetary question is not debatable. Were it not for the gradual marginalization of cash, most of the population would not even notice it.

So if concessions are made, it will not be for the control of money, but rather to whitewash the Reset and gain popular acceptance. At the head of that public relations campaign will be «the fight» for basic income and that other sick joke known as the Green New Deal.

As for the Universal Basic Income, it should be clear without further explanation that:

1. Even with the current monetary system, and without digital money, guaranteeing a minimum income would cost the rich practically nothing, despite the predictable staged saga surrounding its negotiation. Inflation would soon eat up half the purchasing power of that income, to continue indefinitely with the comedy of demands, increases and all the rest. Meanwhile, the assets of the owners can beat inflation by a good margin.

2. In the event that «the governance of digital money» triumphs, since only the banks would know the bottom line, perhaps even inflation could be avoided, but in any case, with or without inflation, it would cost nothing for them, just increasing the burden of debt as usual.

3. The basic income will be nothing more than concealed and postponed debt to help the debt serf continue to meet the deadline, now turned into the new social phenomenon of the serf-rentier.

4. Real wages for labor could be fivefold to tenfold[1] if all the value extraction sustained by debt at all levels was eliminated.

5. The wages of all the workers in the world are only a small, almost negligible part of the money flowing in the system as the sphere of production is completely subordinated to that of financial circulation. Beyond the staged drama about the costs of such measures, what is being negotiated is the servitude of the population.

6. Despite what their advocates say, even now there is no real popular base for these so-called demands, which are completely fabricated and subsidized and which imply the consummation of the servitude to the cybernetic regime based on the regulation and modulation of the circulating flows, and to the anonymous individuals who govern it.

However, it is clear that the destruction caused by absurd and unprecedented measures against just one more epidemic put growing parts of the population at the mercy of the most powerful.

7. The main cause of the impoverishment of the population now can appear as its savior. The new digital currencies of the central banks could even grant rents independently from the governments —like the new emergency program *FEDNOW* to inject money into the households- distributing a small part of what they have stolen with no other conditions than the exhaustive control of all the movements and the personalized negotiation of each debt. A brave new world indeed.

It is not hard to see through this. However, the control of the public debate, tries at all costs to circumvent the monetary question in all its terrible simplicity

*

Why does the state have to ask private banks for their own money, while they reap all the benefits of the trust in the legal tender? This fraud and usurpation are at the very origin of the degradation of the public affairs along the path of democracies that seems to be coming to an end.

If, as it should always be, all the money were legal money, and the fractional reserve that allows us to create almost all the money we manage from private credit did not exist, the main incentive for the creation and accumulation of debt, and with it also for the forced growth to pay it off, would be put to an end. It is absurd to ask for a policy-directed degrowth when the first thing that should disappear is the artificial stimulus to growth.

The radical simplification of sovereign, unadulterated money, still allows two antagonistic options: in one of them the state creates all existing money and also gives credit through a nationalized bank, while in the other the state only creates money and the credit supply is open for all types of entities, including public banks.

One might think that the latter option, which was first outlined in the so-called «Chicago Plan» of 1933 and has been defended recently by economists and even former central bank governors, is much closer to the liberal model of Western countries, while the former would be much more feasible in planned economies such as China; but the truth is that classical liberalism has found its essential engine in the legalized counterfeiting of debt-money and would be unrecognizable without it. On the other hand, the socialist countries have not ceased to assume the model of the Western central banks with similar levers of fractional reserve and a highly selective distribution of the money issued.

There is a curious mixed situation: in Western countries the main economic lever of the state has been in private hands, but in the real socialism, banking, although nationalized, has taken from private banking the essential part of its structure and practices.

But now, with the arrival of digital currencies, such a mix has the potential to create a double contradiction —since digital currencies can be both state and private. Banks and other financial entities have an interest in issuing their own cryptocurrencies, not to mention social media and projects like Libra. Although there is naturally talk of «offering a service», the aim, when it comes to corporate cryptos, is to capture market share. But on the other hand there are attempts to create crypto currencies that seek stability and flee from the speculative precedent of bitcoin or the corporations.

In reality, one hundred percent legal money would have to be synonymous with stability, were it not for the fact that friends of the governance of others have something different in mind. Until now, the United States has made the dollar system and the Federal Reserve prevail through its military power; but already at the past meeting of central bankers in Jackson Hole in 2019, and in the presence of Jerome Powell, the aforementioned Carney advocated ending the hegemony of the dollar in favor of a new modality of digital currency.

Clearly the intention is not to give away the «exorbitant privilege» of having the US debt, which also pays for its military power, sustained by the rest of the world. On the contrary, it would be a matter of keeping under control a transformation deemed inevitable and irreversible, in order to avoid the catastrophic consequences that a flight from the dollar would have for the Empire.

But the issue goes beyond the interests of the American empire, which is still an instrument for the existing plutarchy. And right now the question is whether that instrument is dispensable, and to what extent. Perhaps the United States has been until now «the indispensable nation,» but not for those who serve as mounts, but for those who ride it.

And what really characterizes the Empire, beyond the American contingency, is the continuous drainage and suction of the productive forces through the Great Syphon of debt-money, or parasitism taken to its maximum expression. Since the digitalization of money, as ultimate liquefaction of the capital, allows too many possibilities and threatens to open leaks in the closed circuit and its optimal functioning, laws must be made and rules put in place to prevent this from happening.

Since China is at the forefront of digital money and already competes in economic weight with the United States, the critical point of global monetary «governance» will also depend largely on the maneuvers of this country with its crypto-currency, as well as on the fruit of international agreements.

The crux of the matter is that there are two possible positions of the state with respect to its digital currency and credit —with its radical liberalization or nationalization- but there are also two possible trends for private circulation: corporate currencies of speculative nature that seek to capture users and mutual or libertarian currencies that flee from that model and seek more security or more freedom.

Nations under the umbrella of the Federal Reserve System would tend to protect corporate currencies, and in fact, beyond appearances, within this arrangement there can be nothing but a symbiosis between the state and corporations, since here the money print has already been in the hands of private interests for generations.

The currencies created by groups with a mutualist/libertarian spirit can adopt different strategies: stand alone, associate with other similar currencies, or seek the best convertibility relationship with state currencies when the later are not hostile to them. Now this type of currencies seems to be a residual phenomenon but they tend to fill in the gaps that other currencies cannot or do not want to cover, so in the long run they can play an important and even critical role in the case of a currency war over quotas in the global crypto space. At its worst, they can be a kind of black market, at its best, an alternative to the state and corporations.

Nobody can yet foresee the final result of this competition for the monetary space in the final stage of its liquefaction, since it will depend on a whole bunch of factors such as the present technological wars, election results, maneuvers and alliances of the United States, Europe and China as well as of the rest of the countries, their courtship or prohibition of private currencies and cash, the degree of cohesion or political decomposition of the principal states, the evolution and orientation of alternative currencies, etc.

*

Digital currency may seem to be the ultimate level of control of the monetary circuit with all its movements, but it may also exhibit the maximum volatility and fragility, depending both on its design or purpose, on the other crypto-currencies with which it competes, or on the legal provisions and public reactions that are equally part of this conflict.

It is in this context of potential chaos and currency war that crypto-digital «governance» is sought, but here the clash between the unipolar pretensions of Anglo-Zionism and the aspirations of other great countries for a multi-polar world cannot fail to come to the fore. Also critical are the internal tensions in the main countries, as demonstrated by the deep division between globalists and neo-isolationists in the United States.

The existing plutarchy is by definition a shadow government or cryptarchy, so the global dominance of the cryptocurrency market would fit them like a glove and be their capstone, in the absence of that recognition money can't buy.

Moreover, the advent of digital currency and the demise of cash is projected as the ultimate triumph of transparency and the «distributed economy», although it is not necessary to study the subject to see that it is just the opposite. As it already happens with cyber-vigilance, someone will be able to follow the most insignificant monetary movements of almost everyone, but practically nobody will know where the bulk of money is flowing in the Great Syphon scheme. Not a black mirror, just asymmetric transparency.

Nor will we discuss its usefulness in controlling organized crime, when it would constitute not only the supreme criminal organization but also its enthronement with legal armor included —something just logical in the economy of fraud.

The Marxist analysis that assures that the collapse of capitalism is inevitable due to the progressive reduction of profit, still repeated to boredom, is irrelevant to the fundamental question of the debt economy which is the motor of everything. On the contrary, if rates of profit are progressively reduced, that would only mean that an ever greater part of profitability migrates to credit and financial speculation, with which its ascendancy over labor/services and material and intellectual assets is ever greater, as we observe. Revolution is by no means inevitable, especially if the control of ideological alternatives is consolidated, which is also what we see.

Liberalism has always been a farce, but so has the materialism of the left, so fond of ignoring the formative and mediating role of the monetary system and credit. In this way the supreme arbiter of the manufactured ideological debate remains out of focus while it operates unimpeded.

Marx lived in the London of the Rothschilds, Disraelis and Montefiores, and he could not ignore what was happening; on the other hand Jewish people, and especially its economic elite, didn't want anything to do with productive economy, left for the dull. This was a great opportunity to put a wedge between the 'bourgeois' and the 'proletarian', while the credit that determines their relationship and profit was pushed into a more than discreet background.

Rozanov said that democracy is the system in which a well-organized minority governs a disorganized majority. We can only add that an increasingly reduced minority can only assert itself by increasing actively the disorganization of that majority, dividing and sinking wedges in all its fissures; and that, too, is what we observe.

For the rest, it is worth remembering that the question of the monetary system goes beyond the economy. The circulation of money in society is not only comparable to the circulation of blood in an organism. It is also a system of information and communication as prices are; and it is an essential part of the social grammar, to such an extent that many of the distinctions between the public and the private sphere have become what they are according to its general disposition.

On the other hand, it is good to understand that the tertiarization or rampant hypertrophy of circulation would not have been possible without the impulse given to it by a monetary system in which the bulk of the money is imaginary and based on debt only.

*

In money as in everything else, digitization is not irreversible, and if it were just irreversible it would be insignificant, a mere update on the way to a gigantic mousetrap. What is decisive is precisely the existing degrees of freedom to reverse the situation.

In other words, the only valuable thing about the possibility of introducing digital money is not to dive into some unknown and nonetheless too familiar future, but the one that allows to reset the sovereign, legal money as it always should be and as it was often before the great usurpation.

Some authors excuse the prevailing system of fractional reserve or money-debt by arguing that one hundred percent legal money was impractical in the nineteenth century for purely technical reasons. The argument has some truth to it, but unadulterated systems, or at least way more equitable, have existed centuries and millennia before that time.

A fair monetary system cannot in any case depend on the level of technology, since if it depends on it we are already in a desperate situation. To make justice depend on technical arbitrations is to put it at the disposal of the most powerful, which is the case with the current system. The fractional reserve is already a system of high technical sophistication, a sophistication that seems to be expressly designed to conceal the reality of money created out of nothing.

And here we will also see that the relationship between what seems irreversible and reversibility surrounds the axis of becoming, of the creative and the emergence of everything new, as opposed to that combination of brute force and inertia that only by an interested misunderstanding we have called progress. If technology is more opaque when it is more sophisticated, and one wants to make a «fairer» system depend on higher levels of technology, opacity, and capture of «users», it should be clear that a fair system is possible precisely insofar as its essence does not depend on the level of technology.

This does not necessarily mean outlawing digitization, but rather not making it the touchstone. At the moment, it is digitalization that threatens to outlaw not only cash, but many other physical relationships. One hundred percent legal money, money without massive counterfeiting by its authorized distributors, should be possible regardless of its physical medium, and that independence is a main indicator of its acceptability.

Defending the coexistence with material supports may sound paradoxically idealistic, but since people still want cash for an elementary sense of security —and because they already charge for having it in the banks-, perhaps the result of the currency war for the digital monetary space, which will involve nations, corporations, and collectives, will also depend on how the supply moves on that axis between the reversible and the irreversible, between capture, freedom, and security.

We can then distinguish three axes: a horizontal axe linked to liquidity and the most purely commercial and civic factors, a vertical one related to the legal aspects of the states and the agreements between states, and a temporal axis in depth hidden in the reach of the temporal reversibility or irreversibility of debt, monetary uses and its changes.

That this axis is really alive and that *THE FUTURE WANTS TO LEAVE THE DOOR OPEN TO THE PAST* is clearly indicated by facts such as that both the project of the digital yuan and the crypto-currency drafts contemplated by Washington are considering backing the currency with gold, which in the case of the Chinese crypto is virtually confirmed. It is precisely the volatility of the digital environment, coupled with increased uncertainty and the climate of imminent currency wars, which gives meaning to criteria that the same experts deemed outdated until recently.

*

We are not among those who believe that economy is destiny. Important it is, but it would make only one of the axes of our coordinate system. Even if today it has a hypertrophied importance, it is only because of the vertiginous descent of our attention towards material things in the last centuries; but this reduction of the altitude continues to be a commitment of the intelligence within a relationship that it perceives as vertical.

Economism or the intrusion of economics into everything is only one aspect within a broader logic. Alexander Dugin has distinguished three logos in the crucible of Mediterranean and Western culture: the Apollonian, solar or spiritual, the materialistic or matriarchal represented by the earthly Cybele, and the vitalist mediating between both, personified by Dionysus, sun of darkness. But characteristically he forgets another divine power mediating between the High and the Low that has been much more relevant in the last centuries.

Certainly, if already in the Antiquity relations between gods were extraordinarily rich and complex, with the advent of Christianity it became even more difficult to recognize lineages, potentialities and tendencies. We continue to look at classical Greece as a diaphanous exponent of Apollo's culture, although it undoubtedly had intense contact with the other currents. Babylon became in the scriptures a synonym of material culture. In the painting of Florence the Apollonian element predominates while in Caravaggio, with its double affirmation of life and death, Dionysus rules.

Western culture, which during the long centuries of the Middle Ages accumulated an enormous reserve of spiritual energy, since the Renaissance has devoted itself to invert all that energy into the earthly sphere, something decisively evident in what Weber called «the Protestant ethic» associated with the «spirit of capitalism».

Within Protestantism, it will be the Calvinists and English puritans who will best embody this descent of the spirit into the worldly and material, and at this point it must be said that England, since its rise in the seventeenth century as a commercial and industrial power, has played in Europe, soon to become the «West», a specifically mercurial role.

We say specifically because Mercury or Hermes is a dual god, which like the human unknown in the true Manifesto of the Renaissance, Pico's *ORATION ON THE DIGNITY OF MAN*, mediates between heaven and earth; but here the mercurial activity is clearly directed downwards, to trade and industry, toward the material civilization of Cybele that reaches certain culmination in the time of Queen Victoria.

Of course, this trend towards descent and materialization did not stop then and has continued to this day, driven by new metamorphoses. The era of philosophical materialism is long past, and yet no one hesitates to consider our cultural babel as the multiform flood of materialism. With it, a tamed hedonism has gained momentum, now basically driven by the materialist logic of consumption, not really living but trying to feel alive, as Dionysus regularly officiates in decadence and assists in dissolution.

In this reduction from a culture with spiritual reserves to a neutered, castrating material civilization, Mercury's amphibious functionality played the essential role; but this neutral companion of the human kind assists us in the descent from mind to matter as the other way round. It is not only the numen of horizontal trade between equals —its more civilized appearance- but also the unnoticed trade between the high and the low in a sense that still evades the grasp of our normative, flattening sciences.

In contrast, dialectics use to move around the axis of this other trade, as the symbol of the Caduceus still manages to evoke it. While in its descent towards the cubic stone the spirit tends to be neutralized and deprived of polarity and vitality, in its ascent it uses the poles only to access that which lies beyond and before duality —the unknown common trait of mind and matter.

Dialectics lost explicit contact with the natural sciences long before Hegel, who nonetheless was the culmination of certain mercurial thought insinuated since Heraclitus. The new rationality in Nature gravitated toward the cubic stone, at least since Newton, who also devoted himself to the study of the hermetic science just when it was vanishing as a tradition; because until Lavoisier's times, Nature's science par excellence was chemistry, much more than the newly conceived mathematical physics.

Everything in the ancient hermetic art was a figure of the omnipresence of the spirit in Nature —and that spirit was will and understanding, fire and water, sulfur and mercury, sun and moon, even though they were never exhibited simultaneously, but only successively. In this sense it was a «historical science», even though it had its own criteria of analysis and synthesis which, at least by analogy, could be extended to the most diverse domains.

Marx himself attempted to define the alchemy of capital in arid algebraic terms, while gold diggers in the rivers had used mercury to separate the gold from the ore since ancient times. And this aspect is much more interesting from the point of view of circulation, which is the one that concerns us.

Mercury presides over the inherent logic of dissolution and coagulation, as well as the circulation of heterogeneous elements within a closed process that tends to represent in its own way the great world. A vessel or a hermetic process try to describe the dynamic transformations of a totality in a certain closed circuit; and this is also the cybernetic main concept, something not unnoticed by the specialists of the second generation of this discipline, around 1970.

But the great difference is that Hermes is the master of reversibility even within such manifestly irreversible, open, living processes as generation and corruption. Cybernetics aims to close a circuit between initially open systems, which would then accommodate the reversibility of dead matter. As cybernetic rhymes with hermetic, technocratic does it with necrotic.

*

The horizon of today's world cannot go beyond the cybernetic hypothesis, which, as the subtitle of Wiener's pioneering work says, is all about «control and communication in the animal and the machine», and tries to level them out on the plane of the mind, although we all see their obvious difference even at the most elementary physical level. In fact, «communication» here is the exact reverse of the concept «mechanism», only that the latter was predicated on matter's motion and the former on mental activity. We could equally use the word «commerce».

This inability to see further forces all those who do not want to be left behind, from Davos to China and from Siberia to Patagonia, to bet on the party of technocracy and efficiency, with its perpetual demand that nothing escapes it. It is about playing it safe, minimizing the risks for the control head and therefore minimizing also the reactions, the instability, the vitality of the social body that is the object of control —or channeling its extra energy in safe and profitable options.

So the Great Reset can only multiply the rations of this familiar diet. With such indigestion of titanium and graphene, teleworking, artificial intelligence and cyber-vigilance, one needs some green altruism, concern for nature and sustainability. A winning couple, since the more we deny nature and destroy it, the more we need to light a candle in its honor.

It is not the same to deny nature than to destroy it; first comes denial in the form of science, and only then the implementation of destruction in the form of technology. How deep can be our concern for nature if our theory does not know or does not want to distinguish between an animal and a machine? We are neutralized to such an extent that this no longer impresses us, even knowing that our theories were bent on execution right from the start.

When these guys express their concern about nature, people should rather express concern about their pocket, because even if they create special fees, they will be payed by everyone except those responsible. If they call it today «externalization of costs», tomorrow it will be «cost distribution», and guess to whom will be distributed. Besides, everything in the structure and dynamics of the Great Syphon makes it «natural».

In truth, the Great Syphon is not only the first political, economic and social problem, but also the main issue of social ecology, since in it we would approach the singularity faster than in any other horizon of catastrophe, which in passing are so promoted by the dominant media to make us forget the big one. There is a deeper ecology than that of the external balance of the so-called environment.

Evidently this catastrophe is not reduced only to the distribution of wealth but contains all its consequences, raising the lowest to the highest, intoxicating the collective perception and creating a parallel reality that only in some crucial but distorted aspects touch the ground.

At any rate, for the enclosing technocracy «Nature» is only synonym with resources, and «taking care of Nature» can only consist in taking over its administration, or in a word, in a barely concealed privatization. Privatization of the few things not yet fallen into private hands, such as the use of the atmosphere, freshwater, the seas or mining fields and forests, under the pretext that they cannot be left in the hands of the ignorant masses and irresponsible, «out of control» governments.

«Nature» for them is simply the Geoeconomy of resources. And if the plutarchy no longer has so much to gain in relative terms, in the face of the wild instability that its very existence provokes, tries to transform that overwhelming quantitative advantage into an increasingly iron consolidation of the control of the material aspects they call «natural», and for which it should find a powerful excuse.

That is why the old liberal propaganda is no longer relevant and tends to be replaced by the new jargon of responsibility and sustainability, in a conservative turn towards self-restraint. Yes, they are the masters of self-restraint and self-containment. With so much more to lose than to gain, the essential thing is not to lose all the advantage gained, that should be transmuted into more favorable conditions —for the main cause of the imbalance.

Needless to say that in the grossly quantitative sense, a singularity in the accumulation of wealth is a chimera like any other mathematical singularity. The real thing is to transform and maximize this domain in qualitative terms, negotiating between strain and stress the obedience and stability, freedom and security of the masses.

*

We are the great fundamentalists of the irreversible, and it is not for anything else that we have come so far; and yet the fundamental laws of physics to which we attribute the operation of nature are based on a preceptive reversibility. Is it not strange?

Scientific positivism, which is not science itself, has ended up reducing its idea of nature to that framed by prediction, prediction that plays the same role as profit in capitalism. And this, coming out from our practices rather than our theory, ended up having an enormous impact on what we are willing to contemplate and ignore in our relationship with nature and on the limits that techno-science outlines on our society.

What does it mean that the constant accumulation of scientific knowledge, which we perceive as the irreversible engine of progress running parallel to the accumulation of capital, is founded on the idea of a reversible, lifeless Law?

We will never get to the bottom of this question if we do not manage to reverse the terms.

For the rest, we all know, scientists and laymen alike, that science is a gigantic Procrustes bed that cuts and amputates everything that does not conform to the standards. The question is, how can we get a reasonable idea of what has been left out without reducing it to our current standards? What has to do with the knowledge we now manage? Is it just a series of useless wastes or does it hold the key and the context to understand what now only seems to be a series of increasingly unintelligible enigmas?

Both classical and quantum mechanics are reversible but the irreversibility of thermodynamics and the ordinary processes we observe is seen only as a macroscopic accident, just another illusion. That the most basic property that we appreciate in the real world is characterized as an epiphenomenon in statistical mechanics and its daughter information theory already speaks to us of an inverted criterion and priorities —of the primacy of prediction over description.

On the contrary, what is really interesting, in the most apparent processes no less than in human time and his biology, is to find out how islands of reversibility emerge from an irreversible background, and closed circuits within open systems, not the other way around. This would be the magic ring of Nature, which does not admit comparison with the cuckoo clock of the Law. The day we understand this, we will have overcome the fundamentalism of linear and cumulative time that is the basic assumption of our society.

In a book entitled *Pole of Inspiration —Mathematics, Science and Tradition*, we have pointed out how what is understood today as feedback, the key concept to cybernetics, was already present in old Kepler's problem, in which Newton fitted his theory of gravitation. The same Kepler's problem has already the key to gauge field theories like electromagnetism or the other forces that govern the atom —these also involve a feedback regardless of the mechanism.

Well, this would have to be the least mysterious matter since variational principles were always a teleological expedient, something that conservative scientists like Planck admitted; but most believed that their use was so innocuous, that it was not even worthy of attention. Certainly, for an outstanding mathematician like Wiener this would have been a complete surprise had he ever been lucky enough to notice it.

Of course, theoretical physicists still wonder how the Moon knows where the Sun is and how it «knows» its mass to behave the way it does; and they call this the great riddle of identifying the concrete mechanism of gravity, or its quantization in physicists jargon.

Needless to say that it is not really a question of identifying the mechanism, or, as it is also said, deciphering the form of communication between distant bodies —since a variational problem, whether on a small or large scale, is by definition independent of its mechanism. The Lagrangian of a system is only an exact analogy, and already since Newton we knew that nothing was mechanical, at least up there; but the mere fact of including which is below and which is above in the same Three Principles of Dynamics gave rise to a formidable mirage.

One could say that if we can recognize a feedback in the orbit of a planet now, it is only retrospectively because of the omnipresent influence of cybernetics, but then why no specialist in cybernetics noticed it? Not even physicists, since the times of Lagrange? We must assume that *because it was not useful to the task at hand*, as the planets were already revolving in closed orbits after all, and since calculations and predictions could be made, we could only think that it was a kind of mechanism.

Those whom the gods want lost, they make blind; and in our time men got blind by the way of predictions.

No, the reversibility we are speaking of is not as obvious as the human habit of calling old things by the latest names. On the contrary, one could rather assume that if physicists had realized this in their time, in the century from Newton to Lagrange, cybernetics would probably never have been created, because there would have been no need to create an artificial bridge between living beings and machines.

*

The need for brevity prevents us from going deeper into the subject and we can only refer the reader to the aforementioned book and its obligatory bibliography. As we have already indicated, the true interest of this can only begin to be grasped by trying to see how reversible and closed systems, that we call «mechanical», emerge from an unlimited and homogeneous background — totally reversing the approach that has become conventional since Newton.

It may seem an impossible task to reverse the scientific development of the last four hundred years, but no one intends such a thing. At the institutional level, and despite their vain appeals to originality, current sciences, with their enormous bureaucratic inertia, conform to what power dictates and this has no remedy. But for the individual intelligence it is perfectly possible to turn the glove inside out, and this is what matters, because if one can understand it, so can many others.

There are some very basic guidelines for scientific understanding to come back to life. We are referring to both the theoretical and practical horizon —to principles, means and ends. These basic lines touch directly the principles of mechanics, the definition of equilibrium in different types of systems, the foundations of calculus or analysis, the critical relationship between description and prediction and the no less critical relationship between closed and open systems, reversibility and irreversibility.

Interpretation, which is the end of science as such, is also the principle of technique and technology; so there is also an unnoticed ring in the logic of Technoscience, which so far could not be properly explored because the revision of foundations endangers the monotonous accumulation of knowledge in the various specialties.

*

The forms of knowledge are infinite and so are their combinations, which does not prevent strong redundancies. Far from being on the verge of finding a «theory of everything», Western science has only explored an infinitesimal part of the possibilities of knowledge, even within its own very particular forms, but on the other hand those forms are subject in practice to a rapid evolution, aging and death.

The cybernetic paradigm that dominates us today, no one will doubt it, is the ultimate expression of instrumental reason; but at least since the Newtonian natural philosophy was assumed, and since description was subordinated to prediction, the path of descent was clear.

It is not difficult to change the principles of mechanics; to replace, for instance, the principle of inertia with the principle of dynamic equilibrium, and to preserve the bulk of predictions that physics treasures but changing their meaning, context and interpretation completely. However it is the real inertia of institutions, and of the limited scope of our knowledge, that makes this seem unfeasible.

In industry there can be reversible and irreversible standards. An example of a standard that has been shown to be irreversible is the keyboard layout, which was expressly designed so that the typewriter would not be too fast and the physical letters do not cross their paths; today this limitation makes no sense on the computer, but nobody could change it.

The same thing, and with much stronger reasons, occurs in the sciences, and it is not for nothing that we talk today of a «standard model» in particle physics or cosmology; it is actually the accepted framework of formalisms to make possible calculations and predictions. There are nod doubt good reasons for using these frameworks, but they are still circumstantial —to an extent that we cannot even imagine.

The whole cybernetic reflex is an attempt to close over the open to enslave it; it is therefore not gratuitous to identify it with the death drive. It is also congruent with the interest of a tiny but massively concentrated head that has little left to gain and prefers to clench its fist. But what exactly would be the effect on the cybernetic model of the opposite hypothesis, so extremely plausible, that everything that seems closed and reversible comes from something open and irreversible? Would it contribute to its collapse, or to its consolidation?

The individual evolution of an organism between birth and death tells us in the clearest way that aging is a process of gradual closure and hardening until the break down. And the hardening depends on the inability to eliminate, which in itself is a form of increasing restriction or closure.

If science has not incorporated other possibilities, it is because they were not convenient to their goals; hardly it can try to assimilate them while remaining what it has been until now. Therefore, any development in this sense comes too late for current science as we know it; this could be viable only in a different environment.

However, here we can perceive a common axis that connects our dying logos with others to come.

*

Within environmentalism there is much talk about a new civilization and a new rationality, but their intellectual coordinates continue to be those of the liberal-materialism of the nineteenth century, updated by perspectives such as those of systems theory. That is, basically the same logos of this twilight civilization, but with an ethic opposed to consumerism and in favor of technological disarmament.

Ecologism is an economism with an awareness of limits and scarcity, but economics has always been the science of scarcity, whether natural or man-made, and in the end it is the latter that always prevails in social perception. On the other hand, the administrative jargon of systems theory is employed but the hierarchy that this ubiquitous funneling system we call the Great Syphon imposes on the economy, perception and opinion-making is completely ignored. Just a slight neglect.

A material civilization can make sense only if the idea of balance is duly observed. Babylon could be a material civilization but traits as the periodic cancellation of debts allowed the imbalances not to accumulate. An example of a great material civilization of long duration has been China, where precisely the idea of balance or harmony was so fundamental at all levels. However the attention to balance in processes does not refer only to the physical needs of men, but also to the compensation of material considerations; if this occurs, matter does not exclude the spirit, nor the immanence the transcendence.

But things are very different with the rationality of the modern world, which excludes both the immanence or transcendence of experience, not to speak of its transcendental character, and can only aspire to manipulate and denaturalize it. Thus, everything in our dual liberal-materialism is already the product of an extreme imbalance that nevertheless seeks to perpetuate and aggrandize itself.

Many still wish to believe that modern science not only aspires to rule the world, but also to know it. The Greek idea of the logos, based on geometry, was purely descriptive and prediction was alien to it. But the modern idea of Natural Law is entirely founded on the idea of prediction or calculation in time, and to ask it for a fair description would be like putting sticks in the wheels.

Calculus changed completely our idea of what analysis is. Instead of determining geometry from physical considerations, deriving from it the differential equation, since Leibniz and Newton the differential equation is set up first and then one tries to solve it to get the physical answers. Both methods are far from equivalent, but the same belief in the reality of the differentials follows from the procedure adopted. It is still necessary to reverse this method in order to open the eyes and recover the right perspective.

We use the word «reversing» bearing in mind that what modern science has generally done is to reverse the previous idea of rationality, literally turning many things upside down in order to save prediction. But we have become so used to these procedures, and they are so justified for a definite purpose, that we fail to conceive others.

Of course, even within the absolute prevalence of the predictive, the thickness of the abstract Law in relation to the world of becoming and observable forms still seems quite feeble. Thus, the «hard» sciences, like physics, must be accompanied by a descriptive supplement to fill the abyss between the imaginary quality of our representations and the symbolic or normative value of the Law. Hence disciplines such as cosmology or the theory of evolution, with no predictive value but essential to maintain the believe that science has explanatory power.

However, it is easy to see that these descriptive disciplines are only an ideological supplement to amplify the range of application of the hard sciences to their maximum extent: cosmology to extend the value of our planet's laws to the entire universe, or evolution to reduce all aspects of biology to a potentially manipulable genetic drift.

So, in science there is a double circuit and a double circulation, in a very similar fashion as in our monetary system there is a double circuit of currency, with a legal money issued by the central banks and a much wider imaginary circulation dependent on credit. This is not a vain analogy; although the degree of «stretching» of legal money in science, and the credit we grant it, is incomparably higher. We cannot yet measure the value of gravity on our own planet to four decimal places, but we pretend that general relativity can make calculations to 11 decimal places in black holes and distant galaxies.

This is called speculation, and much of theoretical science is guided by a speculative zeal as intense as that of the financial world, with the great advantage that here there is hardly any possibility of experimental control. Of course, this is all too general. If the legal currency is about 5 percent and the rest is counterfeit or imaginary money, «debt,» the same can be said of practically everything that circulates in the media, ideologies, and all spheres of society —a huge mountain of counterfeit money that nevertheless serves to channel people's efforts and illusions.

Prediction is justified insofar as a law, by definition, expresses a regularity, and this too, by definition, must be predictable. But if prediction is not leveled with description, we cannot even know what the law refers to beyond the prediction itself, as in fact is the case with the law of gravitation and any other. This is how positivism works.

Ideally, description and prediction should be reciprocal, like the memory and anticipation with which we continuously and reflectively create our perception of time. But today, to demand that they go hand in hand may seem a sabotage; something as unreasonable as to ask that imaginary money did not exist, being the bulk of the total. However, this balance defines the plane in which science, technology and men work together.

Whether we find it inconvenient at this point or not, all the real life and the internal logic of scientific evolution, its veracity and its degree of viability over time, depend on it. These are the secondary and tertiary sectors of its economy, and the primary one, the quality and reality of the phenomena and their inclusion in the principle, is too broad to deal with here. Needless to say, here also the tertiary sector has been hypertrophied without the slightest consideration for the most basic requirements.

*

In the second half of the 20th century, just about the same time that the standard models of particle physics and cosmology were emerging, a range of theoretical attempts to deal with complexity flourished, with a vocation for universality and diverse fortune: first and second-order cybernetics, systems theory, catastrophe theory, dissipative systems and non-equilibrium thermodynamics, non-linear dynamic systems and deterministic chaos, spontaneous order and self-organization, computer science and algorithmic theories of everything, artificial intelligence, digital neo-Darwinism and then some.

All of them were postulated both as new disciplines as well as an interdisciplinary horizon; so they avoided questioning the achievements of the oldest sciences and each one tried to approach in its own way all the density of the real world out of the reach of the great «fundamental» reversible laws. If in fundamental physics Wigner exposed his amazement about the «unreasonable effectiveness of mathematics», in fields such as biology, epitome of complexity, these exhibited, according to Gelfand, an «unreasonable ineffectiveness».

Both were wrong, to say the least. The effectiveness of mathematics in physics is anything but unreasonable since what has been done since Newton is a reverse engineering assigning the variables to arrive at the known results, then generalizing the equation and finally declaring it universal. On the other hand, if they intended to apply the same method to biology, the ineffectiveness of mathematics in this area could not be less unreasonable either.

Although there was and there is a lot of space in between, complexity theorists have never succeeded in seeing it because it seemed suicidal to question the application of mathematics to physics in view of its long history of successes and the predictive power of its methods. But it is always power what clouds reason.

Despite the gigantic advances in computing power and how appealing the unlimited panorama was, these theorists of complexity were left hungering for the big bite out of the apple. On the contrary, what we are witnessing is the dizzying rise of a neo-Babylonian ziggurat very much in tune with the ubiquitous proliferation of disorder.

The most complex computers have yet been unable to match the flight skills of a tiny fruit fly, but some talk about the use of AI for the governance of the world. It is hard to tell if this serves more to self-deception, to disaster or both.

In order to see further, we always go backwards, and here as in everything, the first thing that is needed is to question the most fundamental in science, no matter how important its successes may have been. That modern science has an important part of truth is out of the question, otherwise we would not even waste time with it. But what is interesting now is to see the relationship between what has been discarded by fundamental science and what escapes the sciences of complexity. And there is a whole world to explore here.

Differential calculus, and with it the whole application of mathematics to change, has vacillated between idealization and rationalization —between the idea of infinitesimal and the concept of limit. These are the clashing Simplegades rocks against which all ships crash in their exploration of the infinite, and when they are not destroyed, still dramatically limit their field of vision. Yet the middle way of finite differences has only been explored as a mere auxiliary for known results, rather than as the true foundation of the method. This middle way is the axis of reversibility of a development until today irreversible, that of Analysis, which still does not truly honor its name.

The claim that science as today we know it has its foundations consolidated and can only grow upwards is simply ridiculous, and yet it is extremely difficult to change since in it the layers are inevitably stacked both on a theoretical and a sociological level. And yet it is perhaps not impossible that here too the future will open the door to the past, if one is able to understand that only on this axis there is still freedom of movement.

Science can and should be more than a ziggurat.

*

Of course, science is more an effect than a cause. Scientists like to think that they are the ones who really change the world, and no doubt they have contributed a lot to it, but even then all they do is to conquer new domains of application for the prevailing mentality. The religion of prediction is the knowledge of the slave —a conquering slave for that matter.

It is said that we have already overcome the era of the subject and that this has happened for the benefit of, for example, communication. According to this, the subject was an obstacle and, once superseded, «progress» can continue; this would be precisely the assumption of communication platforms like Davos.

However, what they are proposing is not progressive in the least, and less and less people allow themselves to be fooled by the supposed vertigo of technological advances and the industrial production of false novelty. On the contrary, today the only possible progress begins with the realization that the kind of technological acceleration being promoted is neither necessary nor irreversible, and only at that point the subject find again time and space for decisions.

But we are not talking about an impossible return to the past, nor about the recycling or opportunistic combination of new and old things that characterizes postmodernity. The balance between past and future, between memory and anticipation, is the normal condition of reflective consciousness; it is modernity that has broken this balance in the most violent way in favor of anticipation, imposing an idea of progress that can only be unilateral.

Now, this unilateral development, as Hegel could have say, leaves an enormous gap open behind no less than forward. The gold-backed currencies seemed yesterday to be a thing of the past; but the same volatility of digital currencies is pushing its readoption. Needless to say, more decisive than the golden metal itself is the change of orientation in time.

It is a perfect symbol of the moment, whose magnitude is too broad to grasp —and which comes just when the megaphone system repeats to us with the insistence of a fair hypnotist that there is no alternative and no way back. A symbol that coincides and finds resonance with the first clear signs of turning points in other great processes, starting with Globalization itself.

At the same time, awareness of the increasing fragility of this civilization, which relies on technological solutions, is growing sharply; we felt it, and now we even see it clearly: a child will finally come and destroy it, or even something way smaller than a child. Either we keep our backs or we are finished; and the same is true for so many «social conquests» that have been bribed or kidnapped and depend critically on everything going in one direction. We are moving in deep water.

Indeed, if it is really a matter of starting from scratch as they say, and they care so much about the economy, they would have to start by canceling the whole mountain of debt and getting the big looters who have profited from all sorts of scams and wars to give back their loot, and especially the usurped control of the monetary system; but it seems this is not what they are talking about.

Nothing is totally irreversible except death, and those who try by all means to convince us of the irreversibility of any change whatever can only lead us to death, whose party they represent. No one is less interested than they in the fact that consciousness and history can turn on itself.

*

The cybernetic compulsion operates top-down and wants to exhaustively conform the parts from the control of the whole; it is totalitarianism in plain functionality. But we should consider the opposite strategy, which says that in an organism we can always proceed from the apparent effect to a cause that is perhaps no less apparent, and that modifying a portion of the behavior can have far-reaching effects on the perception of the whole. If the tail wags the dog, much better head and tail can put it straight.

it is not ideas that determine our actions; it is what we do and what we want to do what determine our ideas. Used to a direction on autopilot, each line of action that is reversed induce all the others. And when it concerns the unnoticed axis of our trade with reality, it can ignite unhindered like an oil-soaked wick.

Cybernetics and artificial intelligence progress in their closure and enslavement of the system by perception and action cycles. Today our administrators believe they have enough control over the media, money and the executive levers to define these cycles; however, the entire dynamic of material-liberalism or liberal materialism is committed to the descent of the intellect, not its ascent —what is at the root of the extreme imbalance of modernity.

For today's technoscience created in the image and likeness of liberalism, the fact that intelligence is totally separated from nature is an indispensable condition for recombining all her aspects at will: atoms, machines, biological molecules and genes, or the interface between any of them under the least restrictive criterion of information.

Thus, the totality of the cybernetic reason never ceases to be an arbitrary aggregate or collection of elements that attempt to reproduce observable results. The bottom-up movement is the composition or integration of the whole from ideal elements, and the downward movement is the decomposition of the whole into similar parts. It is the same logic of mathematical calculus or analysis, in which synthesis is essential but heuristically subordinated and oriented to results. But it is easy to show that this alleged analysis turns everything upside down and that far from being a neutral foundation it is, not less than in the case of cybernetics, a mere justification of heuristics.

Putting straight what now is upside down will not fail to produce results that are difficult to imagine. This is not a philosophical criticism, but an operational redefinition of principles, means and interpretations. And when we realize that there is an enormous unoccupied space, and every attempt to prevent its exploration can only be encouraging.

The present business of science, like our monetary system and so many other arrangements, is totally oriented towards definite ends and its so-called «pragmatism» ignores the most elemental neutrality. Its objectivity is only a very particular case of objectification. And this strong partiality, which is also at the origin of its enormous vitality in the past, is the same that now slows down its progress.

*

The West owes a rehabilitation to Nature in its very concept and in the very bosom of science and consciousness; as it owes it to itself and to the truth. Perhaps it finds itself too decrepit, or too enslaved by its achievements, to undertake a reform of such depth as it needs even to ascertain that it is alive.

But the world is not only the West, even Europe was not except in this twilight phase, and if people in Europe or North America cannot free themselves from their own yoke, other cultures or civilizations will see the need to consciously use time as the two-edged sword it is —in China, Japan, India, Russia, Iran, Palestine, the Southern Hemisphere or just about everywhere.

Of course, today the first problem for any new culture is how to thrive in the context of a global civilization with such overwhelming technological deployment that cannot leave anything undisturbed. A civilization so comprehensive in space and time that it even seems to exclude any other possibility.

At this juncture, there are two great prospects: either the collapse of globalization is so complete that even our technologies become extinct, and then the spirit is once again free to evolve unrestricted, or the cultures to come end up taking over its logic and overcoming and reorienting it for a completely different purpose. So far that has not happened, but what we are talking about is precisely how that can begin to happen.

And yet, even that overcoming can only happen with a reflexive turn that in many ways is equivalent to a dissolution, but still avoids a traumatic destruction. To turn to itself is not to go back, in fact it is the only non-destructive way of understanding progress.

*

The way in which we understand Nature and our relationship with it will ultimately be much more decisive than any kind of partial measures to «stop the catastrophe» undertaken from the lower order logic of Geoeconomics and its skewed models of analysis.

I have insisted time and again on the importance of freeing Nature from the limited rationality in which we have constrained it. Many unacceptable definitions have been accepted for the mere fact that the intellect was flattered by the feeling of dominance over it.

Today some people laugh at an invented belief in a flat earth, when the truth is that questions of that order could only be entirely secondary to our ancestors; but the fantastic speculative flight of natural sciences just hides the universal flattening of our reality to the domain of efficient causes.

To reduce everything to motion, an unattainable ideal of modern physics, would be the consummation of nihilism, but not because of motion itself, which is as pure an expression of the spirit as any other, but because of the spirit of reduction which for the nihilist must go unnoticed.

I have also insisted on the close reciprocity between liberating our vision of «external» nature and liberating the nature locked in the social construction and within us. Both are spirit within spirit, but not in the way that the mere intellect use to arrogate for itself.

Spirit is the substantial union of will and understanding; and in a very real sense we can say that the will is the innermost of things, «the thing-in-itself», while all representations of the intellect, however elaborate and consistent they may be, remain forever as pure exteriority. Schopenhauer proposed this irreconcilable dualism, but curiously the will he spoke of was also inherently dual, composed of a feminine part, made of appetite, movement and flight —the desire, the imaginary- and a masculine or energetic part made of friction and effort, which wants independently of desire and which we properly call will. Both are like water and fire, or the mercury and sulfur of the hermetic arcane.

Capitalism, like the Satan Trismegistus of Baudelaire, lives evaporating the sulfur of our will by means of desire to recirculate it and infuse it again in the washed metal, thickened by the effort of work, and reaching its maximum concentration in gold, which is nothing else than mercury condensed with sulfur, the true mineral blood. So, the very social circulation of value would be the inversion of the hermetic hypothesis for the natural world; an inversion that seems to correspond well to the logic of the inverted pyramid of wealth accumulation.

This analogy allows us to reconcile Hegel and Schopenhauer in a hermetic key, but the inversion that society implies with respect to nature also entails a commerce between the big and the small; society itself becomes the microcosm of a great world to which it has turned its back, and to which it nevertheless intends to ride.

One can think what one wants of Schopenhauer's certainly limited approach, but the truth is that to think of nature as a series of representations of processes and not to recognize it in our own impulses is an extreme naivety as well as the most finished product of a long dissociation.

Anyone would say that mathematical physics, our highly formalized way of understanding the foundations of the cosmos, has nothing to do with the hermetic dualism or that which results from the crossroads of the great romantic philosophers. However, if in the hermetic key matter is composed of both volatile and a fixed modalities, of something that moves and something that does not, from its very beginning physics finds a crucial and unresolved distinction between scalar quantities like mass and vector quantities like force, between intensive and extensive properties, between matter separated by space, as in electricity, and space separated by matter, as in magnetism.

Beyond dimensional analysis, physics itself hardly raises the extremely complex question of an exhaustive genealogy of its quantities, since in reality they cannot be more heterogeneous and their adoption has always resulted from the resolution of concrete problems, constituting in itself a monumental tower of Babel. It is also worth mentioning that dimensional analysis, today a modest auxiliary tool, can yield completely different and unexpected results when the main analysis is soundly based on finite differences.

*

The individual has been the force-idea of liberalism and it has been liberalism itself, now fading, which has inflated it so much that it has become emptied of content. No matter what the grammar says, today «individual» is more an adjective than a noun, if not the object of the verb «to personalize».

It is not the individual, but the process of individuation, what matters. It is the individuation that makes us feel individual at any moment, and the one who sticks to it can be free from fear. Today, on the contrary, the idea of the individual, from which they still try to extract return by exploiting all the resources of fear and desire, is more exhausted than full.

After the collapse of the I as ordering principle, it seems that we can only understand the individual as singularity, which is itself a theory of collapse. In mathematics and physics, a singularity is the way in which a map or manyfold degenerates, ceasing to be differentiable or giving rise to infinite results.

It is still significant how cosmology has spent more than two generations studying theoretical objects such as black holes that should be seen, more than anything else, as a complete failure of reason; even if this kind of singularity is just another exponent of how a reversible law gives rise to an irreversible process by virtue of its own absolute character.

It is also curious how, instead of trying to see why this type of behavior is impossible, the theorists, not without great resistance at first, finally decided to dive into the funnel of impossibility by stretching their gullibility to Neverland.

Such a singularity is an infinity surrounded by the «normal behavior» of the law. The individuation of a real entity, on the contrary, would have to describe how infinity momentarily sustains the existence and evolution of a finite and measurable being —a particle, an atom, a body or a celestial system.

But for today's physics this is impossible because it has not even been able to go beyond point particles. Special relativity forbids particles with dimensions, while for general relativity point particles do not make sense. But in general, for the fundamental physics of laws, the formation of any body with real size and dimensions, like celestial bodies, needs to assume the conclusion.

Hertz rightly insisted that the Third Principle could not be verified in Kepler's problem because in it there is no point of application for forces. But the Third Principle, understood as simultaneity of action and reaction, is the Global Synchronizer of mechanics, and neither special or general relativity nor quantum mechanics cease to be inserted in its absolute, and therefore metaphysical arrangement of time.

This Global Synchronizer is the supreme arbiter of efficient causality in physics. If it limps, even for the most elementary two-body system as in Kepler's problem, there must be a «proper time», that unlike the proper time of relativity implies an internal principle of self-organization. Inadvertently, the gauge fields that describe the fundamental forces include it, though not explicitly. This principle of self-organization, which we already pointed out in passing, is a particular case of the principle of individuation. And it is the feedback that is already inscribed in the purely relational form of the «law» that precludes the existence of physical singularities.

It is the same logic of modern physics that prevents the proper description of the formation or individuation of entities, or improperly breaks down them in bits of information. And mathematical analysis does the rest, because every individual entity must exist between the infinite and the infinitesimal, but in standard analysis finite differences are accidental.

Thus it is possible to see the mixture of idealization and rationalization with which material-liberalism has ended up dissolving both mind and matter, without clarifying anything about their relationships.

*

If we have not yet succeeded in seeing the obvious, it must be because we have been focused on other objectives —on prediction and manipulation rather than understanding. And of course, on the insistence on making an object of nature, which has to end up turning us into an object.

Today it is people of Jewish origin who overwhelmingly dominate the theoretical discourse in fundamental sciences, and surely this is not a coincidence, but one cannot blame them for the kind of science we have; a science whose spirit really has not changed since Newton, the warden of the Royal Mint in the early days of the Bank of England.

And so we continue with a total dissociation between the spirit of the Law and the way of the forms and becoming in nature; and this dissociation is the basic condition for the disintegration of natural relations which is the goal of liberal-materialism. If liberalism is in a declining phase, this is always compensated by an increase in rationalization, which now acquires strong technocratic-cybernetic tones; the aim is always to dispose of the whole without hindrance.

The gradual transition between liberalism and technocracy still allows to play the card of transhumanism as a vanishing point for the desires of life extension and simulacra of transcendence while the potential for technological transformation is further exploited. And this cyborganic transhumanism can thus pursue and intensify the war against nature both outside and within ourselves, pushing the malleability of the human to new limits.

*

As Israel Shamir says, that Jewry exist and Jewish politics do exist is something obvious for Jews but a forbidden thought for non-Jews —which already says enough about who control opinion. It is not something I am going to denounce here, but some comments will be useful for those who still want to understand.

It was sort of a commonplace, rather forgotten today, that the movement of progress can be understood in the Hegelian key of the being-in-itself, being out of itself and being for itself; and in few cases this dialectic becomes so explicit as in the narrative of the self-called Chosen People, with the drama of the Diaspora and the return to the Promised Land, which should be culminated by the coming of the Messiah. Of course, this messianic culmination, as their scriptures repeat over and over again, would put Israel above all nations, who would be forced to recognize the Lord's people and the Lord of this people.

It is also elementary to equate the people in exile with the alienated spirit of nature, or more generally with a society that can only be constituted insofar as it has to turn its back to nature despite continuing to live off it. Here it is easy to appreciate different levels of discourse, although they all contemplate a horizon of reconciliation.

It should be clear that what must be recognized is the presence of the subject or spirit in nature, and that this coincides with the re-entry of the consciousness of nature into the heart of a society that was thought to be separated from it. It is clear that Judaic theology and teleology rest on the negation of nature, which is only the object of the Law; but not only they, all modern science, though coming from the Greek heritage, has grown under the sign of the same negation.

It is highly probable that Western science would never have taken off as it did without the detachment of the subject and the reduction of nature to the Law; but its culmination can only happen by returning to itself in the sense that we have indicated, from the laws to the descriptions and from these to the single principle in which the phenomena unfold.

The sense of natural law that has prevailed since Newton is pragmatic and external, and therefore superficial. It was the degree of knowledge that could be expected from a culture oriented towards world domination. For the Jewish consciousness, a people within peoples, it was a matter of mastering the dominance and controlling the controller, not of getting to the bottom of this new natural «Law».

Western science has been so far vain and hollow enough to be satisfied with a mastery of the object that relies largely on the rejection of its entity. It has wanted to ride nature, and since it has been content with this, it is not only poetic justice but also pure logic that the rider ends up becoming a beast of burden for a mentality that since the beginning has rejected nature, and not in part but in its entirety.

Messianic teleology in general, whether Jewish, Christian, or Marxist, not to mention the capitalist apocalypse-spectacle, cannot fail to function under the logic of «the worse, the better,» since it requires a worsening of conditions to a critical point for an intervention of another order to become desirable. Only an old Christian could say «the worse for me, the better for me»; but what any political animal thinks is «the worse for them, the better for us».

Playing together the transhumanist and technocratic cards enables to continue exploiting all the possibilities of recombination of nature while humans lose the little north they have left. It also serves to redirect the desires of the public towards trivial objectives, degrading their criteria at every step, while they unknowingly finance the projects and lend themselves to being the experimental pool and database. This continues to present a vanishing point for the liberal imaginary as the means of control become more ubiquitous and invasive.

So, the transhumanist escape would be a controlled destruction and a slow-motion apocalypse that favors the ultimate intervention of a higher order: it is at the service of the most fearful and definitive reaction. The presentation of other alleged planetary crises is also calculated to justify this kind of intervention from above, only in this case the consummation of History would pass through the open moral recognition of the Chosen People as savior. The most tortuous routes would serve so that in the end everyone would walk straight before the Lord.

*

A faith can be absurd and even be at the service of a lie and yet serve very well to a vital interest; on the contrary, if they are not close to life, neither reason nor truth itself are of much use to us.

The unconsciousness that tears ends up finding the consciousness torn. Does the 50/50 share of the American oligarchy have anything to do with the present social divide of the indispensable nation? This fission does not seem to be primarily a matter between Jews or non-Jews but a struggle between the interests of the nation and those of the Empire; however both issues tend to coincide closely. The current Republican misgovernment has made the most embarrassing concessions to the State of Israel in the hope of having its hands free to disengage from the global liberal agenda in return; but in vain.

It is evident that the Federal Reserve is not at the service of the United States, not even of the stock market, but of monetary hegemony and global private banking. The moves to find an alternative to the dollar within the same priorities of the plutocracy are quite eloquent in this sense, although its viability is a complete unknown. Without a doubt, a democratic misgovernment would be preferred in order to undertake the transition, although this is always secondary. That the main donors in the campaigns of both parties are vehement Zionists is not essential either, but telling it is.

The problem is that serving hegemony and plutarchy destroys the country, and even with all the rivers of disinformation it is inevitable that more and more people understand it. This cannot help but cause an increasingly violent division at all levels, even if none of the candidates can do anything about it, and eventually a house divided to such extremes will not be able to sustain the dollar.

For its part, the American Jewish power is far from satisfied with dictating American policy in the Middle East; it would be foolish to do so with such complacent partners. History tells us that Jews have ended up badly in many ancient empires and countries, not without bringing chaos and destruction upon their hosts. It is inevitable to think that this has to do with a invincible propensity to go too far and an inflexible tendency to always use all the advantage gained, no matter how great, to obtain more. It would be another irreversible dynamic where the fear and audacity of those with advantage are fed back.

The infamous «aircraft carrier theory,» which assures us that Israel is just an instrument of convenience for American imperialism —a sort of aircraft carrier to control the oil tap- did not stand up to the slightest scrutiny even fifty years ago; but after what we have seen in the last two decades, I do not think there is anyone who dares to defend it in public.

A nation that depends on its monetary hegemony tends to be eviscerated and torn apart without remedy; and on the other hand the dollar hegemony cannot be maintained without a minimum consensus. Clearly, it would be infinitely better for the United States to end the Fed and create a currency with a simple, fair system conceived for the country's needs; but the fact that this so basic decision seems impossible already says it all.

*

There is in nature a will that scarcely has anything to do with the sick man's will of will, and there is in nature an intelligence that barely has anything to do with the dissociated human intellect; we may suppose that in reality we are dealing with an infinite number of impulses and reactions, but their relative unity is derived from what they are not, and which even in us we do not recognize.

Trying to know that will and that intelligence on its own terms, even with our formal tools, which are still shadows, has a virtue that would have to resurrect even the dead, for it is nothing other than the spirit of life, which is beyond life and death.

Science also ends up meeting the three great questions: why does something exist, why does life exist, and why does consciousness exist. It is more than doubtful that there can be an answer to any of the three questions separately; but if all three coincide, the fact that there is no answer does not even demand a question.

All that modern science perceives as natural Law is just a shadow that only gain new grades of life and understanding by its restitution to the Principle, instead of trying to push the Law toward impossible horizons. But, besides, if laws can be subsumed into principles, this would also mean that, always within certain limits, we can redirect the mediations of technology and society toward simplicity. Our law is hard and the principle is infinitely ductile; but our laws are extremely tortuous while the principle is one of unfathomable straightness. A different, ever present horizon.

*

The insistence by the powerful on the irreversibility of the great technocratic transformation when the signs of its exhaustion are so clear can only produce a reaction in the opposite direction. But this is not just a reactive phenomenon. Insisting that great processes can be creatively reversed gives us an awareness of additional degrees of freedom and this is absolutely vital when the rest of the space is occupied. Those who hold today the vital structures could only access that space by destroying, «cleaning» their own work.

We can always put freedom and security, justice and truth two opposite two; but the selection of their contents and the way they have been opposed today depends to a great extent on «who puts the money» and on the sedimentation of interests during this long period of monetary usurpation. This cross summarizes the real existential crisis ahead, so different from a mere technical crisis.

These three we have: the mind, the matter and the metal, which is matter but in its deadest, more closed form serves to reflect light and see ourselves while eclipsing the secret of its origin. And whoever says the metal says money or the mint, which reproduces an analogous process in social exchange and collective intelligence.

We have barely suggested how reversibility can affect aspects such as monetary policy, science, individuation, or our sense of time and «progress». These are absolutely central points in a civilization, ours, governed by circulation, but in which circulation continues to serve a tendency towards irreversible accumulation, and where that accumulation in turn seeks to transform its quantitative advantage into a qualitative dominance in all orders, a process that can well be called inversion. It is also an investment in cultural influence and hegemony, but primarily an inversion in the general dynamics of social circulation, beginning with the direction of money, from its ascent by extraction to its descent transforming the subjects.

There cannot be singularity here either, but the simulacrum of singularity or emergency plays a leading role and while a change of sign and an exchange of influences takes place, from the material to the mental, and vice versa, arguments for transvaluation accumulate. Still there is no transmutation, but a continuous transformation. And if it is true that this circuit is not perfectly closed, and never can reach that point, also it is true that it works with considerable efficiency.

All this could only work with a strong believe in the irreversible development of history which is equivalent to pull tight the blindfold and surrender without resistance to the double game of growing restriction.

On the contrary, to understand our active and irreplaceable role in the mutual constitution of the past and the future is to access our share of freedom and creativity in the midst of the indomitable current of events. It is precisely because others want to control it that the creative and free element must always elude them, even if they do not fail to grasp it in its reflection.

This would be the axis of the process of individuation for any entity with a recognizable profile in time on an individual or collective level, from living beings to nations, from scientific theories to technological products, from the history of an idea to the idea of history.

It was a common idea of the ancients that the sum of good and evil through time always remains the same. This necessarily made history a secondary phenomenon. There is no doubt that material success can be accompanied by waning intellectual and vital horizons, in biology as well as in biography or the evolution of cultures.

Of course, matter and spirit don't exclude each other, so this should not authorize indifference. There have been phases of material prosperity and transindividual expansion, and there have been times calamitous in all areas, of which it is even easier to find examples.

What then is in the balance in that imperturbable sum across the events and fortune? Does it suggest a double sense of the empty and the full, of what we express and what we keep in reserve? Perhaps there is a sum that remains invariable, and a product that does not need to be so at all. This balance would be linked to the exchange of momentum between the mental and material poles.

The tree of life is not the tree of good and evil; the latter depends on us, but we depend on the former even though we do not see it, and even with all our predictions we cannot stand apart from it. So, let us say there is a vital truth that does not depend on calculation and is certainly the most important.

If the indestructible axis in which reversibility and irreversibility so mysteriously coincide were free of any obstruction, freedom would be unlimited but would also be indeterminate because of the lack of restrictions. As in every human endeavor, even without realizing it, one explores the practical limits of their interaction.

There is wisdom in receding; well understood, it does not imply a setback in the least. A different thing is that the economic elites are in a position to do it. In a climate that calls for the intensification of hatred, I have tried to indicate a way without damage to anyone and available to those who seek it.

Quite debatably, Mackinder said that who rules the Heartland commands the world; with much more reason we can say that whoever adheres to this ungovernable axis can be his own master, which should be enough.

References

Bruce Bogoshian, *Kinetics of wealth and the Pareto Law* (2014)

Laurent Guyénot, *The Holy Hook* (2019)

Jaromir Benes, Michael Kumhof; *The Chicago Plan revisited* (2012

Charles Hugh Smith, *The New Tyranny Few Even Recognize* (2020)

Miguel Iradier; *Futuro y fuga del dinero* (2016)

Una fábula, un enigma y una solución final (2019)

Caos y transfiguración (2019)

La Tecnociencia y el laboratorio del Yo (2019)

El capital y sus amigos (2019)

El pacto de los cacahuetes (2020)

Pole of inspiration -Math, Science and Tradition (2020)

Notes

1 Miguel Iradier, *The little finger strategy*, www.hurqualya.net

STRIKE, POPULISM AND TECHNOCRACY

16 November, 2020

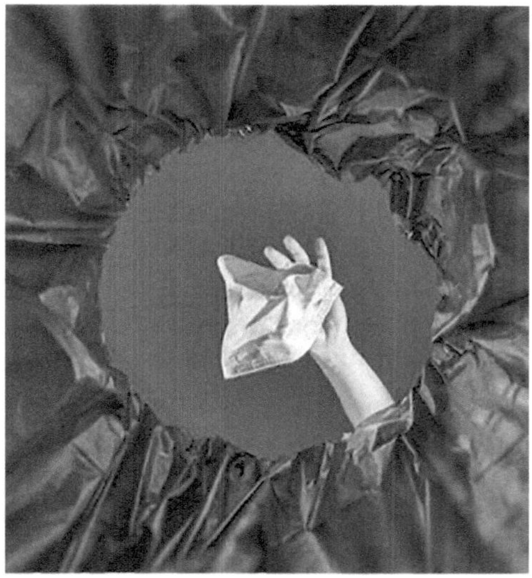

First things first, and rather than speculate about what may happen five or ten years from now, we need to look at the present. What is being tested now is to what extent the powerful are able to shape our perception of reality at will, and were we to believe mainstream media, it seems they are getting it.

We can only trust our instinct, and what my instinct tells me is we have to throw away the masks and go on general strike. Act in unison, not in that social distancing that they want to enforce to its last consequences. Everything else is calculated ambiguity, mediocrity and delirium, a worthless attempt to escape from what is happening right now.

This would be truly cathartic, and the crossing of supposedly contradictory gestures would short-circuit power and the ideological divisions that are at its service.

The future will depend on what we do in the present and not the other way around. Today what we observe is general subjection; each mask is the most effective vote for the party of the technocracy, fear and the multiple levers of repression. We will return to this at the end. But it is also necessary to pay attention to the medium and long term and not give a few the additional advantage of being the only ones capable of playing a long game, since they already have too many.

It has almost become commonplace to put populism and technocracy at opposite poles, and so it is on a superficial reading. What is less appreciated is how they participate in the same lie, and this is always the essential thing if one wants to see beyond banana politics.

Everything encompassed under the umbrella term "populism" has the common denominator of claiming to represent "the people" as opposed to an elite; the unitary character of such a people may be a myth, but the overwhelming reality of a tiny kleptocracy is not, which ends up endowing populist claims with undeniable strength and truth.

On the other hand, it is crystal clear that technocracy is at the service of these minute elites and their facilitators, which exposes a myth much more unfounded and false than that of populism: its legitimization in the name of efficiency. Since the aim is to optimize benefits and structural advantages for a minimum percentage of the population, it is absolutely impossible for technicians at the service of power to work for the benefit of general efficiency, but for a very particular efficiency that is not only different but so often in stark contrast to the former.

No technocracy is more consolidated today than that of the central banks whose archetype and keystone is the Federal Reserve System, a mere consortium of private banks with the official blessing of the Government. And it is clear that the Fed does not work for the general efficiency of the economy —not even that of its own nation- but for the financial interests of global banking with the hegemony of the dollar as instrument.

It is also clear that the economic theory that technocrats force on us is garbage economics at the service of the debt trap, so their alleged technical superiority is simply ridiculous: it is just counter-demagogy at the service of the economic elite.

It is in the interest of this pseudo-elite that everything is, or at least seems, way more complicated than it really is, so that no one but them can be in charge. The best example is the convoluted fractional reserve system, which is ten times more complex than a hundred percent legal money system and incomparably less fair, since it is at the service of and intimately related to the inverted pyramid of inequality.

So-called "right-wing populism" claims to be sovereignist, but today it is impossible to seriously consider sovereignty without a way to guarantee an economic autonomy which begins with monetary independence. In this way any claim to sovereignty is a dead letter from the outset, and national-populism is a neutered cat that is only allowed the mewing, or rather the screaming, of cheap demagogy.

Trump is the textbook example: the base that support Trump is that of the impoverished white worker, but all he offers is more tax breaks for the rich and inconsequential measures against China. His feeble invectives against the Fed are a very small matter because we all know that the consortium has much more weight than the chief executive —he is there to assume quick decisions, distract the attention, and take the hits and the slaps.

Populism could only be something more than demagcguery if it had a place to grasp the question of monetary sovereignty, but this cannot even be posed in the current correlation of forces.

Populism as a "fourth political theory", whether right or left, is an option neutralized beforehand by the prevailing system as long as it is incapable of proposing monetary sovereignty: it continues to be on the same plane and turns within the same wheel as the rest of the parties and ideologies, each with its own ambiguity and indetermination, each with its own demagoguery.

But is it not also demagogy to raise economic sovereignty when it has been totally hijacked and is impossible to access? Demagogy, it is not: it is a vital and elemental question, but totally beyond the reach of modern democracies. After all, democracy itself has been emptied of real content because the usurpation by private banks of such a fundamental public prerogative has also emptied it of effective power.

Geoeconomics and Geopolitics

Except for China, no country today has monetary sovereignty: certainly not the United States, despite what appearances might suggest. And it is this situation that really determines the hostility of the globalists towards the eastern giant, a hostility that they try to transfer to the masses that support them.

The media does a great job, and so we see the corner shopkeeper complaining about the Chinese next door and his fellows, "who are taking everything," instead of looking at the bank to whom he pays the mortgage and who happens to be the engine of globalization.

The Empire's hostility towards China will not be tempered, but rather the contrary, since it is an existential threat. And it is here that the wires of geoeconomics and geopolitics will cross.

The Chinese government never had global hegemonic aspirations and would be content with having a voice in international affairs commensurate with its weight. Understandably, the real hegemon is not willing to yield a single portion of its present power. By increasingly harassing China, it is forcing her to defend herself in all spheres —and today attack is an integral part of defense.

The more hostile the West is to the Sinosphere, the deeper her reaction will be. Previously, Western decision-makers believed that these attacks would leave no wounds and would be reversible since the ultimate goal was to assimilate the Chinese economy into global capitalism. Thus, anything that would weaken the dragon would be good for gaining bargaining power and obtaining concessions more easily.

But, reached a certain point, the hostility begins to take on an irreversible character and the rival moves further and further away from the negotiating horizon, even if it never loses sight of it. It is forced to move deeper. And that is the point we are getting to, because, unable to solve their own problems, Democrats and Republicans alike need to blame China to get returns from both the electorate and the deep state.

In the United States, the fallacy of Chinese guilt becomes the only "floating signifier," the only remaining space left for powerless politicians to the point of pathetic.

But what is a floating signifier in the jumble of lies of the West could end up being a meaningful token for China, and for this very thing that liars cannot meddle with.

There is a much more pressing technological race right now than that of coronavirus vaccines. It is the race for control of the cryptocurrency space, in which China seems to be in the lead.

Digital currency is the ultimate phase of capitalism in the final stage of the liquefaction of money: both a technical culmination and a critical step in the enslavement of the masses and the closure of the system. This step poses a great danger both for those who run it and for those who suffer from it.

It is only natural that in this phase neoliberalism takes a sharp turn towards technocracy and global governance until it is no longer recognizable. But technocracy itself is only a screen to disguise as neutral the unusual power of a plutarchy that also uses the oligarchy to hide.

In previous works we have detected a "double contradiction" at the crossroads of the main monetary decisions. We have seen that there are two basic attitudes of the states with respect to the digital currency, as well as two types of private crypto-currencies, the speculative and corporate cryptos, and the alternative cryptos for all sorts of communities. The latter seem to be a residual option but may tip the balance in the very plausible case of a generalized currency war.

China has no pretensions to monetary hegemony —its currency is still pegged to the dollar- and would be content with some sort of equilibrium; but it is the aggressiveness of those who want to impose the rules that will force her to undertake incursions in the global market even reluctantly, and precisely to preserve the equilibrium.

The coming crypto-currency war can be a great test to find out the truth of monetary theories. The globalist elite wants to impose a "global governance" of the digital currencies so that the subjects have no escape from their system, even if it means the end of the dollar's hegemony, now replaced by a new design and a new consensus. This is something that Mark Carney, ex of Goldman Sachs and former governor of the Bank of England, put frankly at the Jackson Hole meeting in 2019.

But efforts to reach international agreements may come late and badly, at least with respect to a player as important as China; and if agreements were finally reached, today it could only be at the expense of the US dollar, or more precisely of the people of the United States, which would have to submit to a new international discipline for which their internal market is not even remotely accustomed.

Today the different currencies do not have real autonomy because if they turn their backs to the dollar system, the response, whether financial or military, is almost immediate, as we have seen with relatively modest countries like Iraq or Libya. If the monetary hegemony is transferred to a global crypto-currency, blessed by the Bank for International Settlements, they must be certain that the US army will continue to be at its service, or else the always decisive resource of force will be lost.

On the other hand, if the dollar is abandoned, the Americans will see with painful clarity something they should have seen long before: that the United States has always been an instrument and nobody gave a damn about the people. To speak in such circumstances of betrayal of the American people would be ludicrous; but a new kind of humiliation, of plundering and derision, can still await them if they don't wake up. Not everyone will accept it.

The question of the transfer of monetary hegemony, which until now had nothing to do with Chinese initiatives, is a very delicate one and reminds us of a blanket too short for the bed —one has to choose between the feet and the head.

American populism is not going to die with Trump, that could represent only the beginning of a somewhat comical but more than understandable struggle for self-determination. The division in American society is deep and has a dual but unique origin: in the external opposition between globalists and isolationists, and in the internal division of the elites themselves.

It may well be the case that global banking will reach an agreement with China to the detriment of the American people, but this would only further fuel populism and neo-isolationism: one can even imagine soon the vindication of the dollar as a symbol of the lost sovereignty... even if it was through the dollar that sovereignty fled to the Empire.

But on the other hand, a monetary hegemony dissociated from military hegemony and force will never get very far. Add to this the fact that the American stock market bubble is the greatest in history and is yet to pop, and we can begin to imagine the chaos into which the old models and behaviors will be plunged.

Smaller economies and currencies always look for another reference currency to be backed for. After the Last Great Bubble pops, it will be really difficult for the dollar to be the main reserve currency —because the Fed is the Last Great Bubble itself. What is broken, is broken. Hence the urgency to "reinvent" the hegemony by the Great Financial Siphon.

This will make the impending digital yuan, better supported by both a more robust economy and a more solid and balanced state, even more attractive. If the American bullying disappears, the flight to the eastern currency will be much easier —but here we are not only talking about national currencies, but also private crypto-currencies, whether speculative or alternative.

From Chinese dual circulation to the Tao of Judah

Just this fall of 2020 the Communist Party of China has revealed its new five-year strategy of "dual circulation" to reduce dependence on foreign markets and technologies from 2021 to 2025.

It does not take much insight to realize that the dual attitude is a deeply rooted feature of Chinese idiosyncrasy: from yin and yang to the "one China, two systems", to this brand-new dual circulation that admits more levels of interpretation than what commentators have so far discovered.

One of these levels, and a vital one indeed, would be the imminent new digital currency. Let us remember in passing that our own fractional reserve monetary system is also a dual circulation system, with legal money issued by the central banks, barely a five per cent of it, and endogenous money created out of thin air by bank credit, which is the bulk of the total: the debt-money for which we all work, even if we have never asked for credit in our whole lives.

But the "dual circulation" of the Chinese digital currency would have very different implications. The Chinese government could be extremely restrictive about the use of its crypto-currency in its huge domestic market, while adopting a much more "liberal" approach in foreign markets in order to attract long-awaited capital while harvesting new support and a new kind of soft power.

It sounds very attractive, both to the Chinese government and to a growing sector of the foreign population that feels increasingly destitute, not to mention the flow of capitals that ignores ideologies and is always willing to sell out its mother.

This explains why guys like Carney are stressing once and again that "governance is the core pillar of any form of digital currency" —global governance, of course; because it is not easy to empty the sea with a sieve.

China would in no way want to play hard the card of a dual monetary system unless it is strongly provoked, that is, as a defensive countermeasure;

but this game, which would disrupt the global financial corral, would be easily interpreted as an incursion into someone else's yard or 'aggression" in other words. Put another way, if China is winning the economic war, the Western countries will intensify their offensive in the other sectors of the hybrid war, from opinion to military pressure and attempts of destabilization.

It can be assumed then that China will evolve in this arena with great caution and according to the behavior of its rivals, because after all, the official approach to double circulation is to decrease the dependence on foreign markets and to come closer to the Chinese ideal of self-sufficiency, which will never cease to be the ultimate goal of its movements.

The Chinese government would also have an additional reason for concern: the difference in treatment for users of its currency in the domestic and foreign markets could lead to growing discontent at home. It is not easy to harmonize the state-directed economy with navigating the markets, but the Chinese technocracy already know something about it. And they are not alone, because the Fed and the other central banks are familiar with this issue too.

However, this contrasts sharply with the attitude of the Western powers, which still believe they have a special right to rule the world and impose their wills on others. But we should not blame the "West" for what is basically a matter of the plutocracy and its lackeys.

Elite comes from elected, the chosen ones, and it is known that some are more Chosen than others. We have spoken repeatedly of the 80/20 law in the distribution of wealth and its successive powers, and we have also spoken of the x-ray of that x-ray, that informal "law" of 50/50 that affects the share of the booty and the most intimate physiology of accumulation, that special relationship within AngloZionism between leaders in the exercise of violence and deceit.

This share seems insufficient to one of the parties given the notorious advantage they enjoy; the financial elite wants the United States to serve the Empire and not the other way around. The disappearance of the dollar would not at all imply the retreat of the interests of those who manage the Federal Reserve, but its protection against the growing instabilities of the nation and loss of consensus —of which they are the main cause.

This transfer or succession would imply an internecine war that can only be read between the lines, and this conflict is the invisible cause of a social division that is but the ultimate consequence of the projection of power with its control of means and minds.

The transfer of the coercive mechanisms of financial control can leave a vacuum of power and authority that will be favorable to the monetary sovereignty of countries, collectives and peoples, and right now it is impossible to say what will tip the balance, since many of the consequences will be unintended.

China may favor that spring, but so may the United States if it throws off the yoke of empire and creates the first one hundred percent legal tender, as already proposed in the 1933 Chicago Plan. And only in this way they could set a positive example and have the only moral leadership, involuntary, that may be worthwhile.

Social Democrat Roosevelt did a great job for the Fed at the time, which should not be surprising. But it is in the destiny of the United States and of all of us that this unprecedented option —which will not fail to come up at the critical moment of "change of phase" in the state and reality of money. And this will completely disrupt the props of the political spectrum, which, as we see, can no longer stand.

Today there would no longer be globalism or a will to continue with this project were it not for the weight and power projection of "Jewish" money, which is the only will that continues to give a direction to the multiple divergent interests of the oligarchies. The West itself would be dismembered into a heap of disperse interests without this element of cohesion; there is nothing easier to demonstrate, if the question is the fate and concentration of capital. It is good to understand this, and he who ignores it miss the thread of the plot.

Not even the Jews themselves have exhausted the meaning of this reality, since censorship of public opinion must also have a profound effect on involuntary mechanisms of self-censorship. Only at the very end can instinct attain self-understanding.

Too many confuse loyalty to the West with loyalty to Netflix, and there is a point to it, since the West is above all the imaginary of the descent in search of its unfathomable bottom.

As the couplet goes, *'cause the Tao of Judah, neither is tao nor is not nah*, the anonymous poet complaining that the sharp-edged rigor of the Law has little to do with the harmonious roundness of reciprocity. But surely our coplero, whether Benjamin of Tudela or not, had an understandable bias that prevented him from seeing the issue more objectively.

Because yes, there is a Tao of Judah; and it is true that it does not have to do directly with the Law, but with its reversal: we are talking about the dialectic between the Jews and the Goyim —the nations and their peoples. The "tao of Judah" is literally the winding road and tortuous way of the Law among the nations.

Jewish scriptures continuously emphasize the outstanding status of Israel over all peoples and nations; China, on the contrary, was the popular civilization par excellence long before the People's Republic, since there the ideal is not to stand out. The Tao of Judah seeks the extremes with full intention, turning everything into gold included, while the Chinese people, in their instinctive search for the invariable medium, aspire to the golden mean or *aurea mediocritas* by carefully avoiding the extremes.

So the archetype of the Jew and the Chinese are like the scorpion and the frog of the fable: one is always looking for soft parts to nail the sting to the heart, while the other, all defensive instinct, is always ready tc avoid it.

Of course, we are talking of a big frog. And if we are dealing with how the digital liquefaction of money is going to affect the balance and legitimacy of nations —water and earth, after all- it is not hard to spot a curious fight between Behemoth and Leviathan. And in our upside-down world, the earth has to use the liquid element of money to defend itself, while the plutarchs who accumulate the bulk of the wealth seek to poison all the levers of politics to prevent it.

Those who like enigmas may wonder where the Ziz is today —that mysterious creature of the air capable of covering the Sun with its wings.

To avoid being fried in the pan, China would have to combine the three mythical creatures into a single entity: a full-fledged Dragon with brand new claws.

Israel's extremely close ties with the national-populist movements are telling: from Trump to Bolsonaro, Johnson, or the European populist parties. It is as much about controlling their movements as it is about ensuring that their leaders are unpresentable enough to discredit their most legitimate claims. And so far they are having great success, because even if they win, and especially if they win, the expectations about a real sovereignty wear down more and more —especially when they have control of the media and can round out the caricature of these already cartoonish characters.

The current national-populist leader appeals on the one hand to healthy instincts against globalization, while on the other appeals to the most dull and stupid selfishness. Lacking real power, he is a movie loser doomed to ridicule.

The only game changer here would be the new (digital) monetary balance. If sovereignty is able to regain control of money for nations and do so successfully, the myth of the efficient technocracy would be forever discredited.

But this requires not only a window of opportunity, but also another sense of goodness and justice. There should be much more coherence and integrity in speaking about the common good and the ways to achieve it.

There is the erotics of power and there is the necrotics: but now they are barely distinguishable for both politicians and technocrats. The true erotics of power is being able to awaken in the people the desire for good, and the present national-populist has no idea what that might be. If he gets so many votes today, it is only because many people know what they hate. Adherence by negative reaction is already guaranteed by the system; the question is if there is a real offer beyond demagoguery.

Given the nature of the forces at play, the time to rise this question will come.

Saying no

In the coming monetary war we can distinguish three axes: the horizontal axis, linked to the most purely commercial, civil and liquidity factors, the vertical axis, to the political-legal environment of the states and the agreements between states, and a temporal axis in depth encoding the scope of the reversibility and irreversibility of the debt, and of the uses and changes in monetary technology.

As we have already said elsewhere, this third axis is as important or more than the other two, and it is present in all orders and not only in the economy. In fact, it defines the extremes of our sense of temporality, and maybe one day analysts will understand its relevance. Today it is beyond their estimations, and even their ideas, which makes it even more interesting.

The things now being discussed are at a very different level; issues such as basic income, which, by the way, will be one of the star products to gain the assent of the debt-serf without costing the elites a penny. The same people responsible for all this situation have to present themselves as saviors; what less not to end up hanged.

And this brings us back to the beginning of our article, and how to respond to those who care so much about saving us. And for this we do not need to ponder or analyze; we only need to have guts.

They say fear is free but nothing is further from truth. Surely those who seek to frighten us 24/7 are trying to free us.

Many people who often talk about the Revolution and taking over the Winter Palace not only put on their masks but also launch diatribes against the "deniers". It is well known that "fascism" must be fought —but operetta fascism, not the real thing we know so well. Perhaps, by attacking this paper fascism they believe they have fulfilled their political consciousness, a truly revolutionary one. They are the necessary counterpart of all those who say that the Democratic Party is communist.

I am not going to tell anyone what to do, but there is no need to wait for the revolution, or the taking of the Winter Palace, or anything like that. If you wanted a chance to do something against this system, to make a real gesture for once, the time has long since come. It never will be easier for you, because you don't even have to start by doing something, but simply decide *not* to do it.

Not to wear a mask. Not to close your business. Not to submit to disgusting and unnecessary tests. Not to be vaccinated with completely experimental and unnatural vaccines. Not letting them muzzle you or inserting icky sticks and needles in your body, you insert them something bigger by the best place possible, without no need to stain yourself with their shit.

It is way better to die alone than suffer their special treatments, because anything is better than owing them your life. Show yourself you don't depend on them. If there is a time to prove it, this is it.

You don't need to be their guinea pig, let them graciously experiment in their own flesh —this are the only experiments worthwhile. You don't need to back up with actions what you don't believe in. If you stop doing and consuming certain things, you will soon find room to do and assimilate others; if you stop giving in to the all-pervasive pressure to isolate yourself, you will soon find unexpected companions, because you will also be busy looking for them.

Whatever happens, in spite of everything and with all our actions included, I am convinced that this is always the least bad of the infinite possible worlds; but no one can imagine how the others might be. We have enough with just one.

References

Miguel Iradier, *Art and Theory of Reversibility* (2020), www.hurqualya.net

ASTRONOMY, ASTROLOGY AND ASTROPHYSICS — PHASE MEMORY, PARALLEL TRANSPORT AND REFERENCE FRAME

22 March, 2021

Theoretical physicist Nicolae Mazilu observed at the time that if astrology remains a blemished subject, it is because it has failed to correctly address the question of the reference frame and its proper transformations. Indeed, a person who was born on a specific date and time in a certain place continues to maintain the "inner clock" of his birth even if he moves to the most remote place on the globe. And on the other hand, for all those who are not familiar with the astrology complexities, it may seem absurd that the same planets, which in real time are in the same positions in the sky for everyone at the same location, can influence each person in a completely different manner, because each person is an ongoing peculiar microcosm.

Astrology is only one more of the symbolic systems expressing the universal interdependence, but it is the most directly linked to the birth of modern physics, with which it has *almost* nothing to do. This contrast and this proximity has created feelings of mutual grievance. Emilio Saura and Raymond Abellio have drawn a distinction between three levels of astrology: an influential or primary astrology that tries to relate celestial configurations and terrestrial "facts", a symbolic astrology that attends more to propensities and resonances, and finally an astrology oriented towards the absolute that seeks to subsume the apparent multiplicity in the consciousness of the "transcendental I".

I will not pretend to do justice to such a vast and uncertain knowledge in such a brief space. I do want to underline the fact that, although physical astronomy and astrology today have diametrically opposite aims that can in no way coincide, they must inevitably have a common ground that have not yet been properly explored by *either* side.

All physics is built around the general concept of reference frame. A reference frame combines in one two primary elements: the clock and the coordinate system. The clock that regularizes our perception of a physical body in motion, and the coordinate system, which does it with many bodies. But astrology is no different in this either, only that, instead of orienting its whole apparatus towards quantitative prediction, it takes it towards interpretation, and in the very place of the reference frame places the very self, whether the empirical self concerned with the affairs of the world or the transcendental self which is the condition of possibility of experience.

Thus in physics, as a third-person objective science, description and interpretation of processes are relegated to the background in favor of prediction, while astrology, as a subjective knowledge that puts the first person at the center, only prioritizes prediction when it becomes a degraded practice, whereas interpretation should always be its genuine focus. Astronomy sees objects outside and deals with motions of matter and forces, astrology deals with cycles, potentials and even potencies, and tries to finally see the world as the environment of the self.

Astronomy should be concerned with analysis, although overstating the role of prediction has prevented even a complete analysis of such basic matters as the Kepler problem, whereas the touchstone of astrology is the overall composition of a whole and its synthetic perception and judgment, much more akin to the composition and perception of an art creation.

This diametrical difference should suffice for demarcation. However, the I that projects itself onto the world does not want to even contemplate the idea that the world also projects onto the I and affects it. We prefer to think that Nature is always out there, even if it is in the form of genes and molecules, as long as we can manipulate it; but if there is a possibility that Nature acts upon us in ways we cannot control, we choose to ignore it, since it calls into question our standing in the wordl. Moreover, the monster of the social is extremely jealous and rejects the mere idea that its very parts and aggregates may have something to do with things off limits.

*

When physics speaks of forces, it refers above all to controllable forces, since there are also measurable but not controllable forces, as it is shown by something as empirical as the constitutive law in materials science. Those force that are not controllable are, in any case, interpreted in terms of the controllable ones, and so, for example, we apply Newton's three laws of motion to celestial bodies even though, for example, the third principle is not verifiable in an orbit since the force vector points to an empty place, and not to the body of the Sun. This plain fact had to be most embarrassing to Newton himself, but in time people managed to forget it.

A controllable force is a projected force, but not an uncontrolled force. The hole in Newtonian theory was then plugged by Lagrangian mechanics resorting to potentials, and these potentials were thought to be only an auxiliary tool to the full description of forces which remain uncontrollable, but are assumed to be of the same type as the controllable ones. Newton's theory, reasoned Gauss and Weber between 1830 and 1846, amounts to a gravitostatics, not to a gravitodynamics, and it could be that forces, beginning with the electrodynamics then at hand, depended not only on distances, but also on relative velocities and accelerations. This hypothesis also implied a phenomenon known as "retarded potential".

Weber's theory was replaced by Maxwell's, but much later, its heir, quantum mechanics, began to discover, to the general bewilderment of physicists, that quantum potentials were more than a mathematical tool at the service of force. The Aharonov-Bohm effect certified that an electron feels the presence of a magnetic field even when the field strength at its position is zero. Then Berry, in 1983, generalized this type of phenomena to adiabatic processes without heat exchange, and in a few years it was further generalized to all kinds of physical conditions.

The phenomenon in question, known today as "geometric phase", was first detected by Pancharatnam in 1956 in optical interference experiments. Regarding this totally universal phenomenon of the geometric phase, most physicists still circulate two false assertions: that it is exclusive to quantum mechanics, and that it in no way affects the completeness and accuracy of quantum mechanics.

But the fact is that the geometric phase can occur at any scale and is detected in macroscopic electrical experiments as well known as Faraday's induction, in the precession of Foucault's pendulum or even on the surface of water. And on the other hand, its description requires the addition of a bundle over the Hilbert projective space in which quantum mechanics is formalized, which implies that it cannot be traced back to the Hamiltonian dynamics that defines conservative or closed systems. Then, something was clearly missing in quantum mechanics, and the best proof of it was the widespread confusion of physicists for decades, until Berry "formalized" the embarrassing situation.

David Bohm himself, who spoke so eloquently of the *implicate order*, did not seem to have fully grasped the universality of the so-called geometric phase; but the situation is not different now, when this factor routinely enters into the calculations of engineers striving to achieve "quantum computation" or to manipulate atoms at the individual level.

The geometric phase, "global change without local change", is also known as phase memory, since it implies the subjection or "enslavement" of the system to certain additional parameters; consequently it is studied in robotics and control theory, and has also been observed in dissipative systems and in animal locomotion, for example in the righting reflex of falling cats or the undulating advance of snakes.

The interpretation of this totally universal phenomenon is still completely open, although almost all physicists prefer to ignore the question. To deal with systems as having a dynamical and a geometrical phase is just a compromise, but, in any case, of what kind of geometry are we talking about? The answer here is clear: it is *the geometry of the environment* through which the parallel transport is measured. That is, it is an indirect measure of the environment itself, of how the system is not perfectly and ideally closed as mechanics by pure principle pretends.

There is also an extremely elegant mathematical language to describe these parallel transport phenomena: the differential geometry developed by Cartan, in addition to the also very compact geometric algebra initiated by Grassman and promoted in more recent times by Hestenes. Unfortunately, none of these formalisms has much favor among physicists, who, still by inertia, continue to cling to the vector formulation, a poorer and flatter description that often involves less information.

Speaking of information, which as we have already said is usually clipped out by vector formalisms, the geometric phase is often explained as a pure and immaterial transmission of information without any relation to the dynamics. On the one hand we would have "blind forces", and on the other hand, "pure information" supported by nothing; it certainly does not seem a coherent or credible picture, but after all the problem of "communication" has been there forever: no one has ever said how the Moon knows where the Sun is and what mass it has to move as it moves.

Celestial mechanics, in short, still has huge holes, that general relativity, which does little more than complicate the equations, is in no way able to fill. Not to mention Laplace's perturbation theory and its "resonances", which, apart from the fact that we are never told how they are physically possible, force us to believe that gravity can also have a repulsive effect, not a negligible feat.

The geometric phase simply shows us the physical reality of potentials, even if we do not know how to explain their action, and their ties with "controllable" forces, with justified quotation marks because no one has so far controlled gravity. And if astrology is completely unable to justify any kind of planetary influence in terms of force, in terms of potentials and their phase shifts celestial mechanics is in no better situation.

As we say, the purely relational approach to dynamics proposed by Gauss and Weber, which at least put forces and potentials on an identical basis, was more logical. This was the true theory of relativity, way earlier and with fewer postulates and complications, not to mention unnecessary curved spacetimes. But relational dynamics did not even need the principle of inertia and all the scholastic distinctions of inertial frames of reference.

Newton's mechanics is summed up in the phrase "nothing moves unless moved by something else", but with the principle of dynamic equilibrium, which replaces the principle of inertia, nothing needs to be moved by anything, but any state of motion or rest of a body is already the result of the sum of all the forces of the universe on it. Thus, the principle of dynamic equilibrium, key to relational mechanics, is in perfect harmony and consonance with the principle of universal interdependence, and one could even say it is just another way of expressing it.

Moreover, the principle of dynamic equilibrium, while remaining neutral, allows to think that the body itself determines its own motion, as Alejandro Torassa shows, without breaking in the least with interdependence, and without losing power of prediction. So, there is really no need for new theories or new postulates or new particles, but only a new way of looking at the Principle and the principles.

This does not mean that a good principle explains causes or influences, but something surely more important, that it ceases to make them necessary. If even the fundamental physical laws are supported by variational principles, and the Lagrangian that supports them is but an exact analogy that can respond to an infinity of different causal explanations, astrology could never claim to go beyond statistical correlations.

It is also true that even in the field of statistics, physics only see what it wants to see. For example, Simon Shnoll and his school has demonstrated for many decades "the occurrence of discrete states during fluctuations in macroscopic processes" of the most varied type, from enzymatic and biological reactions to radioactive decay, with periods of 24 hours, 27 and 365 days, which obviously respond to a most familiar "astronomical and cosmophysical pattern".

So even radioactive decay is not totally random, as it usually claimed, but such regularities are routinely filtered out and discounted as "non-relevant". On the other hand the so-called "Copenhagen interpretation" of quantum mechanics is still promoted precisely because it is supposed to contain the minimum dose of interpretation, but even that is not true, since according to Copenhagen the wave function responds to an individual system, which is still an unnecessary ontological fabrication, and it would be much more logical to think that it only applies to a statistical ensemble, and that its waves are waves in the coordinate space. This would greatly facilitate the connection with macroscopic classical mechanics and with the arguments of a global nature that make the geometric phase possible.

Such elementary considerations would make the inclusion of cosmophysical factors much simpler, but the experimental evidence for such correlations can only grow consistent very slowly, on a scale of generations and centuries, and on the other hand the demanding autophagy of the Social does not want strangers sitting at its table. Of course astrology has accumulated its own kind of evidence over millennia, but it is not the kind of evidence that modern science is willing to consider.

It is clear that one can legitimately present astrology by elegant projective arguments and global correlation without entering into the question of physical causation, but this is not tantamount to denying its possibility. The physical influence of the Moon, for example, is certainly not difficult to conceive, nor, for many, to feel, but if so many of "the learned" deny even the possibility of such an influence it is also a matter of feeling, albeit in the opposite sense: the I of many feels stronger if it imagines itself autonomous. Of course, it is only society that inculcates such ideas of "independence"; a society that tries to convince us that intelligence is synonymous with dullness, instead of sensibility.

Much more evident is the influence of the Sun. In general, it has been astrologers and cycle researchers who have pointed to the powerful influence of the great planets, Saturn and Jupiter, and especially the latter, on the sunspot cycle and the magnetic activity of our star, so critical to the balance of life on Earth. Apart from the fact that the solar period and that of Jupiter are almost equal, 11 and 11.86 years, there is also a plain astrophysical reason: the Sun's contribution to the total angular momentum of its system is estimated at 0.3%, while Jupiter alone has more than 60%. The other planets would explain the small difference in the cycle. One can also find more specific reasons outside the mainstream theories[1] but we will not go into them now.

And here again we see the overdetermination of the social that wants to refer everything to itself: the media campaign relentlessly to convince us of a climate change of which we have only vague and debatable statistical correlations, while systematically ignoring the absolute basis on which the climate depends, the radiation from the Sun and its well-known variations. But it is clear that trying to convince us of the decisiveness of the "anthropogenic factor" is way better game for the powerful.

*

Most of the "refutations" of astrology are of such a low level that they deserve no comment. On the other hand, every interpretation has its share of speculation and even of folly, and physics is no stranger to this; perhaps that is why it has tried to reduce its interpretations to a minimum or even to dispense with them, which in practice, besides being absurd, is simply impossible. Every experimental or technical application already starts from an interpretation. In astrology, on the other hand, the interpretation itself is already the application, so it cannot be dispensed with either in theory or practice.

The very idea that fundamental physics is made up of "local theories" is pure interpretation, and not in accordance with reality. The Lagrangian of a system is by definition global; a very different thing is that it is systematically used to obtain derivatives —local answers. The same thing happened with Newton's "explanation" of the ellipse; because Kepler problem already contains in itself the whole business of gauge fields, based on the invariance of the Lagrangian. So the "local theories" are already pure interpretation —only in the service of prediction. And the proof is that physical causality, which is what a local theory worthy of the name should really aim at, can never be clarified and it is really out of question.

In physics, then, the local is also the result of the global, even if one is not willing to aknowledge it. Only a few old-fashioned and honest physicists, like Planck, wonder why all fundamental laws depend on action principles, which are global and teleological by its very definition. The assimilation of entropy with disorder, perpetrated by Boltzmann, is another example of an all-too-human interpretation that has nevertheless been passed down for generations as a model of positivist neutrality.

Clausius' original principle of maximum entropy already assumes a tendency towards order, for as Swenson said, "the world is in the order production business, including the production of living things and their capacity for perception and action, for order produces entropy faster than disorder". But furthermore, the diverse Lagrangians of fundamental physics can be reformulated as a balance between minimum change of energy and maximum entropy production, so that we get a totally different picture of mechanics, dynamics and finality.

Any interpretation is partial, but pretending that we do not interpret because we give the last word to the quantitative prediction is worse and even more false. The worst thing that can happen is that scientists and technicians manipulate things without even asking themselves what they are manipulating, but that is exactly what this system expects of them, and they have internalized it to the marrow of their bones. The rest of us are also expected to follow them.

*

There is nothing more mundane or profane than reading the newspaper in the morning: a whole selection of news calculated to check thoroughly your springs. And in spite of its never sufficiently pondered foolishness, one cannot help but react to these news with the same reflexes every day, with only slight variations. The same thing happens during the course of the day, and even in dreams, since the shadows of our planets only stop being projected when the light of consciousness definitively vanishes and we reach the depths of slumber.

In details such as our chronic reactions to the news it is clear that "the stars" are not so much influences as impulses that come from within; there is nothing extraordinary in that these impulses require external stimuli to come to life, that they take the form of reactions. "The stuff of which dreams are made" is the "astral substance," which is but a “tonal or chromatic dimension" of the homogeneous continuum from which we have emerged and in which we still find ourselves. There are astral tonalities just as there are the timbres of the different metals when we twang them. These tonalities, which have many different nuances in each individual, envelop us in concentric layers, like a certain music of the spheres, which an astrology that aspires to be more or less “scientific” would have to recompose.

Astrology undoubtedly has an invariant structure that cannot change over the ages, since it simply connects events in space and time by means of a certain projection. In this sense it is much less fickle than other disciplines that claim to be sciences, such as psychology, economics or sociology, although undoubtedly the almost infinite "plasticity" of these matters gives them a great advantage in adapting to the needs of power. Were it not for the fact that economics deals with the accountable data par excellence, money, and ends up reducing everything to the same quantitative standard, no one would even dream of calling it a science; but at any rate, dealing only with quantities is not something that other fields of knowledge have to envy.

Astrology, on the other hand, was always aware of its great affinity with sociology and psychology, long before these ones acquired scientific pretensions; but its presuppositions transcend completely the reductive and confined vision of such specialties. What is not forgiven is that she has a transparent and first-rate quantitative part, unlike the heuristic and often spurious mathematics of these successful new riches.

The art of Urania could adapt to these times, but that is not a great prospect. If she cares about science, it is not to receive its legitimacy from anyone, but because of her desire to fit some of the most important loose pieces together, out of pure inner aspiration for unity.

The ellipses discovered by Kepler can be interpreted, following Weber and Nikolay Noskov, via the retarded potential and a phase velocity that produces a wave or vibration in the moving bodies—which amount to a direct interpretation of the geometric phase. Let us recall that Weber mechanics gives good predictions but it does not allow us to distinguish between the kinetic, potential and internal energy of the bodies —an ambiguity one should find *natural,* indeed, given that energy is just a human account balance.

Just think about it for a moment. The Kepler problem is not just a matter of celestial mechanics, it is at the basis of quantum mechanics and the atom too. The same Schrödinger equation has a term corresponding to a wave in the body, only that quantum mechanics, because of the limitations of special relativity, is unable to describe extended particles (ephemeral configurations) with volume. Weber's electrodynamics has no such limitations. But it is not only Schrödinger, the electromagnetic waves with which Hertz seemed to confirm Maxwell's theory, are not transverse in a geometric sense, but only as mere statistical average between empty space and matter.

We can then see matter, *when it leaves room for real bodies*, —something that does not happen in the standard version of quantum mechanics-, as a diffuse or statistical cloud traversed by different tones (the inner phase), which are already included within the probabilistic picture of standard quantum mechanics. There are certain *propensities* within the "probability clouds", which would have a layered structure like the atoms themselves —which would explain the discrete states faithfully attested by the Shnoll school for more than forty years. I find this interpretation natural, and it is easy to integrate within the great mass of experimental evidence, even though this is never indifferent to the theoretical framework that contains it.

*

Paracelsus said that without the astral impression man is not even able to mend his pants. This *impressio* is the propensity or disposition that inclines us to do certain things rather than others on our own initiative and even without external coercion. The great Swiss physician and traveler obviously knew what he was talking about, and had probed human nature and Big Nature with much more direct means than our current bureaucratic science, which adjusts its glasses so as not to see what may be on the sides.

The most important aspects of astrology are more of a symbolic than a physical order, and yet that symbolism fits perfectly within the spatial and cyclical determinations of the motion of our planet within its astronomical environment. It is ultimately a qualification of the three axes of space and its six directions, the simplest scheme for our physical and intellectual perception, which Abellio, trying to grant it the highest degree of gnoseological dignity, called "the absolute structure".

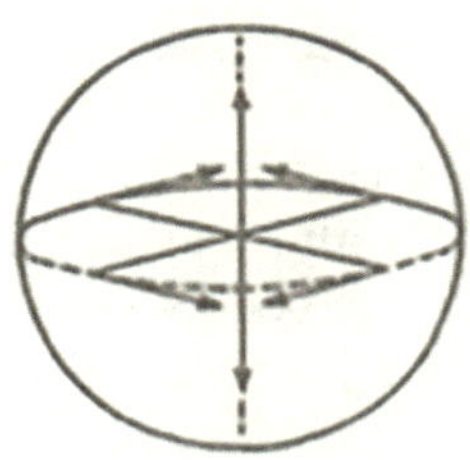

Abellio also proposed a quaternary model of perception and knowledge, in which the mere relation between the object and the sense organ is always only a part of a larger proportion —a relation of relations—, since the object presupposes the world and the sensation of the organ, a complete body that organizes it and gives it a definite sense:

$$\frac{\text{objeto}}{\text{mundo}} = \frac{\text{sentido}}{\text{cuerpo}}$$

The most important aspect of this exacting proportion (object/world=sense/body) is the ignored but ever-present continuity between the extremes "world" and "body", where the world is not a sum of objects, nor the body of organs, parts or entities. In short, the primitive homogeneous medium of reference to which the so-called "fundamental forces" must necessarily be connected, since we already know beforehand that any movement or change of density is just a manifestation of the principle of dynamic equilibrium, either as a sum or as a product.

The body from within is the seat of sensation, the undifferentiated common sensorium from which the different organs have emerged, and without which there would be neither subject nor "common sense." In harmony with this, we can speak of two modes of intelligence, one which seems in motion and pursuing its object, and an immobile one which allows us to listen to our own mentations, and without which they could not exist. Let us try to think without listening to ourselves, and we shall see that this is impossible: the very compulsion to think is nothing but the desire to listen to oneself.

The relation of the motions of the body with respect to its isometric center of gravity as origin of coordinates is similar to the movements of the object-oriented intelligence with respect to the immutable intelligence. Certainly the "space of the mind" does not seem extensive at all, but to verify its intimate connection with the physical it is enough to put into practice any of those isometric exercises in which one remains standing and hollows out simply to perceive the balance in the micro-movements necessary to maintain the posture. If the intimate is the compenetration of the internal and the external, we have here both a physical exercise and an exercise for the intelligence, which allows us to verify the intimate, transcendental relationship between movement and immobility.

As for all the nonsense that says that astrology is contrary to free will, it should suffice with Schopenhauer's well-known phrase: "*A man can do what he wills, but he cannot will what he wills*". Astrology does not deny the deepest freedom, the only truly transcendental one, which can only take place beyond passions and inclinations. In practice, very few reach this place, or do so to very modest degrees. That is to say, as we all know, in practice we follow closely the grooves carved by inclinations and habits.

In other times the passions of human beings had much deeper roots, but great efforts were also made to overcome them. In modern urbanites everything is incomparably more superficial, so much so that people no longer believes that there is anything to overcome. Plants adapt to the qualities of the soil, but they depend first and foremost on their own seed. There are changes in vigor or intensity, but not in the nature of the motives.

*

Perhaps, after all, it will not take centuries to sort out and consolidate a "statistical evidence" of the astral influence on, for example, meteorology. In my land people still spoke of *el astro* to refer to the momentary disposition of the weather, what we now call in a completely empty and abstract fashion *el tiempo* (as in Spanish the weather is "the time"). Astrometeorology was held in regard for nearly two millennia, and still receives credit in countries like India, whose crops depend to such an extent on generous and timely rainfall.

Kepler himself argued for planetary influence on climate in quite a punctual manner that should be easy to confirm or refute: the Sun's conjunction with Saturn would produce temperature drops, and so on. Since these kind of events occur every year, albeit in highly variable circumstances, there is a statistical basis with a much cleaner and defined structure than in many modern applications of data mining and analysis: these indeed are the truly neo-Babylonian practices, in the most familiar sense of the term.

But we all know to what extent statistics are open to debate and manipulation. The data of the Shnoll school, although abundant and solid, have never been taken into account in physics or biology, and we do not envision the heirs of Bohr saying that certain peaks of radioactive disintegration may have been due to a transit of Uranus —but we don't need it either. The same is true of meteorology, which already has its own closed shop of principles, methods and problems.

Cornille, Naudin, and others have developed simple "aether drift sensors"[2] relevant to space weather by simply rectifying the Trouton-Noble's 1903 experiment. Here we have no reservations with these words, provided one has a correct take on the constitutive side of Maxwell's ether problem and its relation to Weber's electrodynamics. Seasonal variations can be measured as with the famous Miller experiment, and if the right kind of test were pursued, we might detect even the equilibrium and reversal points at the solstices and equinoxes. This is a necessary key.

We try to optimize what we choose to measure, which is why econometrics and sociometrics are such useful tools for power. Astrology is transversal to these interests, although not completely, since there is nothing that sooner or later, if it reaches some relevance, cannot be instrumentalized. In any case, astrology, unlike many other branches of statistics, has a solid natural reference and an invariant structure, which gives it a great advantage in the long run in the race of generalized statistical inference that we are witnessing with the new machine learning algorithms.

Said another way, in this New Statistical Babylon the old Babylon could still have plenty of room to grow, but that would only represent the umpteenth denaturalization of a knowledge that should be committed first and foremost to qualities.

*

Newton had no particular interest in astrology, as it is sometimes said without basis, although it is true that he studied thoroughly something way more obscure and confusing in his texts as it is alchemy. There has also been a tendency to interpret universal gravitation as a revamped mathematical version of the principle of universal interdependence, but this could not be more superficial and misleading. Newton certainly reverses the question, which is characteristic of all his researches, but this reversal, far from reintegrating man into the cosmos, isolates him on one of its planes, which can nevertheless expand without apparent limit.

For the universe as a great clockwork as seen in Leibniz and Newton — the two founders of calculus- is certainly different from the great animal of the Platonic *Timaeus* or the organic cosmos of Plotinus. Newton's universal gravity is supposed to operate between *closed systems*, since the Third Principle introduced by the English physicist intends to define precisely a closed system — even if in Kepler's problem such a principle is impossible to assess. Moreover, according to absolute space and time, action and reaction always take place simultaneously, which tacitly assumes a *Global Synchronizer*, something that is otherwise also implicit in special and general relativity even if the application of the Third Principle is hardly scrutinized there.

The traditional idea of interdependence, which was certainly never made explicit because it was taken for granted and because the concept of a "universal mechanics" had not even been raised, the idea, in short, of the great organic animal or cosmos, assumed on the contrary that there are no closed systems in Nature; and since no such thing as a "Global Synchronizer" or Great Clockmaker was conceived, it assumed, by default, that each thing had its own internal organizing principle, which by now we would define as "its own inner clock." Leibniz's monads themselves would not have been far from such a notion had not been defined as "windowless mirrors ".

The difference is that Leibniz's monad is an overtly metaphysical notion, while Newton's global synchronizer, no less metaphysical, has become the tacit assumption of all modern physics. Even the probabilistic quantum mechanics is embedded within it. It is yet another exponent of the inadvertent Newtonian inversion.

Therefore both types of universality must be perceived, with the most profound but unnoticed reason, as incompatible. It is clear that the universal interdependence before Newton did not allude to anything mechanical, but to a resonance or parallelism; only we never quite assume that universal gravitation *cannot be mechanical either*, at least *in the usual sense we give to the three principles and the central forces*.

One could speak of a field theory of absolute parallelism or teleparallelism, not in the sense that Cartan gave to his differential geometry to accommodate general relativity —and which on the other hand has evolved a lot since then- but closer to the parallel transport in the ambient geometry of the geometric phase. This phase or retarded potential is giving us just the "inner time" , and maybe even the inner mood, that is truly the soul of the animal in the macrocosm and microcosm.

In Newtonian mechanics time and distance are functionally identical. But, on the other hand, in the same Newton's equations the mass of a particle like the proton can be written as the acceleration of its radius, according to which mass and gravity are also functionally the same. What does this mean?

Everything suggests that the ground of truth of this theory is purely relational —like the geometric phase itself- rather than mechanical, and that the confusion stems from the unsound assumption that the three principles of mechanics apply identically in our machines and in the fundamental, natural forces. In addition, it is admitted that the real boundary of fields is the conservation of momentum, not the third principle.

This extended conservation principle probably also governs such fundamental organic functions as the blood circulation regulated by the heart and breath and the bilateral respiratory cycle —assuming that their retarded potentials amounts to a geometrical phase. Certainly, this does bring us much closer to the ancient view of the physiology of the cosmos as a great animal, having incorporated only the better part of the modern "geometric spirit" —that which seeks the geometry in the physical problem, rather than reducing the geometry to the reverse engineering of calculus.

Newton's apple would owe its lustre to absolute parallelism, but if this does not seem to have a place in our physical world, it is because we have not given room for bodies either by excluding the possibility of extended, non-ideal particles. Gravity would be nothing other than the weight of absolute parallelism and its flow within bodies, a flow that our theories exclude and block, and which may ultimately be modulated by the body's own self-gravity.

Why? Simply because, unlike Weber's dynamics, all field theories coming from Maxwell, being defined in terms of the particle-field relation, cannot fulfill the third principle directly, and require self-interaction. Noskov's internal longitudinal wave, analogous to the "trembling motion" or *zitterbewegung* of quantum mechanics, would be the self-interaction from within the body of an extended particle. Everyone knows that the drop melts into the Ocean, but few want to know how the Ocean melts into the drop.

The origin of coordinates of a reference frame, as Patrick Cornille demands, should always be located at the center of mass of a particle and has its value. Otherwise the criterion is not physical, but merely mathematical, and yet this is common in modern physics. On the other hand a center of mass is not a purely geometrical question, and if it involves density changes it can also involve a scale transition. Much of physics is based on scale transitions in both time and space, and yet no general principle of transformation is envisaged. In astrology this should be of paramount importance.

*

Tone is the fundamental question connecting physics and astrology, but we do not see how that common ground can be presented in any other way than with Noskov's longitudinal vibrations internal to the moving bodies. His interpretation of the retarded potential, which is also a physical interpretation of the geometric phase without the need to propose a trivial causality, was simply a way of giving content to the conservation of energy which is merely formal not only in Weber but also in the other mechanics now in use.

Noskov further insisted that these vibrations are present in all kinds of natural phenomena, from the stability of atoms and their nuclei, to orbital elliptical motion, sound, light, electromagnetism, the flow of water or gusts of wind: the living actualization of *pneuma*, the vehicle of *logos* in Stoic cosmology. Given the fundamental ambiguity of energy in Weber, it would be appropriate to undertake a careful dimensional analysis of the values of mass and the threefold manifestation of energy in relational mechanics and its implications, which could lead to a number of surprising interpretations and conclusions at the very core of physics.

References

Nikolay Noskov, *The phenomenon of retarded potentials*

Notes

(www.hurqualya.net)

1. http://milesmathis.com/tilt.html

2. https://www.researchgate.net/publication/234879690_Simple_electrostatic_aether_drift_sensors_SEADS_New_dimensions_in_space_weather_and_their_possible_consequences_on_passive_field_propulsion_systems

JEWS, CHRISTIANS AND PHOENICIANS (SATURN, JUPITER, MERCURY & MARS)

31 March, 2021

A different point of view

For once we are going to look for an extra-human and extra-social reference for some issues of our species, namely the so-called «Jewish question» and its relationship with other peoples, and in particular with the great empires. Appealing to a metahistorical element does not mean, of course, that we can dispense with interpretations that will always be highly subjective and human. In any case, I find it more than convenient to take a different angle for this poisoned subject.

Ancient cosmographies were much less complex than the present ones with their vast array of knowledge, but also less reductive. The seven «climates» under which peoples lived throughout the geography also had different influences and were under different poles of spiritual ascendancy; those climates and poles could be represented, in addition, by the seven planets, without the need to be confused with them: «the angel moves the star», it was said to allude to the remote potencies.

At the end of 2020, right at the gates of the winter solstice, the last conjunction of Jupiter and Saturn took place, something which occurs every twenty years and follows a curious cyclical pattern across the four elements of the signs of the zodiac. These conjunctions have always been widely studied in world astrology as major reference points of what they now call «international affairs», and have also been associated, incidentally, with the great plagues and epidemics that recurrently arise. To take just one case, the famous plague of Athens of 430, 429 and 425 has been linked to the also famous conjunction of 424 B.C.E.

If we had to express in two words what the two giant planets Jupiter and Saturn —which account for most of the angular momentum of our solar system- mean, it would suffice to say that they represent the great archons of expansion and contraction, of space and time; a very elementary translation that nevertheless helps to understand many things.

The Jews, «the people of the Sabbath", could only be associated with Saturn, and the truth is that this connection goes far beyond the choice of a day of the week as a period of special observance. Zionist pioneer Leo Pinsker was only echoing a commonplace when he said that the Jews were «the people chosen for universal hatred,» and of course Saturn is the most detested of planets, since it is also the representative of the «hateful old age,» and in general, of everything that deteriorates under the action of time and its scythe. It has also been the planet traditionally associated with downfall and misfortune.

Every empire, while it expands at least, is under the sign of Jupiter; and every power, when the limit of its expansive thrust has been reached, becomes definitively installed in time and involute under the sign of Saturn. These tendencies are as correlative as expansion and contraction; what is of special interest is that different peoples concur in the same imperial endeavor with different functions and tasks.

The descendants of Jacob lived halfway between the first two great civilizations and empires, Egypt and Babylon, and, if we judge by the Scriptures themselves, they soon managed to gain the favor of power, to the point of being given, as in the case of Joseph, freedom of command. The motif is repeated with the Persian empire that ended with Babylon, and then with the Christian project that digs the tomb of the Roman Empire and Antiquity.

The verisimilitude of many of these stories of what the Christians call «Old Testament» is more than doubtful, but the fact that the intrigues for the infiltration and the seizure of power are presented as exemplary stories for the chosen people already tells us enough. But this motif will be repeated in the Middle Ages with the Carolingian monarchs and with the Christian kingdoms in general, all the more so when these gave clearer signs of strength and expansion.

Since the Reformation, with the rise of the northern states, attention turned to the new Protestant creed and in particular to Calvinism and Puritanism, always so impatient to put an end to the bans on usury. Then, the great alliance of Anglo-Zionism was sealed in the times of Cromwell, which has been so successful to this day, and which has shown itself for more than three centuries as the main vector of world affairs. Since this long hegemony is going through a critical moment, and its end is in sight, it would be timely to make some kind of balance.

But as a historical balance never ends, we will try, if that is possible, to shift our perception to other spheres. Even today, the term «Anglo-Zionism» sounds extravagant for many, who are unaware that Zionism was a spawn of Puritanism long before it was adopted by the Jews themselves. Which brings us to the third category of our title, after the Jews and the Christians: the «Phoenicians», who are nothing but the alliance that the richest Jews and Christians have had at the top of power at the expense of the rest of the population, and in the name of free trade, wars, tax extraction and usury or credit; since things so diverse in appearance have too often proved to be under the sign of Mars.

Of course, Mercury being the god of commerce and thieves, one would think that the modern banking system and its way of creating and distributing money, the mother of all frauds, would have to be under the sway of its corresponding planet. But Mercury, being so close to the Sun, makes little difference in the events we call «historical»; its scope is too quotidian, as quotidian is the role of rigging and plunder in the affairs of money, the stock exchange and the global casino.

But to expand markets on a grand scale, looting at its greatest, raise taxes, and skyrocket public debt requires the warmonger impulse of Mars, and if wars do not break out, let them at least look like a credible threat. If fraudulent money is our daily bread, so ordinary that we are already desensitized, war is the great fraud for the critical moments, where there is no choice for those who have already been well squeezed but to pay with their lives, their liberties and their hardships.

It has been a long time since major wars have been fought in the West without the blessing of the banks and the money masters, so to look for other causes is not only idle but blatant distraction. If this has the status of certainty for the last two or three centuries, all the more reason for it to be linked to future movements, now that the economic and monetary war is reaching new scopes.

For example, if with the new digital currencies China, Russia and other countries can circumvent the economic sanctions, apparently the only diplomacy left to the empire, we can expect an intensification of cyberwarfare and cyber-attacks not only on those countries but also within the imperial domains to sow panic and keep the sheeple in the desired monetary line. Even the Economic Forum and Israel's highly trained Cyber Directorate are clearly announcing an imminent "Cyber Pandemic" as "the Next Global Crisis".

Admittedly, calling the select imperial parasitic class «Phoenicians» is quite well chosen: they are neither «Jews» nor "Gentiles», but they are certainly much closer to Jerusalem than to Athens. It is not simply that half of the great fortunes are Jewish, it is that the other half, which has granted them all their privileges and is so often openly pro-Zionist, depends on their initiative. The relationship between deception and force remains that of master and disciple.

Why have the Jews been the initiators and instigators of the local powerful? Because in the West those who came to power by force lacked the talent to exploit their prey to the full; hence the kings left the treasury and tax collection in their hands. This, as far as the economic milking of the population is concerned. And as for politics, they trained them in the art of graduated management, since there is almost nothing that people do not get used to if it is done with the proper gradation —that is the true meaning of the expression «to take measures».

War is largely based on deception, all the more so for the weaker. According to their own scriptures, the Jews are called to lead all nations, but a much smaller people can only prevail by systematic deception, which in this case is not only permitted but practically sanctified, provided it serves the Mission. The rabbis gave the sense of mission to the lie, which in the ears of the powerful must sound like heavenly music. Hence the attraction of Zionism for an ilk that has no other aim in view than to maintain and enlarge their privileges. Naturally, the systematic use of deception presupposes that one is at war with those who are not of the tribe, a perspective also convenient for the Gentile «elite».

I am just summarizing the reasons for this profound symbiosis, which for so long has been pointing to the consolidation of «a state within the state»; because I have no intentions to dig into the details of this abject and mostly unknowable domestic history. The dialogues of Goethe's fatuous Faust and the solicitous Mephistopheles may give a certain literary tone to the question, but the reality must have been way more uglier.

Structure, Dynamics and Fate

In other articles we have spoken of the power law that x-ray the inverted pyramid of wealth, in which 1/5 of the population takes 4/5 of the wealth, and so on in a recursive fashion; we also noted that the fact that in this iterative process of accumulation, there is a 50/50 share between Jews and Gentiles, 2/5 for each half, it is a revealing «x-ray of the x-ray». Then, could it be pure coincidence that the ratio between the period of Saturn and Jupiter, just under 30 and 12 years, is almost exactly 2/5?

Well, perhaps the thing is less esoteric than it seems, if we remember that Saturn represents the continuous action of time —and compound interest — and Jupiter the cyclical phases of expansion, in this case of credit and business cycles. Since we took almost for granted that this extreme concentration, which leads some 3 families to have most of the surplus in purchasing power, had to be related to the structure of debt increased and distilled over many generations.

Credit to wage wars, and the intrigues to instigate them, not only made the usurers much richer but also left the levers of power in their hands. The democratic states were still much more vulnerable, since the creditors could always take over public assets, and this was the real driving force behind the determined promotion of democracy. In this last phase of dollar monetary ultra-imperialism the society as a whole became oversaturated with credit, to such an extent that only by selectively destroying and stripmining the economy new room can be made for excess capital. This is what is happening right now, for example.

Those who already have almost everything may not have much to gain, in absolute terms, but in relative terms they can still greatly increase the dependence and subjection of the masses. The fear of being overthrown sets in motion an implacable logic. Who cannot understand this? Didn't they do much worse things in the last century? And has not their leverage of power increased greatly since then?

That well-known Forum, a mere front for those who do not want to show up, blatantly probes us: «in ten years you will own nothing... and you will be happy».

*

Impossible to come to terms with men. Yet we are doomed to come to terms with much more remote powers, which only by the literary grace of myth have we clothed with human traits. And we are doomed because such powers have already come to an agreement with us at birth.

The Jew is hated because Saturn is hated, not the other way around. So it is... at least to a good extent. However, we know nothing about Saturn, except some hateful manifestations, as if they had stolen something from its origin. The planets are only splinters of a fractured order, and in circling they do their best to indicate the principle from which they emanate.

Santayana said that Protestantism was anti-ascetic and so was Judaism. It is a keen observation; what he did not know is that Catholics would become much more anti-Ascetic than the Protestants and Jews he met. After all, many Jews faithfully observe the Sabbath, and fast at least one day a year; appreciating inactivity and fasting from time to time had to be unavoidable in the «natural religion» prior to Revelations and Laws. But who does so today among Christians, whether Protestant or Catholic?

We abhor the majestic fields of Saturn, we cannot bear even the exquisite boredom of a Sunday afternoon. And he who does not come to terms with Saturn, embodiment of Karma, will have to feed him as an unwanted intruder. The same is true of Shiva, a god so different in many ways. Denying death kills us daily, and it is others who take care of our destruction. One is supposed to take care of these things oneself.

Although I have not devoted time to these matters, sometimes I wonder if the the ritual of the scapegoat, sent into the wilderness and described in Leviticus has not been used in the most abusive way. Everything that has been considered sacred or impure and is used as pretext ends up having fatal consequences.

Postmodern lightness flees from Saturn with the results we observe. But what can we say about Jupiter today? Well, in the last four decades it has hardly been more than the expansion of credit, with the Fed as the driving force worldwide. In other words, Jupiter had already been absorbed by banking before delivering the new joviality.

Have the Jews taken advantage of the favorable moment in each imperial swing or have they been the instigators? For Wiesenthal it was no coincidence that Columbus set sail for America just when the expulsion decree expired, on August 3, 1492; but, naturally, he was not the first to notice it. Whether he came from Genoa, Pontevedra, Majorca or wherever, the only thing that can explain the many unknowns of such a dubious admiral is the very nature of his origin. Of course, Wiesenthal says what he wants, and does not tell us about the notorious Jewish aspect of Eichmann, who even spoke Hebrew and Yiddish.

And that happened after the supposed «cleansing of blood» of Spain and the colonies. A good part of the great criollos' fortunes in Hispanic America, probably half of them, belong to families with abundant Jewish blood. What could have happened within the highly philo-Judaic Puritan environment, whether in England or in the United States? The East India Company and the City have been for long the headquarters of the Phoenician Navy, and even today it continues to be the main nest of thieves in the world. Hilary Belloc already noted that at the beginning of the 20th century there were in England more aristocrats with Hebraic lineages than without them; it was the only possible result of such a devoted admiration.

In all the imperial expansion of the West, the Christians have put force and the Jews, always ahead, foresight and intelligence. This would be a mere truism were it not for the vanity of the former and the justified caution of the later. But my intention here is not to settle accounts with anyone, since these issues have already been dealt with in depth by many authors, but to raise the question of what these two arcane powers of expansion and contraction would be like if we were to deal with them directly and without facilitators.

*

It is totally impossible to understand the dynamism of the West and modernity without understanding the duality inherent to expansive and conquering Christian peoples and Jewish communities with a vocation for exile and diaspora. For it is evident that the synagogue chose exile as a strategy of maximum growth as well as maximum control over its community. If there is a Jewish genius for organization, the distance with respect to all things and the psychology with respect to those who do not keep it, it derives entirely from having chosen these special circumstances, undoubtedly inspired by the Saturnian pole through which they related to the Principle.

Jewish leadership has tried to direct the excess of forces of younger peoples to its own benefit; but the relationship with those peoples has been basically from the top down. It is not these peoples who have welcomed those others who were looking for a living, but rather the upper echelons of power, to whom the rabbis have addressed themselves directly to propose advantageous deals —credit or promises to increase the treasury in exchange for privileges and concessions such as the collection of taxes. As Guyenot observes, it is generally false that the Jews have been allowed no other way of life than usury, and on the contrary they have often been granted privileges forbidden to the bulk of the population.

To spare us arguments, we can only refer the interested reader to the writings of Laurent Guyenot, Michael E. Jones, Israel Shamir, Simone Weil, Douglas Reed, Roger Dommergue, Robert Faurisson, Miles Mathis and many others. It is not necessary to share everything they say, far from it, to realize that they are much better informed than average so-called historians, and on the contrary: if people still swallows massively the bait of propaganda, it is because of their staggering ignorance of history and scriptures, as well as pure masochism and overt feeble-mindedness.

In the traditional iconography, the scorpion has been considered the symbol of the Synagogue and of the Jewish people in general, which is easy to understand as an allusion to the zodiacal sign of Scorpio, the nocturnal and aquatic domicile of the planet Mars. The fact that an igneous planet is in a water sign is directly associated with ferment, poison and venom as well as death. This results in the aforementioned association with Saturn and the deleterious powers; and undoubtedly the Jehovah described in the scriptures more than jealous seems to be a psychopath ecstatic with destruction, particularly that of the non-chosen peoples who are all but one.

In any case it is clear that the expansion of the West is one and the same thing with its loss of internal cohesion, something that every empire is called upon to verify; the United States is just but the latest instance of a process repeated time and again. However, what is characteristic of Western empires is the much more intimate symbiosis or connivance of two different interests in the direction of power, something only mitigated to the extent that the more autochthonous and unthinking part ends up accepting the initiative from the more farsighted one. Needless to recall that Jupiter and Saturn represent the domain of space and time.

*

This intimate «symbiosis» was only possible thanks to Christianity and its acceptance of the Old Testament as a sacred book. No doubt a certain degree of knowledge can be attained regardless of one's religion, but not everything is knowledge in man. Nor is it my intention here to attack any faith, since they are already too much assaulted and offended both by the ambient stupidity of the media and by their targeted campaigns. If religions are so harassed, there must be something good in them, one immediately thinks; for there is no religion more petty or narrow than that of believing that everything is social.

However, this cannot prevent us from seeing the disastrous role that Christianity has played as a Trojan horse against the integrity of human nature and as a moralizing infection; Guyenot is not wrong when he calls it a «holy hook». There can be no greater conquest or more profound colonization than for one to strive to place an anomaly at the center of one's being; from that moment on, one can well say that one's fate is sealed.

It is of no use that in recent centuries people have lost their faith, for since then they have become even more vulnerable to propaganda and bland moralism: to the new secular religions of Marxism, holocaustism and warmism, the first two being Jewish to the core and the third simply Phoenician. If one has doubts about global warming or the gas chambers, one automatically becomes a «denier»; and these undesirables have now been joined by a third group, those who have serious doubts about the present epidemic and its origin.

But in view of the kind of people who use the term, being considered a «denier» is as close to a compliment as can be expected. For it is the kind of disqualification invariably wielded by cowards who rely only on the number and strength of the current of opinion, not on argument or reflection. Just as cowardly and false is the use of the term «anti-Semitic»: it is extraordinary that no one speaks of the existence of a «Semitic race» and yet «anti-Semitism» is continually spoken of as a form of racism.

Evidently, those described as «anti-Semites» do not seem to have anything in particular against the speakers of Semitic languages, who number more than 330 million, mostly Arabic, including the Palestinians. Why then Steinschneider's term has become so popular, not only among Jews, but also among the public in general? Why they never say anti-Jewish, which would be the simple and appropriate term? The only possible reason is that they seek to avoid the term «Jewish» and its derivations to be permanently on the lips of those who maintain controversies.

Since «anti-Semitism» is a foolish, false and coward term —which is now heard and read even in the discourses of the far right— the only thing to do is to call things by their name and not to fall into the conditioning of the masters of discourse. The same can be said about the infamous gas chambers. How is it possible that in all the memoirs of Eisenhower, Churchill and De Gaulle, which fill more than 7,000 pages and were published between 1948 and 1959, there is not a single mention of that chambers?

We will not dwell on topics that have already been covered in detail by some of the authors cited. It is clear that we have witnessed the creation of a new religion to condition the memory of the masses and lobotomize entire peoples such as the Germans. But when we see the biggest political scoundrels in every corner of the planet reminding us that it is «International Holocaust Remembrance Day», we know what to expect.

Figures in the Sky

According to Wolfram von Eschenbach, Kyot said that Flegetanis had found the mysteries of the Grail written in the stars. Undoubtedly more than one secret could be read in the figures and configurations of the sky today, for both above and below demand serious adjustments, in these times as in any other, but each one in its own way.

The extreme slope of the inverted pyramid of wealth, that diaphanous mathematical structure that economists and sociologists ignore so well, can only be maintained on condition of dividing the population equally to extreme degrees. To divide, poison, neutralize, domesticate and castrate it as far as possible, with all the resources of control, the media and science at its service.

This extreme concentration has two basic aspects: dynamics and structure, which also correspond to the respective functions of Jupiter and Saturn. As dynamics, today it translates above all into the regulation of the monetary flow; as structure, it is the same self-similar distribution exhibited by the power law of the Pareto distribution, which entails from top to bottom a whole socio-economic chain of servitudes and favors whose functioning we shall seek in vain among the so-called experts. Of course the dynamics, usurped by private banks for centuries, is overwhelmingly at the service of the maintenance of this structure, with its enormous dead weight projected from above. It is self-evident that the so-called «dynamism of capital» is becoming less and less dynamic, for all apparent movement, and in particular the permanent «technological revolution», is also increasingly at the service of the control of the social apparatus.

Saturn dominates the heights but also has with it the fear of fall, which creates a paranoia proportional to the rate of the slope. In the zodiac the heights are represented by Capricorn, gate of the winter, domicile of Saturn, detriment of the Moon, exaltation of Mars and fall of Jupiter: the constellation of power par excellence. Of the managed power, it is understood.

The greatest vulnerability of capital is its own concentration: it would be enough for the names behind the three or four biggest fortunes to be made public, and for them to be subjected to the same surveillance to which the bulk of the population is now subjected, for the balance of power to change decisively. Everything these people have is due to concealment and lies, and they cannot survive without them; they resemble so much to vampires that would vanish into thin air with the first light. However, lots of people work for this to not happen.

Their chronic dependence on lies at all levels intoxicates the entire society, permanently inoculated with the most varied poisons. Propaganda and psyops saturates nearly every message delivered. It is impossible to recover health without putting an end to this chronic intoxication, and the only conceivable way is to dismantle the leadership as with any criminal gang.

*

There is a fabulous being that combines the characteristics of Saturn, Jupiter and even Mars: the Dragon, especially in its Chinese version. China is today the only great threat to the Anglo-Zionist Phoenician hegemony, and the attempts to infiltrate it and reduce it to vassalage have failed miserably. So far, there has not been a single transfer of hegemony without those great tectonic adjustments that we usually call wars. But how far will Phoenicia be willing to take the hostilities?

It will depend on geopolitical lineaments, although soft and not-so-soft hybrid warfare is already in full swing. The United Kingdom left the European Union, but it does not break with Atlanticism, the instrument of American military domination. The United States will not allow good relations between Russia and Germany, and the German government is compelled to repeat its secular mistakes. The Anglo-Saxons always pitted Germans against Russia to divide and conquer, and by now they should have learned something.

It is proven that the bankers of the United States and England intensively financed Hitler[1], to the point of owning a good part of German industry. Germany never intended to make war on the Anglo-Saxons, but Phoenicia had no doubt about what they were doing. Can anyone be surprised at the final result?

London was a Trojan horse for the Union and even outside it, it continues to be its enemy, now with more freedom of maneuver. Indeed, the mercurial capital of England is still irreplaceable in the Phoenician schemes, for it is not only the historical center of the spider's web, but also the refuge of last resort. But surely, while the United States is trying to separate Europe from Russia and China against its interests, England stands to benefit from it.

To say that the London financial sector wanted to remain in the Union but that the popular will chose to leave is nothing but a joke. There is hardly anything relevant in England that has not passed directly through the Square Mile for over three hundred years; but by claiming otherwise, they can still blame the public for the consequences. Despite intense competition, London is still in fine form as the capital of hypocrisy in the world.

The United States is collapsing miserably, and if its vassals follow it, it is because they have no room for maneuver. If China is a threat, it is above all to the status of the most privileged, who seem to have much to lose. Most people could not be more sold. Only that class interests seem to be far above national interests.

Male, female and neutral

Let's leave the cheap politics and current affairs to touch at least some deeper questions. If we consider the internal dynamics of the West, with its two poles in Athens and Jerusalem, Jupiter and Saturn, expansion and contraction, a good question is who has played a male role and who a female role in this intense relationship.

And the first answer should leave no room for doubt: it is the Jewish culture that has played a masculine role, impregnating the gentile culture, once the pagan culture was destroyed. In other words, the West, beyond its expansive front, is feminine and reactive with respect to the ferment that swells and drives it. Undoubtedly the essence of this complexion is due to the implantation of Christianity among young and uncivilized peoples.

The lowest reading attributes the animosity of Christians against the Jews to their guilt for the death of Jesus Christ, Judas and all the rest. No need to consider such arguments, more typical of folklore than anything else, and that do not even contemplate the other side of the animosity. But I believe that, at a much deeper level, what younger nations have always tried to rebel against is the impregnation of a foreign body, against that unwanted pregnancy that can lead to the extremes of madness.

I am not writing to be politically correct, but as part of a West that is trying to reach a higher degree of self-consciousness. A word as false as «anti-Semitism» is not even designed to stop the blows against the Jews, but above all against their ruling caste and the Phoenicians. Moreover, it is these Phoenicians who are the first to have betrayed the less favored Jews when they saw fit, just as they then tried to make up for it to silence their conscience the only way they know. At least this is what can be deduced from the most recent times and the last great wars.

The infinite but not free stupidity of the media still wonders about the causes behind «the rise of anti-Semitism», but only to someone who does not follow these media could it seem mysterious. For, apart from the fact that all that trash already make hateful enough everything they try to sell, it is also not at all mysterious that a number of things are detested by themselves, such as systematic lying, usury, control of speech, direct and indirect inoculation of guilt, and, to top it all, even the attempt to control hatred itself. Just as it is particularly odious to have to admit that one has been duped.

These people really have no sense of measure, nor do they know what the word «counterproductive» means. Surely the average Jew is incapable of being innocent with respect to the non-Jews, but in their self-interest there is still a huge dose of naivety that never ceases to amaze. Someone might call it their «blind spot,» and it is well known that no one breathes without having one. But as for the Phoenicians who resulted from the mixture, I doubt that there remains in them the slightest trace wether of innocence or of naivety.

If modern «criticism» today so often tries to minimize and relativize the importance of race, gender, or religion, one can be sure that it is not by chance or ignorance, but by mere strategy in the current wars of opinion. If I have chosen here another key completely unrelated to the ones they use, it is not because I do not believe that the others are unimportant —apart from the fact that within such conditioning factors we would have to include things that are no longer even contemplated, since race is not just a question of «genetics», for example.

In any case, for the avoidance of doubt, I will say that astrological matters are not reduced to a lax question of psychology, whether superficial or profound, and that it truly shows a chronological correspondence with the facts of history and the passions and inclinations of men, and even with his motives and plots. As an example, I propose to study the expulsions and persecutions of Jews, whose number composes a «statistical base» of a few hundred of perfectly dated events, in the light of the keys here just pointed.

Let us now return to the gender of certain influences or tropisms, such as that of Jupiter and Saturn. We have ventured that the Jewish spirit, its intelligence, would have played an active role with respect to some passive souls —with respect to that double male-female background of will and desire. This itself suggests that the matter is much more complicated than what a few words can cover and would take a separate study, but here we will only point to something essential.

In reality, the utilitarian intelligence is also a reactive intelligence and therefore secondary; it is the restless spirit at the service of the life that flees, not that spirit that does not need movement to gather the aspirations of life. Between the two there is as much difference as between Earth and Heaven. On a different plane, one may ask in what sense the contraction of Saturn and the expansion of Jupiter are feminine and masculine, and in what sense they act the opposite way.

This is a recurring theme that we already touched upon in a book dedicated to the Pole, recalling that the Chinese and Japanese often gave an opposite meaning to the terms yin and yang in terms of contraction and expansion as well as full and empty. Such discussions of qualities may seem specious and interminable, but in reality they only call for a clarification of the context, since they can be used in clearly different domains.

Thus, as we have already noted repeatedly, Jupiter and Saturn represent expansion and contraction but do so in the completely different domains of space and time. Otherwise it is clear that Jupiter is material plethora seeking expansion, while Saturn is amplitude of space as depleted matter. We have here already three different categories, space, time and matter. In physics and mathematics there are well-known ways of treating dualities in such terms, although we will not dwell on them now.

We will instead give a key more in line with the spirit of this article, which concerns the fundamental tone of each planet. The septenary or week contains, in consonance with the seven traditional metals and their chromatic shades, three masculine, three feminine and one neutral planet. The masculine are, in ascending order of purity, Mars (iron), Venus (copper) and Sun (gold); while the feminine are, in identical order, Saturn (lead), Jupiter (tin) and Moon (silver). Another arrangement has on the male side Jupiter, Mars, and Sun, and on the female side Saturn, Venus, and Moon.

The most relevant of either of these two arrangements is the mediating role it gives to Mercury as a neutral planet and axis of symmetry that connects the others. Let us say that Mercury, which also represents movement and intellect, holds the secret of the transition between the masculine and feminine of the planetary potencies, as well as their reversibility.

Worldly, external aspects of Mercury are money, commerce and the media —the great modulators of everyday life. Of course, the fact that today we talk much more about the media than about communication itself already reveals how far the instrumentalization of the most ordinary acts goes. Another less ostensible but ubiquitous mercurial issue, and always in close contact with the media, is the intelligence services.

It is entirely within the normal order of things that Mercury, as modulator par excellence, is the most unnoticed of the planets; just as nothing is less noticed by the intelligence than its own nature. Both determine and are continually determined, but we are unable to follow the line which describes their true neutrality, and no doubt it is only in that sense that Mercury could be seen as the ageless spirit of wisdom.

The Egyptian Thoth, numen of writing, was a lunar god; latter writing became with the Greeks a mercurial phenomenon, and today we are about to make it an attribute of Uranus, patron of the electronic age and «superior octave of Mercury» according to some, who, however, we already know was castrated by Saturn; and the fact is that for things to acquire their proper form some like to cut out a good deal of possibilities. However, the wise, the neutral mercury, knows the route to the solar Word, and nothing but unconsciousness can obstruct its passage: there is a supravirile power in the herald of the Sun .

Notas

1. https://www.voltairenet.org/article187569.html

METANOIA, CONTINUUM AND QUATERNITY

4 April, 2021

(This is a translation of the final chapter of the essay "Espíritu del Cuaternario" (Quaternary Spirit)

After so many considerations on science we need to return to a more general perspective.

It was probably as a reaction against the intrinsic idealism of the trinitarian symbol that a number of thinkers of diverse orientation turned in the twentieth century, and especially after the post-war period, towards quaternary schemes as symbols of totality. Jung was maybe the first perceiving the need for this shift, followed later by such well-known authors as Heidegger with his earth-heaven-celestials-mortals quaternity, or the Schumacher of the magnificent "Guide for the Perplexed" with his fourfold field of knowledge: inner self, inner world, outer self, outer world, opposed two by two as determinants of experience, appearance, communication and science.

Another one of those thinkers was Raymond Abellio, who also presented a quaternary model *of perception* and thus of knowledge, in which the mere relation between the object and the sense organ is always only a part of a larger proportion —a relation of relations— , since the object presupposes the world and the sensation of the organ a whole body that organizes it and gives it a definite sense:

$$\frac{\text{objeto}}{\text{mundo}} = \frac{\text{sentido}}{\text{cuerpo}}$$

The most important aspect of this exacting proportion (object/world = sense/body) is the ignored but ever-present continuity between the extremes "world" and "body", where the world is not a sum of objects, nor the body of organs, parts or entities. In short, the primitive homogeneous medium of reference for any phenomenon, motion or force, since we already know beforehand that any motion or change of density is only a manifestation of the principle of dynamic equilibrium, either as a sum or as a product.

The body from within is the seat of sensation, the undifferentiated common sensorium from which the different organs have emerged, and without which there would be neither subject nor "common sense." In harmony with this, one can speak of two modes of intelligence, one which seems in motion and pursuing its object, and an immobile one which allows us to listen to our own mentations, and without which they could not exist. Let us try to think without listening to ourselves, and we shall see that this is impossible: the very compulsion to think is nothing but the desire to listen to oneself.

Abellio proposed an "absolute structure" of space to which the "movements of consciousness" would also be referred, and which would be nothing but the axes of ordinary flat space with the traditional six directions.

The relation of the motions of the body with respect to its isometric center of gravity as origin of coordinates is similar to the movements of the object-oriented intelligence with respect to the immutable intelligence. Certainly the "space of the mind" does not seem extensive at all, but to verify its intimate connection with the physical it is enough to put into practice any of those isometric exercises in which one remains standing and hollows out simply to perceive the balance in the micro-movements necessary to maintain the posture. If the intimate is the interpenetration of the internal and the external, we have here both a physical exercise and an exercise for the intelligence, which allows us to verify the intimate, transcendental relationship between motion and immobility.

According to Abellio, "the perception of relations belongs to the mode of vision of the "empirical" consciousness, while the perception of proportions is part of the mode of vision of the "transcendental" consciousness": this would apply to the four parts or moments of the quaternity already expressed. As an heir of Husserl, Abellio makes a great effort to go beyond conceptualist schemes but still remains within the bondage of knowledge. Global awareness or "consciousness of consciousness" is not the object of an infinite regression only by virtue of a sort of ultimate "passage to the limit" that keeps reminding us of such things as Peirce's "ultimate interpreter".

The ultimate dilemma of understanding, as Siddharameshwar puts it, is that without detachment there is no knowledge, and without knowledge there is no detachment. Now, this detachment is not the mere separation that we take with respect to the object, but something more deeply connected with our will. We suppose that we want to know, but know what? We do not even know that, and perhaps we do not even want to know that. Scientists strive to find the solution to inherited problems, but it is way more important to know how to put aside the questions that one has not asked oneself.

The empirical self cannot be the operator of global consciousness —consciousness without object-, and the "transcendental self" is only a name for that which never says "I" and doesn't need it: for that body-world continuum within which sense objects appear. That continuum sometimes gives us some knowledge, if we aspire to deserve it, perhaps with the sole object of bringing about the detachment of the intelligence and the self that are normally conscious of adhesion and separation and live by their alternation, but lose ground in the middle.

So it could be said that all global knowledge is simply a grace of which the empirical self is the object in order to facilitate its detachment and give it some aptitude; and it is perfectly within logic that it can only arise beyond the desire for particular knowledge, as grace is identical with being, the non-particular par excellence. But there is also around this neutral word, "being," a transubstantiation of intellect and will.

Abellio's fourfold proportion points to a succession of thresholds, of ever-widening perspectives, but which we should not see only as angles of knowledge, but as degrees of participation in being, arising from a double movement of assumption and incarnation. Of course, this double movement also occurs in scientific knowledge, but as long as Nature is just an object we are talking about two types of knowledge that are not even comparable.

In this context, the *metanoia* or metacognition can in no way give rise to an infinite regress, because what it actually implies is a repeated transformation of the mutable or apparent with respect to the immutable that by definition cannot be seen.

On the other hand, the "turn towards the body" of the most recent philosophy is too partial not to be easy prey to instrumentalization —much has been said about sex and "desiring machines" but the truth is that desire, which is a feminine agency, acts more in the soul or even in the spirit than in the body; while the will, which is much more literally confined in bodies and is a masculine agent, is systematically ignored. Basically, the body continues to be seen as an object, even if on the other hand science prevents us from considering an external desiring, naturing Nature.

Considering the bodies it is obligatory to mention the fundamental ambiguity of the relational mechanics inaugurated by Weber with respect to the three energies —kinetic, potential, internal— which, although a mere consequence of the equations, seems nonetheless natural, and which should be taken into account with respect to certain balances and proportions.

Noskov's longitudinal vibrations internal to the moving bodies, postulated precisely to justify the conservation of energy which in Weber was merely formal —no more and no less than in the other mechanics— are at the core of the issue, and it is easy to see how they should "fit" within the data of classical mechanics, quantum mechanics and relativity —and in the parallel transport of the so-called geometric phase. In general, in any field when distinguishing between particle and field we have a *self-interaction problem* under accelerations that quantitatively coincides with Noskov's oscillations. This view in terms of resonance is in harmony with the ancient, and much more timeless, conceptions of Nature.

As we have already seen in different places, the retarded potentials and the corresponding oscillations would not only be present at the micro level, but also in complex organic systems such as respiration and blood circulation.

The Solar Verb, the arouser, *Savitr*, is totally incomprehensible in our objective representation of Nature without the twin notions of internal vibration and external resonance; and these notions are naturally connected when we put force and potential on the same basis. To get another view of the physical Continuum, we would have to deal yet with other questions of relevance, such as scales transitions in length and time, fractional dimensions and fractional operators, but this would lead us back to complex matters; in any case we have already suggested some relationships.

"Help me and I will help you", today as always, this is what Nature tells us. Modern science, as imposing as it is, ignores this connection, which passes immediately through one's own body but extends to the limits of the world.

*

Greek science was a science of observation, while to arrive at experimental science, in which man unites physical and mathematical manipulation to take possession of Nature, we must wait for Arab science with authors such as Alhazen, whose great work is now a thousand years old and who clearly prefigures Galileo six centuries in advance —though it is clear that it was in the West where this experimental stage reached its peak.

Although today the vast majority of scientists remain confined to this stage, the only one of interest to the social apparatus and the powers that govern it, there are some for whom mastery over Nature expressed in Laws that allow prediction represents only the outermost possible limit of knowledge.

Today we are in a position to arrive at a third stage of scientific reason founded on self-observation, which reintegrates mathematics and the variable observation of phenomena in the universality of the Self, the undifferentiated space that is the basis of any intellection. The self-interaction present in that field clearly points to the reflexive aspect in the principle of equilibrium from which all mechanics can be derived.

This third stage of reason will be able to manifest itself to the extent that we understand that there is no law of nature that does not depend on self-interaction, just as there is no consistency or tautology or circularity in current physical theories that is not expressly elaborated. The concept of self-interaction should help us to see instead how and to what extent a physical system is effectively open or closed.

On the reflective path of this self-observation, mathematics, as "pure form", can still find another way of endowing itself with content when mathematics and physics seek a balance between description and prediction. And it is to be hoped that this will tend to happen *spontaneously*, to the extent that the obstructions created by the unilateral justification of a given purpose are removed.

Calculus is the best example of this, and since it is analysis itself that makes quantity take command to its present extreme, combining the idea of infinite division with the construction based on indivisible elements, we have counterposed two alternative ways for calculus based on the other two extreme forms of the indivisible: the unit interval and division by zero. Both would return us to a sort of continuum somewhere between the current conceptions of the physical and mathematical ones.

On the other hand, it could not be more significant that, now that the cybernetic control model takes command over the scientific and social rationality, we find the principle of feedback at the very basis of natural "systems" governed by "fundamental forces". This brings us face to face with the other essential aspect of differential equations, the boundary conditions, which are also interface conditions of the system.

All modern calculus since Kepler's problem is a reverse engineering on the global contour of a system, which also involves an inversion of the relation between what is conceived as the "core" and the "periphery" of it —the internal domain of the function and its boundary conditions. But, from a strictly descriptive point of view, it is the boundary that has shaped the dynamics, and the integral, what may well be called a pseudo-differential; and only the predictive power and the magic of variable assignement have led us to forget this fact.

The same logic of usurpation operates in the now indiscriminate application of the "cybernetic reason" to which we are all subjected. However, applying relational logic to a system such as the blood circulation and the heart according to the Weber-Noskov retarded potentials, the difference between margin and center vanishes —in a case where such a distinction could not be more relevant.

It is then of particular value to try to bring the problems of dynamics back into the reflexive domain of an evolution detached from a homogeneous medium but still dependent on it, since any feedback can only work by the contrast of something heterogeneous with its homogeneous background. Although the crucial question would be to check how the mathematical function is connected to the the regulatory principle in its direct functional aspect, for example, through biofeedback.

The case of the pulse wave could serve as a perfect illustration of something much more general. The essential thing about biofeedback here is not its capacity for modification, which is always so limited; in fact, ideally this capacity would have to be nil, since what is involved is *the interpolation of the subject himself in an objective functional reality*. If the modification of organic functions can be induced, even then that serves above all to see that its principle of action or regulation can be qualified neither as conscious nor as unconscious, neither as voluntary nor as involuntary, since any of these characterizations contradict itself.

We also speak of a system traversed by a retarded potential that is open but tends as much as possible to stabilize as a closed system, and should therefore exhibit asymptotic features.

It is usually assumed that asymptotic behavior is the simplification of a more complicated particular case; but we can also see it from the other side. For every asymptotic aspect, as an exponent of a discontinuity, defines more limited boundary conditions with respect to a more general case, from which it follows. In this sense, asymptotic analysis unwittingly lies in the trace of the universality of the Self.

Applied mathematicians and physicists as competent as Kurt Friedrichs or Martin Kruskal have argued that asymptotic description is not only an appropriate tool for the analysis of Nature, but that it points to something more fundamental, and we cannot agree more with them. Undoubtedly, interest in this field is growing steadily in the age of computation, since asymptotics is the natural mediator between numerical methods and those of classical analysis; numerical analysis being more quantitative and asymptotic analysis more qualitative. But even if we are apparently talking about a method of applied mathematics, we cannot fail to notice in it a symbolic component deeper and more evocative than in the so-called pure mathematics.

The object of asymptotics, as a heuristic, are the limit cases; but the whole analysis is permeated by heuristics insofar as it is application. It would then be of great interest to see what new developments the asymptotic analysis can acquire between the two poles of the indivisible we have commented. The constant differential calculus, as a method of finite differences based on a unitary rate of change, seems much closer to numerical analysis, but allows, together with the relational criterion of homogeneity, to scope out way better the dimensional analysis and boundary of problems. It also presents a much more basic consistency than standard calculus.

Then "asymptotology," as Kruskal called it, still has much to gain in consistency, depth and unity, and not only by virtue of the new perspectives that alternative forms of calculus may open up. There is good reason to believe this, beyond the fact that the consistency of classical analysis is one of results rather than of procedure: because it points as directly as possible to simplicity from the point of view of the limit, and the simplest use to hide the most folds and unnoticed dimensions; because it lies at the basis of field and potential theories besides thermodynamics, because it tends to connect different theories and to define the transition between different scales, because it connects with the more general projective arguments of geometry and is at the very heart of pure arithmetic as well as the most elementary calculus, and because it is a signature of equilibrium, stability and self-interaction.

"Asymptotic analysis" is practically everything the applied mathematician does when he is not doing numerical analysis. For historical reasons asymptotics is associated with perturbation theory —first in celestial mechanics and then also in quantum mechanics- as well as the study of singularities. However, constant differential calculus makes it possible to treat celestial mechanics without the perturbative approach, and the velocity-dependent forces in relational gravitodynamics make black holes unfeasible. Like many other branches, the emergence of asymptotic analysis in perturbation theory has much to do with contingency, reflecting the punctual fact that it was the first important question in mechanics outside the direct scope of the conventional methods of calculus.

Throughout the eighteenth century, in the wake of Newton, absolute space was seen as the element of contact between the conditioned and the unconditioned, between the things of the world and their Creator. In contrast Leibniz, the other founder of calculus, expounded the idea of space as a set of relations, and both notions had a decisive influence on the Kantian transcendental doctrine of the ideal character of space and time, which attempted to propose a synthesis. However, Pinheiro has shown that neither of these two positions explains satisfactorily something as basic as the whirlpool of water in Newton's bucket experiment.

The asymptotic features would be essential, and not merely incidental, insofar as every observable system is a separation from the continuum or homogeneous medium in which space, time, matter and motion cease to be discernible. In the absence of a complete characterization, they shows us some of the most notorious aspects of its relative differentiation, some of its "boundary layers", to use a very general concept from fluid mechanics.

Thus, certain aspects of asymptotic evolution would be both an index of individuation of systems and the most hyperbolic symbol of the absolute that is in the midst of everything and has no contact with anything. Gaudapada spoke in his Mandukya Karika of "contactless" union, *asparsha*, although it seems clear that what does not admit of contact does not require union either. And yet Nature seeks in its own way and in every way the unattainable that resists nothing —surely to compensate for the fact that it has separated itself from it. Venis' morphological sequence also echoes this evolution as an individuation process, and it is no small matter that the asymptotic enters the projective hypercontinuum of appearances.

Let us return to the circuit of the blood pulse seen as a delayed potential with an internal wave a la Noskov. Let us recall that for Noskov these oscillations permeate everything, very much in the style of the stoic *pneuma* that conveys the *logos*: "*It is the basis of the structure and stability of nuclei, atoms, and planetary as well as stellar systems. It is the main reason of the occurrence of sound (in particular, the voices of people, animals, and birds; the sound of wind instruments etc.), electromagnetic oscillations and light, whirlwinds, pulsations of flowing water, and gusts of wind. It, at last, explains elliptical orbital motion…*"

Certainly these are great claims for something that has not even been recognized —except for, alas, those de Broglie's matter waves so faithfully verified at any scale. The much more massive predicament of relativity, much more than the irreducible ambiguity of the three forms of energy in these equations, suffice to explain this overt non-admission. However, they seem to fit perfectly into the known process of the arterial pressure wave, which is also amenable to subjective interpolation via biofeedback.

As we have already said, the geometric phase is an index of the "geometry of the environment", of how the system is not reducible to the conservative idealization of the theory; it is not by chance that Berry is a specialist in asymptotic methods and semiclassical approximations, which he has also tried to apply to the central problem of number theory. The idea emerges that the ambient geometry of the retarded potential traverses as an oscillation precisely the effective aperture that the system has with respect to a closed conservative system, and that it is due to this aperture that the system tends to appear conservative or closed. At least in a manifestly open system such as our organism, with the dependence of the heartbeat on the effect that respiration has on the circulation, the idea of self-induction acquires its full meaning.

Since it goes without saying that in a closed system self-induction is null and needless: this is the tautological element presupposed in a concept such as "mechanics". And yet Maxwell's equations, the proverbial exponent of a tautological symmetry, give rise to self-induction. It is only with Noskov that the "closure" of the system take place inside the bodies themselves, that what has been ignored outside, and relegated to the convenient nebulosity of the field, is literally embodied.

But we have already said that electromagnetic induction can legitimately be considered as a mere particular case of mechanical induction —even if the quite reproducible experimental evidence has not been properly addressed. The heart then is a "pump", but not a mechanical pump: it is not a "vacuum pump" either, but a pump effect in the ordinary vacuum, which is nothing but the ambient environment. We are talking about the fundamental physical vacuum —there is no other one-, which obviously has nothing to do with the wild guesses of energy levels of theoretical physics.

And since we will never measure this vacuum except in bodies, all mechanics must, by definition, be the average of what happens between bodies and the vacuum. There is no possible escape from this, and yet we are incapable of assuming it. The word "mechanics" has acquired undesirable tones because of the restricted domain we have imposed on it, just like the word "automatic," which Aristotle still used as a synonym for spontaneous, for that which moves "of its own accord." It is by leveling force and potential that mechanics and dynamics become equivalent terms, as the study of the relationship between bodies and the vacuum that can never be reduced to mere motion. This is the main caveat to be made to a poorly understood "relational mechanics".

The pulsations of the Sun and the other stars, an ever-expanding area of study, also raise the same question as the heart: that its entire body is traversed by the non-uniform resonances or oscillations of the Noskov potential. Here we can very clearly identify the main sources of variation of the potential on the planets —even if the assessment of interplanetary density raises other kind of questions. In any case, and including a number of factors such as the relation of the solar sphere to the barycenter of the system, these resonances would be associated with the back-coupling of the central body with the global field; some will be surprised that an interpretation that cannot be more straightforward is not even considered.

The modern idea of mechanics, in tune with the overall trend, consists in reducing "the other" to "the same"; here we want to indicate at least how "the same" that is not contemplated is the genuine principle of differentiation, of directly appreciable and measurable forms and profiles. A sameness, an aseity that, far from reducing anything, neither to analytical conditions nor of any other kind, would be the condition of openness by which any entity breathes.

Thus, the "interpolation" of the subject in a self-induced mechanical system is only a detour to embrace something already fully present and operative, since the entire mechanics, and not only that of biological organisms, is a reflexive balance that includes the environment at any given moment. Conservation symmetries refer in the penultimate instance to the Third Principle, but ultimately to the First Principle of inertia, the same principle that proposes "an isolated object that is not isolated". By simply dispensing with the idea of inertia, and replacing it with that of dynamic equilibrium, all this would cease to sound strange.

Truly, that the order we observe, from light to atoms, to complex biological molecules, cells, organisms, planetary systems and galaxies, can subsist even for a single instant without this reflexive principle of equilibrium that is present in everything, still seems to me a chimera. The idea of inertia, or that of accumulation over time of biological events synthesized in inheritance, are mere ghosts without an immediate principle of actualization, which obviously cannot consist of "blind forces". This spontaneous balance would already include statistical mechanics and entropy, as implied by Pinheiro's thermomechanical reformulation of dynamics.

Studied as a biomechanical subject, the dynamics of the circulatory system allows us to short-circuit between our mechanical notions trained by habit and the consciousness prior to thought. This would already be quite an achievement. In addition, there are many open questions to keep thought busy.

For example, the seemingly anecdotal fact that both the ratio between diastole and systole time intervals in humans and other mammals closely approximates the ratio between maximum systolic and minimum diastolic pressure, the continuous ratio 0.618/0.382, allows us to directly connect asymptotic and numerical analysis, the physics of discrete events and continuous values, a feedback loop for the numerical and continuous methods themselves, or an optimization with a particular recursive ratio for algorithmic measurement theory. It further invites us to connect all these aspects with a thermomechanical description including entropy like Pinheiro's.

Since the heart functions here as a regulator rather than as a pump, and its action is the effect of global motion rather than its cause, it can be clearly seen in what sense it is the "internal clock" that rather than marking indicates the system's own time: such is the sense of the proper time of dynamic systems that would have to replace the "global synchronizer" of the old mechanics. This return of the local measure to its effective configuration is something generalizable and far-reaching, because to perceive the creative power expressed in Nature one needs to see beyond the global synchronizer.

Such a description shows that the two usual proposals to explain the observable order, or the metaphorical "watchmaker" of complex biological systems, be it the random mechanics cut over time by natural selection, or the "intelligent design" attributed to an ultra-mundane creator, lead us away from the essential, the spontaneous principle of organization which is also the principle of immediate actualization. It is always from the immediate openness with the environment that islands of organization and stability are reached.

This same biomechanical subject allows us to define as far as possible the relationship between the arterial pulse wave, a palpable and concrete biological illustration of Noskov's hypothesis, the effective openness of the system with respect to the conservative case and the corresponding asymptotic approximation. It should also make it possible to define the internal tone of the system, the essential element of the nonexistent definition of health, and to investigate the quantities that allow its characterization.

We have also seen that the three *gunas* of samkhya, that peculiar coordinate system, seem to correspond also to the three principles of mechanics extrapolated to open systems with conservation of momentum. Their most tangible aspect, even if reactive or derivative, the *doshas* of pulsology, can serve to investigate the relationship with the threefold manifestation of energy and what we call its "relational ambiguity". What is the logic presiding over the recursive scaling of the three qualities of material nature within a precise but open quantitative domain? What interface conditions are they defining?

Of course samkhya, insofar as it is a "dualistic" system, has a genuinely asymptotic character: the *gunas* or modalities of Nature mark a tendency, a path of ascent and descent without in any way coming into contact with *Purusha*, "the pure spirit" or consciousness, the same described in the Rig Veda as a giant from whose dismemberment the parts of the cosmos emerged. Only the total, non-dynamic equilibrium of the three modalities would produce their fusion with the absolute. Under an entirely similar logic, also the joint relationship of the three feet or letters of the sacred syllable points asymptotically to the fourth foot.

Asymptotics, like so many other forms of analysis, has proposed its own "uncertainty principle" between simplicity and exactness via localization, pure approximation itself. On the other hand, classical mechanics is the first asymptotic approximation of quantum mechanics, but the latter cannot be defined without reference to the former: a fine example that there is something fundamental in asymptotics. It would be opportune also to investigate in depth the connection between the indistinguishability of the three energies in relational mechanics and the various uncertainty relations —for there is not one, but many- of quantum mechanics, developing Noskov's arguments.

*

The relativistic continuum, in either of its two versions, is notorious for its asymptotic profiles, to which it owes much of its fascination and speculative impulse. We only need to contrast how the Weber-Noskov mechanics (not to mention Pinheiro's here) deviates from such aspects, to begin to see the other side of the question: the asymptotic not as an approximation to infinity but as a relative, partial detachment from unity.

The same can be said of the so-called "holographic principle" originating from entropy considerations in the boundary conditions of a singularity or black hole, which would reduce the entire universe to its projection on a mere two-dimensional surface. This principle cannot but be true insofar as *we know nothing except through light*, the universal, unavoidable mediator. But, obviating the fact that such a principle is due to the very pertinence and universality of the geometric phase, it is clear that the question does not beg any singularity: as Mazilu puts it, even the surface of any extended particle such as the electron, the day one tries to describe its ephemeral configuration, will have to reflect these limits, as intrinsic as the very spin of the particle.

In this sense it is idle to think of what may occur inside bodies, inside any body. And yet light faithfully reflects the interaction of each body with the void in which the other bodies are placed —its innermost pulsation, if that is how we want to call it, if the void and the bodies cannot exist separately.

Let us enjoy for a moment this supreme irony of the history of physics. Physicists who have insisted that the geometric phase is only a small appendage of quantum mechanics without the slightest fundamental entity, now appeal to this same "phenomenon" in its last frontier to wrap the whole universe in a thin film and set it up as a great principle with which otherwise one does not even know what to do. So much has been discussed about the limits of information density, about quantum fluctuations of position, about quantum gravity and holographic noise; so many attempts to verify it, including tabletop experiments, have been proposed.

And one may ask, why look for "quantum black holes" to verify a phenomenon that by definition has to encompass absolutely everything, from the beating of my heart to my perception of colors? "Because a global argument is of little use if we cannot predict new things", any expert so well trained to make predictions would immediately tell us. Of course, all field theories, and not only the electromagnetic field and light, have been built from the outside to the inside —from the boundary conditions to the "corpus of the theory", the equations that allow us to calculate. To then separate the corpus of the theory from its conditions, to turn it into Natural Law, leads us to forget their mutual dependence.

Predictions are blinding if this circumstance is ignored. Why is there so much discussion about the range of energies to experimentally verify this holographic principle? Because there is not the least certainty about the background that determines its scale, the Planck scale itself being a gigantic extrapolation. Even the uncertainty of the energy of a photon has nothing to do with it, as the most elementary analysis shows, since it does not change at all if we change the value of the constant; which does not prevent from using it to make dimensional analysis of the entire universe. If the estimate of the energy of the vacuum has turned out to be such a tremendous nonsense, no one will dare to say it is written on stone. This reminds us another infamous global principle, deserved object of so many jokes: the so-called "anthropic principle" proposed to answer the enigma of the fine-tuning of the magnitude of the great constants. Neither estimates nor predictions serve to shed light on the context. The homogeneity principle tells us that constants with dimensions are not universal, but the by-product of a conventional cutout from the background, an "emancipation" from ambient conditions —which is precisely what the geometric phase, mother of the holographic principle, reflects. There is no other "fundamental physical vacuum" than the ambient environment.

The geometric phase and the very structure of gauge fields suggest a sliding knot in energy constants and scales, and several mutually ignoring theories seem to point in this direction. Not to mention that the black hole hypothesis pretends to give us at once "ultimate boundaries", singularities, and processes beyond the singularity, which is like wanting to give a cake, have it and eat it: no longer two, but three incompatible things at the same time.

"There is plenty of room at the bottom", but surely not the way it is expected, stretching the Planck scale to the limit. On the other hand, if today we know from the gauge theory of gravity that we can dispense with curved space-time to describe its field, all the more reason we can dispense with the elementary Minkowski continuum. This should make us think more about the theories following Weber, which become field descriptions just by integrating over the volume and do not require neither a continuum with two flavors nor additional dimensions.

It should be clear that current prediction-centered theories throw into the sea the key to describe scale transitions, which are the very "background" that the word "fundamental" would like to appropriate. Nottale, and later Mazilu and Agop, have proposed a meritorious theory of "scale relativity", but one can also dispense with relativity itself with much more straightforward arguments. Mazilu and Agop's development is expressly neoclassical, and yet goes into the jungle of fractal geometry and the non-differentiable continuum: a good example that one can have simple ideas that nevertheless allow to envision the superabundant complexity of Nature.

"Nature cares naught for analytical difficulties": few words more certain than these of Fresnel's immortal phrase. If the non-differentiable continuum is possible, one can be sure that Nature uses it with real profusion; and the same path integrals of light would be the best exponent of it. Mazilu employs it directly to try to create an operational model of the brain based on the inexhaustible geometry of light and de Broglie's wave mechanics, which vividly recalls Bandyopadhyay's "fractal mechanics" devised for the same purpose.

In any case, Nottale's or Mazilu's use of the Planck scale deserves some examination, for it should be clear that the common use of it as if it were a rule is nothing but the extension of the global synchronizer to all the more amorphous domains of physics. Nature simply evaporates where this chronometer rules, which entrusts us with the task of trying to imagine it free of this equally imaginary constraints. A belt suits her much better than a rigid ruler.

The sad thing is that the holographic principle has even been seen as a "confirmation" of the idea that the universe is a gigantic computer, to bring the empire of the global synchronizer to its apotheosis. And yet, one can be quite sure that were the world a computer it would not even have lasted long enough to explode into pieces, let alone overheat. If this thing we are involved in and participate in "works" to any extent, it will have to be to the extent that it is not a computer and is not based on our idea of computation. And yet "quantum computation" itself, understood as the most exquisitely minimalist modulation of that which they now call "individual quantum states", even if there is nothing individual in that domain and precisely because of it, is at the very crux of the matter at hand, doing an extraordinary job to ignoring it. It is enough to carefully follow the thread that leads from local prediction to global description for the geometric phase to go from being a control parameter to begin to impact and resonate in the entire sphere of knowledge.

If there is something "automatic" in the universe, it must be precisely in the sense of spontaneity, of that which moves by itself in the open: the principle of dynamic equilibrium already guarantees it just by dispensing with inertia. And yet this trifle, apparently a mere game of definitions, takes us as far as we can walk, for even if we can dispense with inertia in absolute terms, nothing prevents us from still counting on it in relative terms. The "three and a half principles" can walk the walk.

Scale relativity physics moves between two asymptotic scales invariant under dilation —the Planck length and a maximum cosmological length associated with the Lorentz transformation- so that resolution requires explicit variables. Reworked by Mazilu and Agop, it also becomes a theory of the infrafinite, the finite and the transfinite without leaving the domain of light. It is at least an idea of great interest with strong reminiscences of Leibniz's monad, though it is clear that it has not been sufficiently developed. And if it is undoubtedly speculative, it is still much less so than the black hole sagas, which enjoy the greatest hold even if they scandalously violate the fundamental convention of the genre, claiming among other impossibilities that physical reality can go beyond mathematical singularities. The Riemann zeta function has often been used to regularize the vacuum energy levels and divergent series at the event horizon of these holes, including the light path integrals, by an asymptotic expansion of the temperature evolution under scaling transformations of the background metric. This may sound like pure scale relativity but it is actually its antithesis from the perspective of global synchronization.

The daunting scale differences between particles and the size of the universe, or even between it and the Planck length, are insignificant compared to the difference between any number that can ever be computed by a machine and the numerical infinity that constitutes the integrity of the Riemann zeta function. One could even say that they are *absolutely* negligible, but that would be to disdain the evidence accumulated with so much work by mathematicians. In any case, if the minuscule evidence of the computed zeros of the function has *any* meaning, it would be precisely because the entire function and its underlying "dynamics" involve some kind of scale relativity within it, although to approach it and find its resolution would require taking into account a number of factors that we cannot even list here.

It is a theorem that, if the Riemann hypothesis is true, the zeta function allows to approximate any analytical function of the infinite possible ones with any degree of resolution. In fact, the function would be a concrete representation of any text and accumulation of knowledge that the human being or any intelligent being can achieve, which would also be repeated in it an infinite number of times. Since the non-differentiable continuum also contains "all that", but the zeta function itself is infinitely differentiable and incomparably more structured, it can be assumed that this function and its large family of associated functions would constitute the widest and narrowest bridge between the differentiable and the non-differentiable —although what is differentiable and what is not, also depends crucially on the criterion of the calculus we use. Since the major problem of this function is to relate the local information to the global condition, if there is any way to gradually bound its dynamics, even if it is an indefinite process, it would have to be by reversing the relationship that physics and calculus have always posed between the derivative and the boundary conditions. At least the turn in orientation should not pose so many problems, if one admits that modern calculus as it is used is already the product of a thorough inversion.

The so-called scale relativity is a very general principle using present-known physics as a guide, but it is incomparably simpler to begin with the contrast between relativity and the Noskov equations when we take them to the supposed limits, starting always from the Lagrangian in Kepler's problem, the real keystone of modern physics; it is not very advisable to try to swallow the Whole when we do not reasonably understand any of the infinitely more modest totalities appearing everywhere.

There is already something singular enough in the evolution of any entity in relation to the background from which it emerges, in which it is maintained, and to which it returns: if we are able to see this, even with just the imagination, we will have already achieved a great deal. In fact, it is for not understanding the unobstructedness of this ephemeral and self-sustained "singularity" that we plunge headlong in search of crossing singularities that are non-holes by definition. One speaks of "the wave function of the universe" without blinking an aye, and the same holographic principle makes one think of a wave front of inconceivable complexity; but the simplest wave in three dimensions is already a challenge to the imagination.

Think again of the awesome ingenuousness of Venis' wave-vortex, who has not made the slightest use of physics or mathematics to unveil the most "symplectic" of morphologies. Although his transformation sequence also seems to show a wave front as a characteristic surface phenomenon, its tremulous boundaries fluctuate between full and empty with more delicacy than the best brushstroke. But how is it that a wave that is supposed to exist in an infinite number of dimensions, including fractional ones, still exhibits such a recognizable profile in six dimensions, and in three, and even in two? The only conceivable answer is that the totality that escapes our perception is still reflected in a point of dynamic equilibrium, which is a line or a plane of equilibrium, and so on.

Venis's sequence shows without profanation in what sense Nature is always equal to itself: fleeting like a whirlwind and immutable like the sphinx. Between one aspect and the other there are infinite layers. What is that amorphous bulb that insinuates itself everywhere, shows in the contour and hides in its center? These surfaces and profiles are beyond measurement —they are purely projective in nature- and yet they are reflected in all kinds of phenomena that can obviously be measured. Following the curves of the sequence one can guess when this imaginary flow speeds up or slows down, gets slower, and slower still, until the wave front roars and breaks and bursts in the midst of silence.

Light between space and matter; between the full and the empty, the light. We have already indicated why the categorical separation between the vibration of sound and light is due first of all to a misunderstanding —although it is unquestionable that light is vibration emancipated from matter and sound is not. But, can light be heard? It is a good question, although from the above it would be easy to judge rather the contrary: light would not only be a principle of expression or actualization, now it looks like a global organ of perception and an all-encompassing tympanum in which matter itself resonates. But, as vibration, who or what could listen to it? Nothing that can be exposed, nothing that is manifest; which may concern the bodies, the void, both, or neither. There is a point from which all that recedes to mean nothing, and yet there is still a long slipknot not only related to the scale of energy but also to the ambiguity inherent in its threefold manifestation.

This "relative detachment from unity" is an omnipresent question and by no means confined to physics or mathematics, which can nevertheless reflect it. For the formal logic that seeks self-consistency, unity only reaffirms us in the tautological; but from the point of view of applied mathematics, to which *all* calculus belongs, unity is never a given formal question, and in pursuit of it much old knowledge can be transformed into something new. From the very moment it is assumed that calculus or analysis necessarily proceeds from the whole to the part and from top to bottom instead of the reverse, the general disposition of the sciences will have changed.

*

Heir to a large extent to de Broglie, David Bohm made an interpretation of physics openly challenging the prevailing reductionism and spoke, between marked extremes of eloquence and vagueness, of “wholeness and the implicate order". Bohm cannot be reproached for being theoretically conservative, for he was already heterodox enough among his contemporaries; and yet such a discourse as his finds much less echo today even if in all this past time we all have gained an invaluable perspective.

Bohm was not really aware enough of the universality of the geometric phase and its relevance to classical mechanics —though we have already seen that present-day theorists are no better off. Nor did he conceive of the effect of potentials as a vibration or resonance that runs thoroughly traversing matter, but, following in this the spirit of the age, as a non-dynamical information field. The latter aspect is important for interpretation; the former has innumerable consequences yet to be explored. On the other hand, Bohm saw the importance of the measurement problem but he could not question its general relations with calculus or analysis as can be done today with utmost clarity. Finally, his work lacks a real discussion of the fundamental principles of physics as such.

In proportional terms, the universe is almost entirely empty; not only in the intergalactic immensities but even in our own body. Of course, this physical vacuum is still far from being mere empty space, revealing itself to the closest inspection as a tenuous sea of radiations. For general relativity, space tells matter where to go and matter tells space how to curve; but the space-time continuum is not empty, unconditioned space, but a totally different dynamic animal subject to curvature within a specific metric discipline.

We have already mentioned that Poincaré, the first clear proponent of the principle of relativity, preferred to think of the bending of light rather than of space, and at least it is now accepted that field theories, including gravity, can be formulated in flat space. After all, light was always seen to bend in water, but space never, and there was no need to artificially condition what had always been perceived as unconditioned.

The density of the physical vacuum with radiation is almost zero with respect to matter, but the density of really flat space, without the slightest trace of energy in it, is strictly zero with respect to the radiating physical vacuum. Waves attenuate indefinitely, and on the other hand, even a particle of matter such as an electron tends to dilate without limit when there are no other bodies in its proximity, as has been proven time and again. We would thus have three asymptotic echelons, from matter to radiation and from radiation to space without curvature and even the least metric restriction.

For the Vedanta, pure space has no qualities or even dimensions; in modern geometry the closest we have would be the primary projective space. Pure space is the perfect symbol of spirit in many traditions, but it is clear that space is not a sentient being that perceives itself. So the sense of aseity that is inherent to me must come from something other than that space prior to concepts of which I nevertheless still have some notion.

Today it is customary to call field excitations "particles" —in accelerators, for example- but the relaxation of an electron that can extend meters or kilometers, or to infinity if the environment does not prevent it, is not usually called a "wave". There is thus a very human prejudice in favor of high energies that leads us to overlook a powerful fundamental tendency. Moreover, this tendency of matter to relax as a function of its surroundings beats in unison with the most elementary interpretation of entropy, since molecules also tend to spread out as much as they can, going from the hottest to the coldest regions and regularly occupying the available empty space.

In this sense, so different from that of relativity, *space is telling matter where it has to go, but matter tells space nothing, because space, both in the limit and in absolute terms, has no curvature*. Just as I know my body but my body does not know me, there is no possible reciprocity here. There is however an unquestionable reciprocity between matter and radiation, which is something completely different. In reality, relativistic space-time encompasses the dynamical relations of the physical vacuum based on interactions, and its potential theory remains on undefined ground: that is why there are alternatives in flat space such as the modern gauge theory of gravity or those derived from Weber line of thought.

In other words, relativity dispenses with absolute space but maintains the absolute time of the global synchronizer —when with the theory of the "retarded potential" the opposite can be done: maintaining absolute space without further qualifications and allowing everything to be governed by its own internal time, which implies the environmental variation of the forces and constants that now are postulated as fixed.

Absolute space can still "speak to matter" with two different languages: that of the asymptotic tendency of matter and radiation to expand as a function of the environment, which implies a mediated and dynamic mode that also includes thermodynamics, and the immediate language of the potential, which is simply derived from position. And this last language supposes the intimate vibration of matter.

Let us recall that almost all the production of radiation is dispatched as "spontaneous emission", which is the most opposite that can be proposed to a mechanical explanation. And yet this spontaneous emission would be nothing but the manifest form of the internal vibration of the particle's body. In this sense, one could also speak of "spontaneous absorption", even if we say simply absorption.

We have then a circularity between the immediate or instantaneous, completely independent of motion, manifested by position or potential, and the dynamic or mediated aspects of the interaction between matter and radiation. This resolves the aporia presented by both Vedanta and samkhya: how the absolute and the conditioned can coexist without the slightest problem. The dualism of the modern age can also be seen in a different light.

Absolute space without qualities or dimensions corresponds to *Prajña*, the state of deep sleep in which all determination disappears. However, one knows that the aseity is even closer than that, that there is something even more general than the most undifferentiated, which logically implies that it is also not separated from the rest, from that which manifests itself. That is why we have speak of a primitive homogeneous medium as something somewhat different from undivided space. In the spontaneous response of matter and radiation to something that by definition does not move, we have a very simple way of grasping in what sense the cosmos is not creation but manifestation, the manifestation being nevertheless something truly spontaneous and creative.

If physics is an inexhaustible matter, it is because it can never be reduced exclusively to motion or extension. To want to explain everything by motion or extension is, besides being the ultimate expression of nihilism, completely trivial. The Cartesian *res extensa* was conceived expressly as a theater for motion, at the same time that for Galileo rest ceased to have proper entity. We are speaking, precisely, of the two fathers of the principle of inertia. And yet, when we return to the original perception of space as the immobile by definition, independent of any dynamics, the internal and the external once again interpenetrate.

It is only a matter of perceiving rest in motion, in order to see the activity in the rest. To achieve such feats, which many will believe to be miracles, it is enough to duly probe the theory of the potential, which will take us into as deep waters as we wish. Most may protest and argue that a potential is never independent of dynamics, but physicists called the geometric phase by such a name precisely to separate it from the realm of forces or interactions, and there was a closed consensus on this. As we have already pointed out, even if we wanted to insert it into the principles of dynamics, it would have to be through the internal articulation of "the three states of rest" with the principle of dynamic equilibrium at its center, which, depending on the case, can be the same or something very different from the multiple and complicated criterion of the relativistic principle of equivalence.

Indeed, from the very Kepler problem on down, there are all sorts of important cases in macroscopic physics that are not well covered by current field theories; but even ignoring that now, we have the fact that quantum entanglement exhibits instantaneous correlations, and the geometric phase, which appears at all scales, is only a lower resolution expression of the same order of correlations. Gravity itself, if it is not really a force, could perfectly well be found in this case. Then, even general relativity, which also does not consider gravity as a force, would be describing only interactions from the point of view of the holographic principle —which we already know where it comes from. It seems that physicists do not know what to do with this reasoning either.

Relativity would be a pure theory of surfaces, but so is the picture of interactions in quantum mechanics. Instantaneous correlation has to be the true and genuine global synchronizer —a synchronizer independent of interaction, but echoed by any interaction. This reveals the extent to which comparisons of the universe to a large computer are irrelevant. The true global synchronizer is as ungraspable as predynamic space; its local translation in wave mechanics is the de Broglie hypothesis of the particle "internal clock", an oscillation no doubt, which does admit of indirect experimental confirmation but attracts little interest. It may well be said that the only global synchronizer is the one that does not need to synchronize or take measurements of anything, but in which everything resonates.

I have shown, in the most general way, in what sense "that which is above" can guide "that which is below" without the need for contact; but that which we imagine above is also always below and at the bottom of what appears. I have also indicated what kind of changes make it easier for the scientific mentality to feel comfortable with this idea, which has its roots in something prior to thought. The problem of finality, which physics always eluded as much as it could, and which it misinterpreted by projecting its own shadows from above, has a very simple solution that nevertheless has two sides.

*

According to what has been said, time is a more superficial phenomenon than the space without qualities, and if it seems subjective to us, and even if it could seem to Kant to be pure form, it is precisely because it flows continuously through that domain independent of motion which is at the basis of subjectivity. It is only that almost all of our attention is pending on the innumerable variations of the contents of experience in time, while the other, though invariable, determines the immediate vibration of novelty.

For the persistent dream that is wakefulness, subjectivity must reside in the variation and novelty of its elements; but deep sleep not only does not dream, it does not even sleep. It is already fully present in the midst of this busy wakefulness —except, of course, for my lack of attention. Attention is then enough for that third stage which is always in the background to become the fourth which is nowhere in the sequence.

In physics the demarcation of these superimposed planes of experience and time would have a similar pattern; although the incompatibility between the two great dominant theories, in addition to their own internal contradictions and paradoxes, makes everything much more difficult to arbitrate. The universality of the geometric phase makes it a natural bridge between the micro and macroscopic, but the global synchronizer, which with the principle of inertia shapes the whole account of physics, prevents the local translation of what the conversions of scale imply.

To get the slightest idea about this, it is necessary to leave the prevailing theories. We have already pointed out several by way of contrast and now we will bring up another one, which is not even based on physical or mathematical, but purely morphological considerations. The wave-vortex sequence in Venis' hypercontinuum is based on projective arguments but cannot avoid being reflected in planes as varied as living beings or the great cosmic formations.

Following the compelling logic of his vortices, Venis concludes that the cosmological redshift does not indicate a general expansion of the universe but only a local contraction by a gradual shift in the dimension of our local group: if we were to shrink we would not notice it in ourselves, but would only see the size of the environment increase. This displacement would have a speed that can be estimated, but it would involve an internal time that could only be verified by leaving the group in question, be it the solar system or the galaxy, and passing to another time branch to return again to the place of origin.

This is simply speculation —as is most of cosmology- but it serves perfectly well to illustrate the gulf between the exigences of the global synchronizer and a physics that takes the circumstances of the environment seriously. Surely the stellar immensities do not exist for everything to be as it has been measured in the corner of a laboratory; but, even so, for a science as mathematical as physics, overcoming its own yardsticks represents an extraordinary challenge.

But we don't need to leave the galaxy to see how the environment modifies the ruler instead of the opposite: the "retarded potentials" already tell us this at any scale. The spiral traced by the planets of the solar system only deviates from an exact profile by the same order of differences as the value of the Lagrangian of each orbit, which is already eloquent, and this spiral approximates the plane of a vortex. And so we can continue down to the atoms and the orbits of the electrons, vortex-waves themselves. At all scales we find entanglement of the potentials, be they classical or quantum.

The anthropocentrism or immodesty involved in speaking of universal constants might well be called comical were it not for the difficulty of getting out of both the logic of global synchronization and the legitimization of appearance in the name of the irreducibility of the observer's position; and in this case both are mutually justified. Even those who are most aware of the unfoundedness of this claim have the greatest difficulty in imagining what the implications would be if there were no synchronization in such an obviously anthropocentric sense and what might lie on the other side.

Today the famous redshift of galaxy light is directly associated with the microwave background radiation to reinforce the big bang narrative. What the story does not usually tell is that Gamow's 1952 prediction was neither the first nor the most accurate, and that a good number of physicists had already estimated its temperature with greater accuracy, many years in advance and with much less data, from Guillaume in 1896, to Regener, to Nernst, to Herzberg or Max Born, among others. All of them had in mind a universe in dynamic equilibrium.

Venis also presupposes a dynamic balance between local expansion and contraction, but invokes the third principle of action-reaction in justifying the balance. However, the third principle alone does not amend the principle of inertia, which is the background that dominates this whole narrative. Indeed, if what is observed everywhere is motion, and with Newton's three laws nothing moves without having been moved by something else, we are forced to believe that everything proceeds from an original impulse, no matter that it is the most grandiose violation of the principle of conservation of energy.

To become aware of the extent to which principles determine the plane of contact between physics and metaphysics, it suffices to recall the contrast that Roger Boscovich, the great precursor of field theories, proposed for the postulate of absolute magnitudes: "*A motion which is common to us and the world cannot be recognized by us —not even if the world as a whole were increased or decreased in size by an arbitrary factor*". Even if the whole world were to shrink or expand in a matter of days, Boscovich continues, with an identical variation in forces, there would be no change in our impressions or in our mind's perception. Joint change has no rank of experience, only differences count.

Of course, this would have nothing to do with the big bang if in it there is no joint evolution of the constants. But if the constants modified their mutual relationship in the past, why don't they do so right now? In fact, we have been assured for many years now that the "expansion of the universe" is accelerating.

Venis obviously distinguishes velocity of motion and velocity of flow in time, but he does not know how the correlation could be established, nor what other factors would be involved. Even after more than a century of relativity, that there can be different temporal velocities determined from the outside — which has nothing to do with "time travel"- will seem impossible to many. Now, what has been said here is that the background of all experience, including the experience of time, has nothing to do with motion and its infinite possible relations. A velocity in time does, whether it is a measurable physical time or the subjective time that in dreams can contract or dilate without the slightest apparent rate.

In short, the background of experience is completely formless and common to all beings, be they gods, humans, galaxies, planets or atoms; there is nothing individual about it, but it is what makes the subject subject. That is the great difference between non-duality and the variants of modern idealism. On the other hand, time itself is part of the individuation of beings, indeed, it is incorporated in their form and that of their body and its "resident matter", as Venis calls it.

It can be seen then that Venis, almost unwittingly, is proposing a far-reaching strategy for seeing time beyond the fiction of the global synchronizer and its universal flattening. And it does not matter that the connection of its morphology to physics and mathematics is entirely unexplored, since its foundation is in projective space itself prior to metrics and determinations. Even as pure phenomenology, it provides a connection between understanding, imagination and vision that the present sciences would like for themselves.

Moreover, Venis sequence of transformations poses fine and profound problems for calculus and mathematical analysis, since it exhibits a smooth and differentiable evolution in the hypercontinuum, which can take any real number among the integers of the dimensions. Today they use to work with "fractional" dimensions in countless areas of applied analysis, and they are commonly known as fractals; however these fractals are not differentiable. On the other hand, only very recently a bridge is beginning to emerge between the fractal geometry so common in Nature and the so-called "fractional calculus" which is not restricted to operators with integers. Fractals belong to space, but fractional operators affect the time domain.

In short, fractional calculus covers intermediate time domains in processes of all kinds, including wave processes. Fractals are nonlinear, while fractional dynamics governs linear processes, —and yet it poses a disturbing anomaly for calculus: it exhibits nonlocal dependence on history and long-range spatial interactions. And so, fractional calculus itself raises a huge problem of interpretation that neither physicists nor mathematicians have been able to solve. Igor Podlubny proposed to distinguish between an inhomogeneous cosmic time and a homogeneous individual time. Podlubny admits that the geometrization of time and its homogenization are due first of all to calculus itself, and warns that intervals of space can be compared simultaneously, but not those of time, since we can only measure them in sequence. What may be surprising is that this author attributes non-homogeneity to cosmic time, rather than to individual time, since in reality mechanics and calculus evolved in unison under the principle of global synchronization. In his reading, individual time would be an idealization of the time created by mechanics, which is to turn everything upside down: in any case it would be the time of mechanics that is an idealization.

It is clear that mathematicians also fail to rise above these aporias since calculus is not just an accomplice of synchronization but its principal agent. Fractals are a geometrical expression of equally ubiquitous power laws of scale, but the operators of fractional dynamics would govern their evolution and what is called their temporal "memory". Now, Venis vortices evolve smoothly and linearly through fractional dimensions while describing a temporal evolution that drives logic, intuition and imagination, showing a common thread amidst an impenetrable jungle of figures. The enveloping boundary layers of their profiles also invite a competition between the various modes of calculus to see which one best fits their contours.

Given that Venis' concept of vortex encompasses not only whirlwinds, but things as amorphous as a spherical bulb or drop, the formless is emerging as logic and process within something much more formless still. This is why it is possible to think of a universal morphological scope, even assuming that the forms do not cease to be fleeting ghosts, apparitions in the domain of the impermanent. And if Venis has been undoubtedly inspired by Far Eastern philosophy, his phenomenology is, as he himself calls it, an "infinity theory", heir to that unmistakable Western impulse to see beyond any limit. Here, however, there is a tendency to perceive the infinite in the very limits of the concrete forms.

Most real processes, for example the stresses and deformations of a material or its fracture, are highly non-polar, as they cannot be characterized like fields in terms of vectors, displacements and forces, nor do they show any detectable axis in their evolution; fractals, on the other hand, are suitable for describing them. In contrast, fractional dynamics may retain the mathematical polarity of fields, so that the connection of the two is a way, still very abstract today, of exploring the transition from chaos to order. Vortices, instead, are present in all phases of matter and light, and define the most visible transition between chaos and order, turbulence and form.

The zeta function, with its pole at unity and its abysmal critical line, is associated with the so-called "quantum chaos" which would be softer, but as in any other circumstance, it has never been possible to specify the boundary of its transition. It is extraordinary that such a universal "mathematical object" as this function maintains such isolation from any visible geometry, not to mention logic; the multiple analogies with advanced physics or the innumerable analytical graphs of the function cannot make us forget its rigorous isolation in arithmetic, the pure domain of time. This "pure time" is nothing but time absorbed and without any external relation, the most opposite we can conceive of to the time we experience. Since Venis wave-vortices are also of extraordinary universality, and have a deep and countless-layered link to time and entropy, it is impossible for them not to have a connection to this function, and a close connection at that, too, even if no one has yet explored the subject.

The intellect that perceives objects never perceives itself; that is why it is not the self, precisely, otherwise both would inevitably tend to merge. This irreducible fact is also reproduced, of course, in the simple act of counting, at the base of the whole arithmetic. True geometry is always synthetic and elementary; arithmetic, when it is not merely trivial, always has analytical depth. In this sense one may well speak of an explicate and an implicate order, in quite a different sense than that given by Bohm to these words; Poincaré preferred simply to say that geometry is a posteriori and arithmetic a priori. The geometricians show a good hair, and purebred analysts were bald, such was the infallible rule in the old days. But those who turn to number theory have their hair again, and today even algebraists and logicians can wear a long mane; the interconnection between all branches of mathematics has reached such a point that one no longer knows what will happen with the scalp of its practitioners.

Venis vortices are inevitably nested within one another at very different scales, making it no simple task to identify the "substance" of the temporal flow. In this sense, they would be as evanescent as the forms themselves, even if they are not confused with them in any trivial sense. If we think a little about this circumstance, we realize that a time is emerging from here that is both material and formal, in a sense different from those that physics and philosophy have so far handled. Even if we understand that the subjectivity of its flow does not ultimately depend on motion at all.

It is quite possible that the Riemann hypothesis will never be proved, as it is quite possible that many other incomparably simpler conjectures of arithmetic, such as the Collatz conjecture, will never be proved. Nevertheless, the zeta function has already intrigued mathematicians enough, and will intrigue them much more, and in that sense it will have fulfilled its mission; but for the moment it has only summoned algebra and analysis, which have hardly offered any real contrast to ti. Instead of continuing to grope about in the darkness of the mouth of the cave, they could confidently advance backwards beholding the light at the entrance, geometry at any rate. Even elementary geometry is infinite, when it knows how to find its problems. Modern calculus, on the contrary, is a grinding machine that gives predictions, but hardly knows anything about the physical geometry of the problems.

Something much more probable but much more disturbing than the demonstration of this hypothesis can then be foreseen: its contact with geometry and its entry into the human imagination, and not only into the mathematical imagination. Some analysts have already detected spiral waves in the domain of the function and exercise themselves trying to imagine their morphogenesis, but such abstract patterns still require many degrees of decompression to make contact with the genuinely geometric. This requires a radical change in the orientation of calculus, along some of the lines mentioned above.

It is just reasonable to think that the Riemann hypothesis is only the consequence of the asymptotic behavior of the prime numbers tending to infinity, yet this has very little to do with its demonstration; however this is not the main interest of this function. There is, in any case, an absolutely elementary geometry of the primes within the line, described by the periodic curves from the origin intersecting each number and its multiples, and the primes being cut only by their own curve and that of unity. Even from this very simple representation surprising patterns gradually emerge, and the complexity in the superposition of the curves increases as we plunge into the depths of the number line. The most basic patterns can only be seen by extracting them from the line and displaying them in at least two or three dimensions. It does not take long for the first spiral motifs to emerge. It would be necessary to continue in 4, 5, or 6 as in Venis sequence, and in the intermediate domains, modulating the possible dynamic components. After all, there are also infinite dimensions that fold and compress in the number line and the zeta function.

The time-form that accompanies time-matter is not only in the vortices that describe particles or galaxies, but a little everywhere and nowhere. When a wave is broken a spiral wave appears that persists and excludes all concentric rings; this was precisely Arthur Winfree's simple observation as he began to model the geometry and resonances of biological time. Spiral waves can be found even in the motion of the heart.

Venis transformation sequence is a game of balance between contraction and expansion; these are the "three feet" of his vision. The fourth foot must necessarily be beyond the forms but in the middle of them. In any case, equilibrium seen only in terms of action-reaction does not take us out of the limitations of the principle of inertia: it gives us the explained order of a closed system, not the order that an open system implies. For, indeed, dynamic equilibrium is implicate order by its very definition, "holomovement" in Bohm's parlance, even if its implications can never be extracted. The geometric phase, on the other hand, can be described as a torsion or shift in dimension, which thus acquires not only a geometrical but also a morphological connotation; although it is understood that it is the dynamic component that is deformed, not space or potential.

Astrophysicist Eric Chaisson has observed that the density of the energy rate is a much more decisive and unambiguous for a complexity metric and its evolution at all scales than the various uses of the concept of entropy, and his arguments are very simple and convincing. Contrary to what one might think, this rate, measured in ergs per gram per second, averages 0.5 in the Milky Way, 2 in the Sun, 900 in plants, 40,000 in animals and 500,000 in human society; the relative value in an atom or molecule is missing here. This density of energy flux, as we say, indicates not only the complexity, but also the individual evolution of an entity, its transformation throughout the process of birth, maturity, aging and death. It must therefore be linked to what is commonly understood as "subjective flow of time" although it is still just a transfer, an exchange with the environment within something alien to motion. No doubt this flow rate can be applied to Venis' vortices if physical values were assigned to them, since they too reflect the evolution of individual entities and their appearance. As in the whirlwind of life, the increasing restriction in entities, the essence of aging, is itself a matter of great subtlety and multiple concentric layers that only gradually unveil themselves. In-depth study of the subject would make it possible to see to what extent we are talking about an incarnated time, a time in matter and form. Returning to the physical world, another question would be how this rate relates to the principle of maximum entropy or to the thermomechanical equilibrium that can account for the various Lagrangians of atoms, planetary orbits and galaxies.

Increasing restriction is the touchstone of the individual evolution of an entity and its aging, which so little attention has yet deserved despite the fact that its profiles are always before our eyes. Even in the almost insurmountable abstraction of number theory can be found multiple domains and orders of increasing restriction, counterpoint of the various measures of numerical entropy, much more irreducible and singular than what in calculus is meant by "approximation" and "boundaries"; for calculus only uses arithmetic as an instrument, while number theory seeks in it its own physiognomy. Entropy is an extensive property proportional to the logarithm of states; the quantity of prime numbers is inverse to the logarithm of the integers, and the same nontrivial zeros of the zeta function on the critical line with exact real value of 1/2, would reflect the principal domain of an increasing constraint that wraps as a product the total sum of the prime numbers up to infinity. Fechner's law tells us that the intensity of sensation is proportional to the logarithm of the stimulus, and the time of life and memory unwinds with logarithmic spiral rhythm. As Tolstoy wrote, there is only one step from the five-year-old child to the one I am now, but from the five-year-old child to the newborn there is an immense distance, an abyss from the newborn to the embryo, and the inconceivable between the embryo and the non-being.

It may seem counterintuitive that the energy flux density is much higher in a blade of grass than in the incandescent atmosphere of the Sun; but this is closely related to the aforementioned fact that entropy tends to the maximum and order produces more entropy. Both facts give away that there are very fundamental things that we are looking at from the wrong side, not to mention the more than possible contribution of entropy to the very order of dynamics. Also in all this, one would have to stand with one's back to the mine to appreciate the light that floods the cave. In any case, these considerations allow us to see that even aspects of time that we believe to be highly subjective can have physical and mathematical expressions with their own entity; and it goes without saying that in Venis' morphology there are many more motifs than spirals.

Considering special relativity, it has been said since Pearson that an observer traveling at the speed of light would not perceive any movement and would live in an "eternal present"; but it is clear that light is transmitted and moves, and even pulsates with more precision than the best clock, so this is simply false. We also know from Huygens' principle that its propagation implies a continuous deformation at each point, the very opposite of timelessness expressed in the most graphic way. Precisely for relativity, there is no other way of conceiving time than motion, so this claim contradicts its own assumptions, although it expresses superbly how the principle of global synchronization would like to be above all contingency and even above the very time it imposes over everything else. Presumably, physicists have given up on such illustrations, which only expose the theory. If anything physical is to be outside of motion and time, it certainly cannot be this.

The multiple inertial frames of reference of general relativity only succeed in creating a permanent confusion, even about the very meaning of the word inertia, which already speaks for itself. To see and consider, with eyes wide open, that there is no inertia in this world: there can be no better suspension of the mechanically learned, no deeper meditation. The most intense, too, that the permanent agitation of thought does not bear for a moment. Is that hard to assume that there is no inertia in this world, to understand that the dancing principle of dynamic equilibrium is telling us that everything is interpreting its position at every instant spontaneously and with all its being, beyond freedom and necessity?

Venis readily admits that we still know almost nothing about how, if there is only a hypercontinuous infinite-dimensional space, our perception is so severely limited to three-dimensional space. Also the Hilbert space of quantum mechanics is initially infinite-dimensional, but are we talking in this case about abstractions? Venis' inferences can be very risky, and yet they have the peculiarity, today more than ever exceptional, of being guided almost exclusively by the specific spirit of form. The infinite dimensions of an abstract space leave us indifferent, but what is implied here is that the dimensions of concrete forms in physical space are only a minimal section of entirely formless domains, much closer to us than we think; of course, physical reality has never allowed itself to be reduced to the visible either. In any case, any supraformal or infraformal domain must be connected to our familiar world of form by the same thread and principle.

Is there a physical process in the reduction of dimensions, or will it never cease to be for us something purely imaginary? Both things are not that incompatible, but they depend on the starting point; physicists need measurements, imagination is an individual exercise that does not require instruments. Finally, we can always count on that which does not even need imagination.

Our hyperkinetic culture cannot conceive anything outside of motion; thus, the geometric phase is given a highly abstract and derivative status when it reflects in the most immediate way the place from which we look, which is certainly not an inertial frame. However, it is the description of the dynamics that has become rarefied to the point of representing practically nothing. Since for the physicist the immediate is not worth anything either, there is no choice but to bring in another type of argument.

The dynamics of a system is supposed to control the geometric phase, which plays an entirely passive role; and this is the part now routinely assigned to it in the laboratory. The unavoidable question is, how can something that merely follows motion be independent of it? But, on the other hand, an instantaneous correlation cannot be associated with motion at all. How to decide the question? There are multiple ways of approaching this new abyss; all that is required is to change the idea of what is to be measured.

Nor there is any need to smash particles in gigantic accelerators, but only to get in tune with the exquisite sensitivity of potentials in field theories and beyond. In principle, the difference between an ordinary potential and an "anomalous" potential such as that of the geometric phase seems very clear; in practice, as always when we talk about energy, the question is much more delicate and subtle. Heisenberg's misnamed "uncertainty principle" has been disproved countless times and its various relations are subject to successive corrections depending on the experimental framework. Not only high-precision experiments, even phase-sensitive frequency meters of color TVs show an accuracy many times higher. But, following Binder's suggestion, and beyond this ailing principle so in need of the simplest dimensional analysis, all kinds of phase-locked loop devices can be designed to create a feedback or back-coupling between the dynamic and geometric phase, and study where is the equilibrium and what is the boundary condition. Binder interprets that it is the geometric phase that controls the strength and sign of the coupling constants of the fundamental forces, and even conceives of scenarios to shift the Planck scale to the nuclear range. According to Binder, what would vary and "curve" would be the dynamical space-time; but the instantaneous velocity of the geometric phase cannot curve at all. Velocity is the fundamental quantity of physics; *physical* space and time are notions derived from its measurement.

We know very well how little attention is paid to proposals such as Binder's, and we know also why. But, quite simply, that which is instantaneous cannot be secondary with respect to that which takes time to operate; that which is pure act cannot be passive with respect to that which takes time to actualize. This simple reflection makes one think that physics has been inverting all this time the notions of act and potency, but it will take a long time to recalibrate their relation, and perhaps even longer to assimilate their consequences. All matter is interpreting, spontaneously recreating that pure act. Light is vibration emancipated from matter, but to transcend the global synchronizer amounts to step out of any temporal order linked to motion, be it that of the solar system, our galaxy or the entire cosmos, observed or unobserved.

Physicists, scientists in general, can still choose between control and wisdom; between closing themselves over man and Nature, or opening for them the vastest of horizons.

Everything that moves is but a reflection of that which does not move, a ripple upon its waters. It is not the potentials that are retarded, it is the non-fundamental forces that are varying. We do not know whether gravity has a speed limit or not, nor do we know how other completely different forces that may exist at other scales, in galaxies or beyond, would behave; but we do know that what is instantaneous is not retarded and is the reference for anything showing speed limitations. As some like to say, "you can't beat that", though you can ignore it. And, after letting the question sink in, one might add: "you are that".

Referencias

Henri Poincaré, Nicolae Mazilu, *Hertz's Ideas on Mechanics* (1897)

Nikolay Noskov, *The phenomenon of retarded potentials*

Nikolay Noskov, *The theory of retarded potentials against the theory of relativity*

K. T. Assis, *Relational Mechanics and Implementation of Mach's Principle with Weber's Gravitational Force* (2014)

A. K. T. Assis, *History of the 2.7 K Temperature Prior to Penzias and Wilson* (1995)

Alejandro Torassa, *On classical mechanics* (1996)

Koichiro Matsuno, *Information: Resurrection of the Cartesian physics* (1996)

Nicolae Mazilu and Maricel Agop, *The Mathematical Principles of the Scale Relativity Physics I. History and Physics* (2019)

V. E. Zhvirblis, *Stars and Koltsars* (1995)

Mario J. Pinheiro, *A reformulation of mechanics and electrodynamics* (2017)

Miles Williams Mathis, *A Re-definition of the Derivative (why the calculus works—and why it doesn't)* (2004)

Igor Podlubny, *Geometric and Physical Interpretation of Fractional Integration and Fractional Differentiation* (2001)

Salvatore Butera, Mario di Paola, *A physically based connection between fractional calculus and fractal geometry* (2014)

Juan Rius Camps, *La nueva dinámica irreversible y la metafísica de Aristóteles* (2010)

Patrick Cornille, *Why Galilean mechanics is not equivalent to Newtonian mechanics* (2018)

Yuri N. Ivanov, *Rythmodynamics* (2007)

Anirban Bandyopadhyay et al., *A Brain-like Computer Made of Time Crystal: Could a Metric of Prime AloneReplace a User and Alleviate Programming Forever?* (2018)

Bandyopadhyay, Tam Hunt, *Resonance chains and new models of the neuron* (2019)

Jan Bergstra, Division by Zero: *A Survey of Options* (2019)

Okoh Ufuoma, *Exact Arithmetic of Zero and Infinity Based on the Conventional Division by Zero 0/0= 1* (2019)

Raymond Abellio, *La structure absolue, Essai de phénoménologie génétique,* París, 1965

K. O. Friedrichs, *Asymptotic phenomena in mathematical physics* (1955)

Peter Alexander Venis, *Infinity-theory —The great Puzzle*

E. J. Chaisson, *Energy Rate Density as a Complexity Metric and Evolutionary Driver* (2010)

Bernd Binder, *Berry's Phase and Fine Structure* (2002)

Miguel Iradier, hurqualya.net

POLE OF DESTINY

16 June, 2021

Effects of climate change? We are the effect, or at least one of them. But we are not talking about the "greenhouse effect" or other concoctions forged and stirred up by the media. Instead we will talk about pole reversal in the solar system, in our planet, and even in our non-society; and most important of all, about reorientation with respect to the Principle.

That the "greenhouse effect" is just baloney, we already knew that a long time ago. It would also be necessary to find out how it is that a large group of scientists allows themselves to be the object, not only of discredit, but of outright ridicule.

But it's not that hard to figure out, given that these people believe they are beyond ridicule, or at least safe from it. It was in the interests of those in power to create an alarmist scenario within which to manage new arbitrariness and interventions, as well as to divert attention from more pressing matters; and by serving those interests, a collective of experts could in turn gain ascendancy and obtain more funds. Modern science has always been close to power, with the results to be expected.

Of course, it's not all about power and collusion; to go far you have to have good intentions, and here the well-intentioned reached new heights. There is always more that can be done, in any matter, and for some it must have been very flattering to see themselves as saviors of the planet.

In general, I prefer not to talk about such manufactured issues as global warming, and since children are used to "raise awareness among the elderly", the subject is already irremediably infantilized —which is also another way of shielding it from debate. However, reading a Spanish translation of Paul de Métairy's book *Climate Change: the truth of what is coming*[1] has given me some new ideas on the subject.

The original title literally reads "*Climate: not guilty!*", and its subtitle "*The revelations of those who do not have the option to speak*". The Spanish translation covers only the scientific part, but that is enough. It is a very short text and is easily read in one sitting. It is highly recommended for anyone looking for clear ideas and a broader perspective on one of the biggest hoaxes in the century of the most blatant and unbelievable lies. Only towards the end does it have a few blunders, such as saying that Galileo ended up at the stake, which in any case do not affect its diagnosis. Needless to say, one can find in the web many other good sources of information about this great scam.

Métairy undoes in a few paragraphs the hoax of the greenhouse effect with elementary logic, but he still contemplates scenarios with rising temperatures, although for very different causes. In contrast, there are serious climatologists who consider we may be heading towards a new glaciation, and adduce arguments worthy of attention; but any real debate is impossible within a discourse and an entire discipline hijacked by the Intergovernmental Panel on Climate Change and its supporters.

The fact that this Panel is headquartered in Geneva tells us enough, although, with a little less disguise, it could have been in Basel, Davos or Wall Street itself. Is not there a lot of trafficking in "carbon credits"? But Geneva is not bad, and apart from being the capital of diplomacy, it is also home to other unspeakable rackets such as CERN's hadron collider.

I will not repeat Métairy's very basic and conclusive arguments against the greenhouse effect because that is what his book is for. However, the main purpose of the French author is not so much to disqualify a laughable theory as to give us a more probable picture of the evolution of the climate in the medium and long term. Of course, the geological and climatic timescales have nothing to do with those of the hysteria cultivated by the media, but they still affect us enough.

*

Métairy seems to be writing in 2012 or 2013, when the European Space Agency launched the three satellites of the SWARM mission to study in more detail the Earth's magnetic field and its apparently increasing instability. SWARM is an example of a project relevant to our species and with a cost, 236 million euros, reasonable for today's Big Science.

The scenario envisioned by this author is that of the probable reversal of the Earth's magnetic field and what would entail its inevitable zero crossing. This would be a full-fledged Great Reset, and not what the plutarchy wants to establish. Now it is common wisdom that without the protection of the Earth's magnetic field, the solar wind would riddle us with high energy charged particles and unleash all kinds of cancers and mutations with the greatest celerity. As for the electronic devices that constitute the nervous system of this civilization, one can already imagine their fate.

But of course, Métairy is not saying that this zero crossing will happen tomorrow. The average frequency of such events is about 200,000 years, but the last one took place as far as 780,000 years ago, which would make a relatively close reversal more likely. At these time scales, "close" could still be tens of thousands of years; however, there seem to be indications that the current instability is becoming more pronounced.

Geomagnetism specialists speak of an average period of between 1,000 and 3,000 years of magnetism around zero before the field reversal consolidates. A period in which the magnetism is almost null, or in any case so weak as to let pass almost all the wind of ions that reaches us from the Sun and other stars and that is normally stopped by the magnetosphere. The estimates of these specialists are based on multiple analyses of rock samples and fossil trees.

However, long before this "zeroing" of the system takes place, there is a gradual weakening of the planet's field which acts as a protective shield for living beings. Do we have any solid indications that such a weakening is already taking place? For Métairy and some experts, this could have been happening for about 300 years.

Any minimal perspective on so-called climate change should consider that we are just emerging from a "little Ice Age" that peaked around the mid-17th century but lasted until the late 19th or early 20th century. In this mini-glaciation, temperatures plunged significantly below earlier or later averages, which should always be kept in mind when speaking of "alarming temperature rises". The subsequent rise in temperatures may have as little to do with human activity as the immediately preceding decline. This is without even contemplating the inescapable physical and chemical considerations that Métairy so rightly reminds us of, and that the media avoid systematically.

As always, one cannot maintain a hoax without appropriating the evidence and creating indissoluble associations between patent facts and empty ideas. Skin cancers, and not only skin cancers, are increasing at a striking rate. Is this related to the destruction of the ozone layer? It could also be linked to the rapid raise in chronic intoxication due to an increasingly artificial diet, to mention only one of the many carcinogenic factors of modern life; and yet it seems certain that sunlight is becoming more and more harmful. But, in any case, the creation and destruction of ozone has its own physical and chemical cycle with hardly any relation with human activity or the famous chlorofluorocarbons, but with its balance with the ionosphere-thermosphere. Why else is the main hole in the South Pole, being by far the area and hemisphere with the least pollutants?

It is even embarrassing to recall these things, but confusion, however crude, is manufactured on an industrial scale. And as for the political exploitation of the staged confusion, it is simply what could be expected.

*

Let us try for a moment to see a bit beyond this all-pervading confusion and superficiality. The retreat of the glaciers had begun to be observed in the Himalayas as early as 1780, which obviously could not have had anything to do with the Industrial Revolution, which was then just beginning in the small, remote England. Métairy lists a series of phenomena, from the melting of Arctic ice to deformations or mutations in birds and insects, which are accelerating but did not begin yesterday, nor are they due to human activity.

It is perhaps striking that this author traces the beginning of these changes back about 300 years, which coincides with what Polanyi called *the Great Transformation* of the social fabric that set the stage for the Industrial Revolution. However, given that some of these often associated and confused phenomena are of natural origin and cannot be attributed to man, it would be worth asking the question that Métairy has failed to raise.

Indeed, it may well be that there is a climate change independent of "anthropogenic factors"; in fact, it is comical to ignore the geophysical role of the Earth and its relationship with the Sun, when these are the only major factors. But, going a little further, it should be asked whether the "Great Transformation" of human society is but a concomitant effect of the transformation of the physical circumstances of the Earth, and in particular of its magnetic field, which is its aura or subtler aspect.

This is certainly not intended to deny the specific reality of human culture and history; rather, it is a matter of conceiving the weight of the imponderable, the medium, as unnoticed by us as water is by the fish. There is a certain parallelism between the electrical activity of the cerebral cortex and the ionosphere and thermosphere; although the increase in brain activity in modern man seems to us much more vertiginous, quantitatively, than the possible overheating of these regions so tenuous and distant from our heads.

But we are in our own skin, not in the thermosphere; and much more than in our skin, in a technological "extended reality" that looks like a progressively amputated reality. It is clear that an overheating in the upper layers does not induce bees to deploy 5G antennas all over the planet, nor does it induce an archaic human society. But do we realize that the more material development a civilization has and the more it relies on "knowledge", the more reactive it is? We see it every day; our hyperactivity is already pure reaction to large overdoses of almost always misplaced knowledge.

Of course there is a correlation; in fact it cannot be other way. But the mutations that a global change induces in birds and in a civilization with material hypertrophy, even if they have a common denominator, are practically incommensurable.

*

Exponential increase in cancer and genetic mutations, growing sterility of the population, the obligation to live at night, the return to caves and salt mines, life in underground cubicles with no other window than a webcam as long as it still works, a surge in the demand for lead and lead shielding, spectacular climatic and hydrographic alterations... Métairy is not prone to horror movies, he only reminds us, with very few words, of the incredible trap in which we live and our dependence on factors until today almost unthinkable.

It has been speculated that the eruption of the Toba volcano in Sumatra 75,000 years ago, with a power equivalent to the explosion of many thousands of atomic bombs, could have reduced the human population to less than one percent. Even if this hypothesis was true, such an eruption would not in any case have affected the gene pool of the species, as magnetic field reversals should.

Perhaps this is the mechanism used by the Earth, or rather the Earth-atmosphere-Sun system, for its periodic renewal; or in any case this would be one of its effects. According to various traditions, there have been on this planet many Adams before Adam, and many other "humankinds" before this one with which we now identify ourselves and which could be writing the last paragraphs of its script. On the other hand, from Heraclitus and the Stoics to the Puranas, fire has been spoken of as the agent of that renewal; a fire that does not seem to be of flame, but cosmic in its origin.

Adversity undoubtedly has a value for biological, and not only biological, evolution. Adaptive mutations exist, and one only needs to look at the passion with which neo-Darwinian orthodoxy combats this idea to know that it holds an important truth. On the contrary, it is the so-called "neo-Darwinian synthesis" that indecently exhibits its own impotence to find the meaning of novelty; but its promoters are not interested in understanding life, but in turning it into a Meccano. In any case, beyond chance and necessity, there is no activity that is not adaptive, the local expression of a global imbalance. One of these local expressions would be our present civilization.

Human activity itself is a way to unwind the pressure that comes to us from the environment, of deflecting it instead of suffering it. Someone really capable of doing nothing, like some vigilant or contemplative souls, may be able to perceive those layers of pressure, which extend far beyond the thin barriers of society, without the need for satellites and instruments; even if no one has ever told him about solar winds and electromagnetic fields. That could have been the case for many human beings before the Neolithic.

*

No one will think that the SWARM team is hiding information, since the uncertainty of the data gathered so far is not compromising to anyone. However, it would be worth watching the reaction of the various bodies, centers and agencies if at some point in the future the red alarm light comes on. Politicians and executives have invested heavily in other narratives, precisely those with which we are now being hammered, and it would be a long time before they shifted their bets. But these elements always arrive too late and to use the situation to their advantage; they are clearly part of the problem, not the solution.

Perhaps the most interesting point about a polar reversal is that it would have no solution, even if it is hard to imagine us resigned. But we seem to be conceding that the scenario outlined by Métairy is inevitable and even practically imminent, which could not be more premature. The intervals between geomagnetic reversals do not show the slightest regularity, and the average duration is just that, a mathematical average deduced from rock samples. There have also been recorded "excursions" of the poles, i.e., large shifts that did not end in reversals but nearly eliminated the magnetic field, reducing it to a small fraction. The last major excursion was the so-called Laschamp event 41,000 years ago, when it dropped to 5% of its usual intensity for several centuries. However, it does not appear that this significantly affected the biosphere or the human population.

There is really no pattern for the pole reversal. Are we doomed to total uncertainty on the subject? We would, if we had only the mainstream theory of Earth´s magnetic field, known as the "dynamo theory" created by the convection of the planetary liquid core; but this hypothesis, besides being highly speculative, has very little explanatory power and raises more problems than it solves. Modern sciences, increasingly bureaucratized, gravitate towards standard models that impose a forced consensus by eliminating opposition, thus depriving themselves of the indispensable contrast for their ideas. Let us see a good example of this.

Miles Mathis has been proposing an alternative model of electromagnetism and the fundamental charge field for a number of years. Earlier, several authors had insisted once and again that the idea of an "elementary charge" attached to electrons and protons as if it were a label can only be a useful convention, or, if one prefers, just another of our priceless fabrications; but Mathis has been connecting over the years more and more levels of evidence with a greater scope.

For Mathis, electromagnetism and dipoles are only an epiphenomenon of a fundamental charge field composed only of photons. This charge field is recycled by the poles of particles and celestial bodies like the Sun or the planets. We know of very close planets, such as Venus itself, that have no magnetosphere but stop the high-energy particles of the solar wind —which can only be explained by a charge field similar to that proposed by Mathis. But Mathis' model also explains many other things that are completely unintelligible in current standard theories.

For example, the solar magnetic cycle and its relation to the major planets or the galactic core. Mathis acutely observes that the so-called "periodic reversal" of the Sun's magnetic field every 9-12 years is not such, nor does it have to do with its core, but with the coupling or reconnection with the charge field of the great planets —Jupiter, Saturn, Uranus and Neptune-, inversely proportional to the fourth power of the radius (1/r4). These circumstances are highly predictable.

On the contrary, the true pole reversal affects simultaneously the Sun and all the planets of the solar system, since it does not depend on what happens inside their bodies, but on the regions through which it passes on its journey around the galactic core, with its unpredictable differences in the local flux of photons and antiphotons, that is, of photons coming in one direction or another, with their spin orientation changed.

The axial tilt of the various planets can also be explained by just three numbers: mass, density and distance. The present "explanation" by collisions cannot be more contingent and sloppy; but this can be extended to the overall picture of celestial mechanics and its perturbation theory.

Mathis obviously prefers to ignore any possible connection with astrology, but this major difference also allows us to understand the "double standard" of the first encyclopedia of humanity, the ancient zodiacs: the tropical, based primarily on planetary relationships, and the sideral, focused on the stellar influence and the cycle of precession of the equinoxes known as the Great Year —although the cycle around the galactic core is about ten thousand times longer, some 240 million years.

One cannot ask of Mathis' theory, the product of a single individual, the same degree of "formal sophistication" than the standard theories developed collectively by thousands of them with much more time and dedication. And yet Mathis makes very concrete predictions and beats the mainstream model time and again in the most diverse areas. To give just another example, it has never been explained why the solar wind acts so differently for positive and negative ions.

Very recently, the so-called "climate change science" has outdone itself and reached new heights of stupidity by claiming that faster ice melting under "global warming" is the most likely cause of the pole shift[2] observed in the mid-1990s of the last century; even when the phenomenon of polar wandering was known long time ago. Mathis relates it to the charge field[3] and the libration of the four major planets observing that it follows the same cycles as the Solar cycles and, extrapolating on the known data, estimates that the shift would have a variance on the order of... "300 years or so". Incidentally, this oscillation reached its maximum around 2005.

The SWARM mission is by no means among the most expensive scientific projects of these decades, but much more could still be known and at much less expense had the experts the courage to contrast different theories and fearlessly sought to falsify them all; that is, if competition were real and not just a mere word. There is already a huge and varied mass of data; what is decisive in order to draw conclusions is to be able to contrast and view the problems from opposite sides. Otherwise we are still groping around in the dark.

*

So, probably, the fearsome Lead Age will never come into effect. But how could we know, if the ban on hunting theories is not lifted? Any expert in geomagnetism will admit that the dynamo theory is tentative, and even highly tentative; but to question the classical and quantum theory of electromagnetism, on which all our technology rests, is quite another matter. And yet neither are Maxwell's equations or those of quantum electrodynamics anything other than a heuristic development.

Many argue: how can a theory like QED, which achieves accuracy to twelve decimal places, be heuristic? Precisely because the theory has been optimized all along the way to make increasingly accurate predictions, disdaining everything else. In fact, the key numbers of this theory must be entered by hand and one has no idea how to justify them; not to mention that quantum mechanics is not even able to predict the collapse of its very wave function. And what about the prediction of the vacuum energy, an error of 120 orders of magnitude? But even Maxwell's classical theory has huge holes in it.

Mathis, for example, finds a very straightforward justification for many of these enigmatic numbers; but he does so within a frame that has nothing to do with the standard model —which prevents his arguments from being taken into account, whatever their value. The dominant theories are neopositivism in action, which says that the descriptive framework is ancillary and it is the prediction that is decisive; however, when an alternative theory obtains better predictions, its description is said to be totally inappropriate.

The standard model of particle physics knows about four fundamental forces, but even if we knew forty or forty thousand forces to twelve decimal places of precision, our science would still be a prodigious display of ignorance. On the contrary, it is clear that it would be much worse, because it would only increase our confusion. It would be better for us to know *one* single force well, than forty thousand in the present way. Why? Mathis believes that physics can and must know the causes of phenomena, but I think this is just another illusion, albeit a very useful illusion.

It is not that we can know the causes, but that a theory with more balance between description and prediction puts in evidence the theories in which both parts are more dissociated. The balance is the needle of the scale, the compass that allows us to navigate in the seas of our ignorance. And since Mathis' model is much more balanced, even if it can never identify the causes, it can expose the false causes of other models that pretend to be beyond causality but in reality identify themselves with all kinds of empty words. Nothing is more unbalanced than the instrumental vision of science that prevails today.

Mathis finds no use for field potentials, since everything would be due to the direct effect of the bombardment of large, ordinarily spinning particles. Certainly, talking about photons as if they were marbles may seem very crude in the world of rarefied abstractions of mainstream physics, but this "naive" view serves to narrow down problems, sometimes with completely unexpected results. For it is not a matter of being highly "original" and "creative", as is demanded today of theoretical and even applied physicists, but rather the opposite: no trick pulled out of the hat can compare with simple rectitude.

There is something in which the great American author does coincide with the prevailing model: there are no forces at a distance, causality should be based on local description. Newton himself did not accept the idea of a force at a distance, and, had he been able, he would have strongly promoted the concept of field that was later developed. But quantum electrodynamics claims to be local and yet it has very little of causality in it. We are talking about mirages. Action at a distance *is* the immediate given; fields, locality and causality are just our mental elaborations.

We have not been able to remotely understand even electromagnetism, the only force that we can and do manipulate at will in all kinds of devices. This speaks volumes about the *level* of our science and the idea we have of it, as it also speaks of the abysmal difference that can exist between manipulation and prediction, with all its cohort of "wonderfully symmetrical" equations on the one hand, and understanding on the other. And we will have to remember once again that the interpretation of physics is not a philosophical luxury, but also marks the limits of technological applications; which of course gives food for thought.

*

Some physicists have proposed that terrestrial geomagnetism and data collections such as SWARM could be an interesting way to search for "new physics". For example, to find a "monopole moment" in the field that would allow to better define the conditions of existence of the famous magnetic monopole, that unicorn of theoretical physics that came out of Dirac's imagination. It is well known, for example, that all modern supersymmetric theories require monopoles, which have never been found.

But, as Mathis says, if the electromagnetic field is not a dipole to begin with, there is no need at all for a magnetic charge unit. Nicolae Mazilu, following E. Katz, reasons that there is no need to complete the symmetry, since in fact we already have a plain and more significant symmetry: the magnetic poles are only separated by portions of matter, and the electric poles are only separated by portions of empty space. There is no polarity of any kind without these conditions. But the relativistic Dirac electron equation, precisely because it is relativistic, only works for ideal point particles with no extension, hence the postulation of this perfectly unnecessary particle.

Actually, the only possible "monopole" is just Mathis' charge field, so one should know where to look to find fabulous animals; and yet I keep thinking that fields are nothing but metaphors. Physicists call the excitations of the field "particles"; I call the excitation or actualization of the potential a field —but in any case, and as we have already mentioned elsewhere, the difference between the kinetic, potential and internal energy of a body should be analyzed more carefully.

No, I do not believe, like Mathis, that it is possible to do physics only with forces, or that the problems of wave mechanics can be solved with particles with "stacked spins" and without resorting to complex numbers, which seems frankly impossible. But Mathis' model allows a much more insightful and less blind interpretation of multiple phenomena today attributed to fields.

According to Mathis, all spherical bodies, from stars and planets to extended particles, are absorbing this charge field of light at both poles and emitting it at the equator; and presumably our own organism does something similar. This bears a strong resemblance to the ancient ideas of prana, chi or pneuma, although Mathis' predictions and descriptions are strictly physical and do not explore the coincidence. As Harald Maurer already said, it is not light that is part of the electromagnetic field and the straitjacket of Maxwell's equations, but on the contrary this field is to light as the water in the pool of a cruise ship in relation to the surrounding ocean.

If the fundamental field of light already has such relevance in the formation of atoms and molecules, it must be equally relevant in the more complex biological molecules, such as globular proteins or enzymes, which are already a form of life on their own, albeit unacknowledged; and even more so in organelles, cells and complete organisms.

Mathis attempts to give a "local explanation" for phenomena, such as elliptical orbits and fields, which have always been global in nature. Poincaré already observed that any law expressed with a variational principle admits an infinite number of mechanical explanations, which by the same token cannot be but futile. But the same variational principle can be a scaling factor, a sliding knot, and this brings us to another kind of considerations.

If the monopole is a fabulous being, it is nonetheless of great interest, being closely linked to such ubiquitous phenomena in Nature as vorticity or the geometric phase initially detected in the polarization of light and the magnetic field of the electron. The geometric phase, "global change without local change", is probably best known for the Aharonov-Bohm effect, which shows us how a charged particle "feels" or reflects the potential *even where the electric and magnetic fields are zero*.

Interestingly, the geometric phase was first detected in 1956 and 1958, while the discovery of the magnetosphere and the idea of the solar wind also date back to 1958, at the dawn of the space race.

*

In the earth's magnetic field, the cosmic and the telluric coincide, which also makes us think that the geophysical can also disrupt the geopolitical. We all know where the compass was discovered, and what consequences it finally had for the world when its use in navigation became widespread; but we have yet to discover a very simple principle that will help us navigate the ages to come, whether the sky falls on our heads or not.

Although it is impossible to verify with certainty, it is estimated that the peaks of exposure to electromagnetic radiation of human origin are today between fifteen and eighteen orders of magnitude higher than those of the natural electromagnetic field of planet Earth —a ten followed by fifteen to eighteen zeros. Comparatively, human air pollution and CO2 emissions would only have increased by a ridiculous fraction over that generated by the natural forest fire cycle.

As if this were not enough, with the implementation of 5G this exposure will be multiplied tens or hundreds of times. Elsewhere we noted[4] how the geometric phase can be used to investigate the impact of electromagnetic pollution on organisms, since although we cannot directly measure potentials, we can measure their effect on cell movements and on the shortening of the telomeres of DNA strands, which also exhibit geometric phases. This phase being equivalent to a torsion component, it can be seen as a measure of the degree of forced contortion of the system with respect to an unforced fundamental state, making it robust to various kinds of noise. Even in our bilateral nasal cycle such a phase memory can be detected and compared with variations in the electromagnetic potential, to establish a more akin biophysical connection.

We have already spoken repeatedly of the geometric phase and its uncertain status in modern physics. Since its discovery is more recent than the development of quantum mechanics, theoretical physics tries to see it as a mere appendix, although it really does not respond to the conservative, Hamiltonian mechanics of closed systems. The geometric phase precisely reflects *the geometry of the environment*, the "parallel transport" of that which has been left out of the idealizations on which field theories depend, no matter how little or much. Therefore, its strategic importance in theoretical physics would have to be evident, were it not for the fact that, beyond the horizon of closed systems proper to "fundamental physics", everything is considered secondary.

It is a serious mistake that theoretical physics has paid dearly for. Just as the geometric phase adds a sort of "fifth equation" to Maxwell's four conventional ones, it can also be said to add a "fourth principle" to the three principles of Newton's mechanics. We have recently seen that this geometric phase allows us to speak of a purely mechanical induction which is more general than electromagnetic induction, and which can be easily verified in various devices and even in our own body but which has not yet found its proper place in the theory.

The geometric phase is a universal phenomenon at all scales and is not exclusive to quantum mechanics as is falsely claimed; it should therefore have already allowed us to cross the threshold between classical and quantum mechanics were it not for the inaptitude of the assumed models to do so. And on the other hand, it allows us to step outside the box of electromagnetic theory and see what lies beyond it —both in theory and practice.

This non-dynamical displacement of the geometric phase, at the basis of the so-called "holographic principle", and interpreted by Bohm as an "information field", is also directly linked to the so called "retarded potential theory" of Weber's electrodynamics, appeared well before Maxwell's, which can also be used to explain the general occurrence of ellipses. And although a theory based on the potential is at first sight incompatible with a perspective such as Mathis', one need not recall that the degree of polarization and the entropy of a beam of light are equivalent concepts, which in fact adds a connection of the broadest scope with thermodynamics at the fundamental level.

The "fourth principle", still to be properly defined, involves the other three as a term of self-induction: this is the sliding knot of which we spoke before. This self-induction is linked to the relation between bodies and the ordinary ambient vacuum. In modern field theories there appears what is called self-energy and self-interaction due to the acceleration of charges. The geometric phase is so called to distinguish it from the phase arising from interactions or forces, but that does not mean that it is merely passive with respect to them. How could an instantaneous correlation be passive with respect to something that takes time and motion to actualize? It would have to be the other way around, and it would seem obvious were it not for the fact that physics is already founded on motion, and on a certain way of calculating and determining it, which we have called the "global synchronizer".

All that today we call "physical laws" of Nature are but pale and very restricted reflections of a guiding principle that evades us the more we cling to the Law and the letter instead of the spirit from which they emanate.

For the modern theoretical physicist, “nothing more practical than a good theory", since here theory has been proposed right from the start to fit predictions. But what we will soon understand, on the contrary, is that nothing has more theoretical depth than a good practice; and this, in today's technoscience, also means that it has to become concrete in the form of a machine, or, if one prefers, in a new relation with machines.

All our mechanology depends tacitly on the three principles of mechanics, but up to now these principles have served us to enclose and recirculate things that were alien to it. Already that first classification made by Jacques Lafitte of passive, active and reflexive machines was very directly related to the three principles, and it would be enough to return these to their original background to be able to break the circle of the spell and see everything in a different way.

Since mainstream theories prove incapable of overcoming the inertia of their triumphs and have already died of success, it will be in the machines that the synthesis will take place, never so consciously against the tide of that same theory that confiscates a posteriori the explanation of what it cannot predict. This time it will not be so, because the machine will be expressly conceived to destroy the theory and the mad spirit that inhabits it.

*

The so-called geometric phase, identifiable in animal locomotion and today also used abundantly in control theory and robotics, makes it possible to establish interfaces between machines and humans at many different levels, and certainly also allows a conversion and transition between these different levels. In this sense, it poses an extreme danger, since it is obvious that today Control wants to close its claw definitively on man and Nature. The cybernetic hypothesis had and still has a big hole at its center, but this discreet assistant is able to cover it grade by grade without us hardly noticing what is going on. Is not the geometric phase used as a routine parameter in state-of-the-art technology with hardly anyone wondering about its meaning? Nothing could be more disturbing than this thoughtlessness.

To put it more clearly: what is used today as a rectifying factor in control theory, and even in the so-called "quantum computing", is something that Nature has always had built in, from the ellipses of the planetary orbits to those of the electrons; nevertheless, we are still told that all this is governed by "blind forces". That which today is systematically used to bridge the gap between the human and the natural by tightening technocratic closure is the very condition of openness that connects the purported "closed systems" with their source.

We have already seen in other places that the three laws of motion can be profoundly transformed by substituting the principle of inertia by that of dynamic equilibrium, while preserving the mass of data and predictions of current physics. But the geometric phase, perfectly compatible with the principle dynamic equilibrium, also allows us to profoundly modify the idea of the "global synchronizer", founded on the concept of simultaneity of action and reaction, introducing a proper time for each system. The true global synchronizer cannot be inside the dynamics, but outside it, precisely in the instantaneous correlation of the geometric phase. It is the dynamics that reacts, and it is everything that moves that has a time of its own. The ideas of locality and causality are due precisely to this circumstance, not the other way around, as is thought today.

We cannot yet grasp the consequences of this transposition because our own subjectivity is embedded in the mechanical scheme of things, and the Global Synchronizer is the guardian of all the accumulated material-symbolic capital. Going beyond this demon is tantamount to liberating man and Nature; but do we really want liberation? It is not that machines can liberate us, but that we can liberate ourselves from machines, from our instrumental compulsion, by more gradual or more direct routes, though following the same orientation. However, this whole route is full of dangers and deceptions, starting with the illusion of being able to have everything at once without the slightest sacrifice.

Although it should be self-evident, it is worth remembering that in practice it is impossible the liberation of man by man if our idea of Nature does not transcend the instrumental threshold, since the techniques of the use and exploitation of Nature end up being the same as those applied to the use and exploitation of men. Machines, apparatuses, dispositives and institutions are "coagulated spirit", to use Weber's image, and the spirit does not release its prey without other dispositions. Now, the purportedly "neutral" three laws of motion define the general disposition and limits of our world, the symbolic economy in the relation of our civilization with the unknown natural background. Neither relativity nor quantum mechanics has changed this in the least.

*

René Guenon wrote somewhere that, in time past, man was receptive with respect to the Principle but active with respect to Nature, whereas modern man has turned his back on the Principle while becoming reactive with respect to the natural world. This is not the image that the moderns, in search of the mastery of the external world, have of themselves, but Guenon's judgement is more faithful to reality. The nature in relation to which that man was active was none other than his own, and that was first and foremost his garden. And, indeed, to make nature something external leads us unfailingly to our interior becoming reactive with respect to our increasing interference with the external world. Thus the Principle becomes an iron Law for those who ignore it.

If we were to balance the predictive and descriptive parts of physics, which is not even remotely the case now, we would soon see that the question of the polar axis and the orientation of matter leads us by the shortest path from laws to the elusive but omnipresent Principle. For this, no gigantic scientific projects are required, but only a change of mentality. For a large number of crucial experiments, satellites are not even needed; some very basic sensors are derived directly from tests such as the Trouton-Noble experiment of 1901-1903, and the like. It's not a question of cutting-edge technology, but of another orientation of intelligence. The biggest obstacle, of course, is the enormous inertia of Big Science, its crushing bureaucracy and the miserable, self-serving narrowness of criteria it imposes.

The three principles of dynamics and motion, even in their conventional version, are closely linked to those "three asymptotic steps" in the structure of the physical world and the way we perceive it: matter, light or radiation, empty space. Between matter and radiation there is a certain reciprocity, which we now ordinarily interpret in field theories with metric space and curvature. Beyond flat or empty space is the primitive homogeneous medium, with unit density, which has nowhere to go because any motion or determination does not apply to it. Neither can be said if this medium has zero or infinite dimensions, whether it is full or empty, conscious or unconscious.

The relation of these three steps or stages to the fourth defines the connection of the measurable with the immeasurable Principle. These three stages have a non-trivial correspondence with the Indian threefold idea of the dense, subtle and causal bodies, superimposed on the Self. Indeed, what we call "light" and "photons", with their absorption and emission spectra, and with many other peculiarities that we have not yet investigated, are not simply particles traveling through space, an idea that is already quite comical. They are part of our interiority, and of the stochastic noise integrated with the signal at the levels of wakefulness and sleep, although they are not yet that which we call "subjectivity", which cannot be related to motion or time, but with their background.

*

The all-pervading electromagnetic radiation coming from our devices is "non-ionizing", and it is clear that it has nothing to do with the deleterious high-energy charged particles of the solar wind. In our current scientific jargon, this radiation of human origin is made only of "photons", massless particles with very low energy. I can hear the laughter if I say now that these photons are already part of our interiority and mental background; but to believe that photons are traveling from one part to another "out there" seems to me even more laughable and inappropriate. To decide what is "out there" and what is "in there", if such things make sense, one would have to carefully analyze the relationship between the three states of manifestation, of these with the principles, and of these with the Principle.

There are still all kinds of doubts about the effect this radiation may have on organisms, not least because impartiality has become impossible and we know where the money goes; so everyone will have to judge from his own experience and careful observation. But it is clear that it is an aberration to saturate the atmosphere to such extremes and that we should oppose it head-on. There are clear alternatives and technical solutions, and neither do we need so much stupid data traffic whose main use is to increase surveillance of the population ad nauseam. Those who push the new wireless generations at any cost will pay dearly for their mistake. There is still time to redefine these non-strategies.

Beyond that, issues presented here such as the fundamental charge field and the multiple levels of manifestation of the geometric phase are also of great relevance to biology: let us recall that this phase denotes the coupling of "a system" with the geometry of the environment, or as contemporary jargon has it, with its information.

The current state of the sciences is anything but accidental and few things are more carefully directed from above. Their extreme closure and inertia, their marked necrophilia, also respond to a situation of threatened global hegemony, as well as to a historical evolution and a gradual process of takeover by public and private bureaucracies. In the colonies the laggards make pathetic efforts to do "competitive science", but there is nothing less competitive and open to debate than science today.

All countries would do well to disengage their objectives and priorities from this globalized "science", which is really going nowhere, except where a few want it to go. Since the current situation responds to a geopolitical impasse, let us hope at least that here and there, the most discerning heads refuse to do the dirty work that is being demanded from very few centers of power. That in countries such as China, Japan, India, Iran, Brazil, Australia, Germany, France or Spain itself, there is enough personality not to be fooled by a model of research, without competence and increasingly incompetent, that others manage to impose for their own prestige and interest. Or even in the United States itself, which is treated by its media as another conquered nation.

Not everything depends on money, and neither does science, otherwise it would already be dead; but it is evident that the more funds are poured into it, the deeper the truth is buried. It seems almost a foregone conclusion that around 2026, the four hundredth anniversary of the publication of Bacon's unfinished fable, we will witness the collapse of this *New Atlantis*, and with it also the decomposition of its entire network of influences. Science has been a political phenomenon of the first order at least since 1703, when Newton assumed the presidency of the Royal Society, the first think tank of modern times.

The science-power collusion cares very little about Nature, because if they did, the first thing they would do would be to radically change their idea of science and their idea of the natural. And the same goes for all of us: one cannot want to "save the planet" while embracing a science of death. Claiming moral superiority without questioning the prevailing worldview is just cheap politics, since the prevailing idea of Nature and our nature is totally unacceptable to begin with. Today "Science" serves above all to legitimize the passage from a democracy that is hardly formal anymore to the most opaque and technocratic of tyrannies.

Where there is a barrier there is a way to cross it. A single "device" can short-circuit forever the arrangement of mechanics and what we believe to be mediated and immediate inside and outside of us.

*

But we cannot stand idly by and wait. The only pole reversal within everyone's reach is the one between dominator and dominated, with an increasing use of all sorts of pseudo-scientific arguments. And the only way we can alter that relationship is by not giving credit to their ubiquitous lies and refusing to obey the mandates of a leadership that is not only unworthy but clearly criminal.

Let us recall once again the power law or 80/20 rule that shapes the distribution of wealth in the world: One fifth of the population owns four fifths of the properties, but in turn one fifth of that fifth owns 4/5 of the 4/5, and so on. The iterative succession of this rule leads us to conclude that three individuals or families own as much as half of the planet, and more importantly, most of the surplus purchasing power, which serves precisely to buy the subjection of those below them in the hierarchy.

This inescapable power law, which economists and sociologists have the great merit of ignoring, is precisely the fundamental fact of economics and sociology within our non-economy and non-society, the structure and dynamics that define the real concentration of power, its hierarchy, and the real intricacies of subservience and favor in actually existing techno-feudalism. This extreme concentration can only be sustained in anonymity, for public exposure of the top of the pyramid would make it extremely vulnerable —and it goes without saying that the core of the plutarchy, necessarily a cryptarchy, is not constituted by the vicarious potentates who show their faces in the media and appear in the lists of great fortunes.

Nothing of what is happening now with the plandemic or the nefarious mass vaccination campaign can take place without the will of that tiny leadership, the keystone of the whole structure of world debt. Just like capital, power is calculatedly distributed to evade any responsibility; which has allowed a sustained increase in the scale of crime, up to what we are now witnessing. Yet the structure and concentration remains, while we are unable to identify three elephants in the laundry room; although there are certainly no laundries like banks, corporations and funds.

There is a direct proportion between this accumulation, induced and pumped by the debt system, and reflected in the successive iterations of the power law, and the cycles of contraction and expansion of debt and the markets which in turn echo astronomical cycles. Cycles that concern Jupiter and Saturn, and therefore the Sun and our planet, and which have been decanting and gaining momentum over the last three centuries. The structure of debt and of the usurpation of the power of creation of public money is a sculpture in time, punctuated by widespread fraud, corruption, war and crime.

A more than superficial parallel can also be drawn between the standard theory of electromagnetism and its "intrinsic" electric charge, on the one hand, and the idea that still prevails today about the creation of money, which is attributed to central banks instead of private banks —although in the final analysis it is the borrowers and repayers who really create it. Will we ever overcome these mystifications and deceits?

Money creates space for its indispensable expansion by destroying the small wealth: it is Schumpeter's "creative destruction", which in the present imbalance has acquired a new dimension. But those who do not have the money can only create space for themselves by not believing, not obeying and not assenting. Only from that created space and that freedom not to do can an initiative worthy of the name arise.

References

Paul de Métairy, *Cambio Climático: la verdad de lo que se nos viene encima*

Miles Mathis, http://milesmathis.com

Miguel Iradier, http://hurqualya.net

Notes

1. https://phys.org/news/2021-04-climate-shifted-axis-earth.html
2. http://milesmathis.com/wander.pdf
3. Miguel Iradier, 2021. *Stop 5G, medida por medida*. https://www.hurqualya.net/stop-5g-medida-por-medida/
4. Miguel Iradier, 2021. *Judíos, cristianos y fenicios. Saturno, Júpiter y Marte*. https://www.hurqualya.net/judios-cristianos-y-fenicios-saturno-jupiter-y-marte/

THE 80/20 GAME AND THE ZION/BABYLON CODE

6 July, 2021

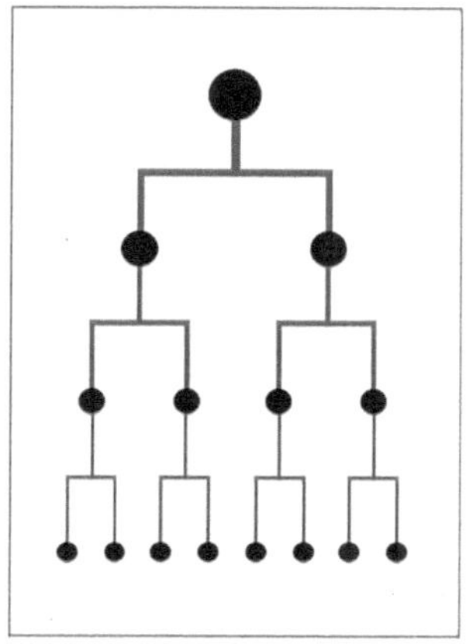

Most have heard of the 80/20 power law or Pareto principle that governs the distribution of wealth in the world: one-fifth of the population has four-fifths of the assets, but in turn one-fifth of that one-fifth owns 4/5 of the 4/5 (4% has 64%), and so on.

Most have heard of it, but only to forget it soon, not least because most of the topics with which the mainstream media try to fuck up our minds, from climate change or coronavirus to transgenderism and critical race theory, are reheated every morning to make us forget this fact.

So we have no choice but to get our heads around it keeping the scope in focus until we fully understand what this distribution means. And although I cannot fully unravel the issue here, far from it, I will have already done my part in duly drawing the attention of independent spirits.

Last year a kind reader commented to me that, more disturbing than the successive iterations of the Pareto rule in the distribution of income, was for him the process of informatization/digitalization of being as a "project of consummate control and domination". I suspect that many people share this same opinion, apparently justified, which nevertheless reveals a too superficial understanding of what this law of distribution implies.

And what this law implies is much more than the scandalous but well known wealth inequality, so often summed up in the phrase "three fortunes own as much as the poorest half of the planet" — they use to put 60 or 62 instead of 3 or 4, but the latter is more credible if one understands the gist of the law. If we are witnessing something that looks so much like a project of consummate control, and which some now call "biodigital convergence", it is precisely due to the fact that this huge accumulation is not something amorphous, but highly structured and integrated into the social fabric and the instances of power. There is an accelerated project of domination because there is already a situation of overwhelming domination that at the same time supposes an enormous imbalance, not the other way around. This double circumstance has a mathematical pattern reflected in society and it is precisely this distribution.

We touch this point time and again on purpose and with the best of reasons: if economists and sociologists systematically ignore the most remarkable feature of our non-economy and our non-society, or treat it with a frivolity unworthy even of entertainers, we can be sure that it is still more important than it seems. Their silence is doubly informative, and my order of priorities does not depend on what they say is important, but rather the opposite.

I wrote about this rule here[1], here[2], and in many other posts. We all know intuitively that an inverted pyramid of wealth is potentially more unstable as the slope of inequality increases; however the graphs and illustrations of this inverted pyramid, provided by the banks themselves, hide much more than they reveal.

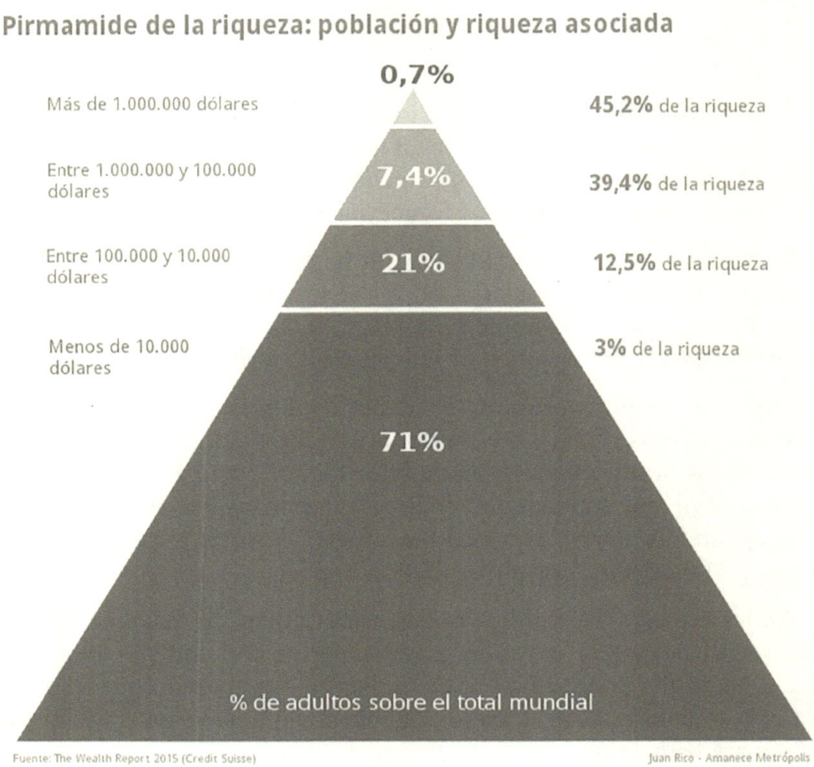

The graph above, for example, says absolutely nothing about the underlying dynamics, and one would think that wealth has a natural tendency to rise upwards, while the big beneficiaries at the top, magically isolated, want nothing to do with those at the bottom. There are two half-truths here that serve to cover up the real whole truths.

The 80/20 rule has a self-similar, recursive structure, like that which exists for example in the blood vessels of our body: it is therefore a structure that responds to an optimized function with a high degree of organization. You don't need to scratch your head to guess what kind of "flow" is being optimized here, for in effect, we are talking about a large suction pump that reaches down to the capillary level. Only money, and in particular debt-money, can achieve in our society such a simultaneous degree of concentration and diversification.

The Pareto distribution has much more to do with a fractal hierarchy like the one below, although this one is simpler and without due proportion; not to mention that the structure alone tells us nothing about the time evolution.

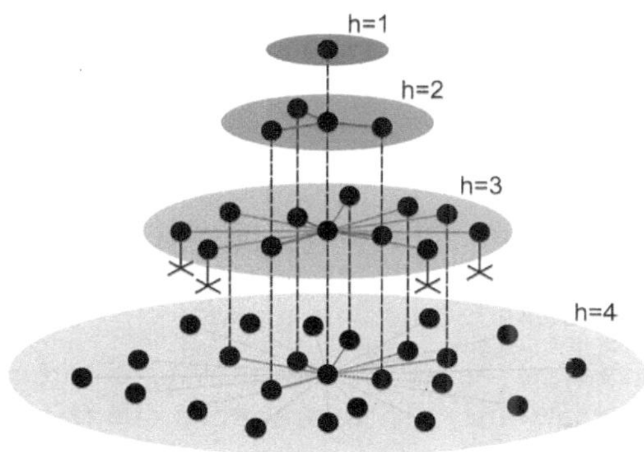

This inverted pyramid is above all a great hierarchy, a highly selective filter in both directions, upwards and downwards, which determines positions, priorities, favors, and obedience. We are told that we live in a “liquid”, horizontal world but they hardly talk about the conduits that use the front of "the markets" for the benefit of the Great Siphon, the vertical hydraulic apparatus that works at all scales.

This "parallel reality” explains the very gradual transition from the fictions of the free market to the actually existing techno-feudalism. What appears to the unwary as a new phenomenon has been steadily taking shape for centuries. This slow forge has so far been the guarantee of its stability, just as it may be the guarantee of its collapse beyond the point of no return.

*

The Pareto principle remains a mere fact of observation, hanging like a picture in the sky without any apparent reason to justify it. This make it even more interesting to the eye of the naturalist, for this rule governs not only the distribution of wealth, but of many human activities and natural processes, from the size of cities and corporations, to earthquakes, grains of sand or stars.

There are two remarkable things about this principle: on the one hand, it has no explanation, on the other hand, it is ubiquitous in the products of both Nature and Culture, which we assume are separated by a yawning gap.

But the number of iterations of this law, the slope and the spectrum of inequality, depends enormously on the context, or as analysts say, on the constants, variables and parameters of the "system". Some have wanted to wield the Pareto principle as irrefutable proof that human inequality is just a reflection of natural inequality and diversity, and to some extent, that could be true. But to what extent?

The inequality of gifts and talents is undeniable, and only the most mediocre could dream of universal equalization. However, it is clear that the distribution of wealth is incomparably more unequal than that of talents, because no one is a billion times smarter or stronger or faster than the most impaired, and very often it is found that those who have the most barely reach the average.

The weight of human beings has a normal or bell-shaped distribution, because there are obvious limits to the gain of weight. Power laws can occur at all scales, as there are no known limit on growth.

However, there are limits on the growth of fortunes, and some are approaching them dangerously. There are limits due to the size of the world and due to the fact that monetization is relative: if you have everything, you have no one to sell to, and your properties have no price, even if they have “all the worth”. Needless to say, this system relies more on price than on value.

On the other hand, the 80/20 rule is neither the 70/20 nor the 80/10 rule, and says both that 20 percent has 80, and that 80 has 20 percent; that is, at least ideally, a curious asymmetry in reciprocity and a round reciprocity in asymmetry, so as to please everyone. It is all about the whole, and the question is how many times the joke is repeated and how the balance of the imbalance grows up.

It seems obvious that the "freedom of scale" for capital growth, the condition for its permanent accumulation, is, rather than inheritance, the appropriation of the mechanisms of money and credit creation —for it seems also obvious that the inverted pyramid of financial wealth has as its counterpoint the accumulated mountain of debt of our economies.

Business schools cannot do enough to give us a different idea of the Pareto law. Nassim Taleb asserts and even demonstrates that this is the expected outcome from highly "dynamic" systems; for example, most of the companies on the Fortune 500 list are not the same as they were a few decades ago. Therefore, the law gives us "big winners", but these big winners change over time, as it is always the case in societies where equal opportunity is paramount!

Some of us have a different picture. By 1835, a single family of bankers controlled the public debt and the finances of the main European powers, from the Bank of England down. This family took control of the levers of the national banks, and on the other hand, it is not known to have squandered or divided its inheritance, nor was it subject to expropriations.

So it is highly unlikely that this fortune would steadily diminish over time, while it is highly probable that it would do the opposite, especially considering that they had no need for risky bets, and that the name of the game is not "innovation", but the coldest and most ruthless exploitation of any advantage gained.

On the other hand, it is trivially true that in a world in perpetual "creative destruction" new opportunities are always arising, and that those who exploit them by obtaining sufficient financial backing can grow much faster. Therefore, large financial surpluses are always looking for new investment possibilities to diversify and minimize risks, seeking to support monopoly or oligopoly situations as they are the most profitable while diverting the attention from themselves. This kind of "synergy" makes it much easier to explain the irresistible rise of corporate empires like Standard Oil in the 19th century or Amazon in the 21st.

Risk engineering experts like Taleb should know it better. The stock market casino is mostly a front, and large companies, especially since they have the direct support of central banks, are mainly intensive fund-sucking devices to make insiders richer at the expense of the unwary outsiders; under these conditions it is normal for company names to change, as they are not made to last. But with personal fortunes things are very different.

*

Statistics reveal time and again that about half of the great fortunes in the United States, Western Europe or Russia belong or are allocated to individuals of Jewish origin, a truly remarkable fact considering that they make up about 3 percent of the population. If the 80/20 rule and its successive powers provide us with the most revealing x-ray of the state of affairs in society, this officious 50/50 share or law gives us an x-ray of the x-ray, an additional penetration into the complexion of that which, with very convenient anonymity, has always been called "capital".

Structurally, it does not matter whether the holders of these fortunes are Jews or not; but historically it does. Especially if one takes into account that most of them, Jews or not, align themselves with the cause of Zionism or use it as a binder, trying to hinder it as little as possible. As the adage goes, the disciple always follows the master, and anything that moves what does not move. The "50/50 law" reflects the internal physiology of this creature, and the delicate balance between violence and deception that has allowed it to impose its rules. What appears to be a mere result gives us, on the contrary, the genealogy and the cause.

The power law is not relevant to our society because 3 have as much as 3.5 billion. After all, in the overall scheme of things the dispossessed count for practically nothing, while the tip of the pyramid counts all the more the less it is visible. What is decisive is not the relationship of the top to the bottom half, but to the top half, and more so as one moves up the ladder. The arch keystones counts with most of the surplus purchasing power, which cannot belong to those who are in debt but to the creditors.

On the other hand, the front men who continually appear in the media and whose names we all know are undoubtedly among those who owe favors, and showing up for others is part of the price they have to pay. The mere recursive structure of debt, operating at all levels and scales, makes the conclusion easy; as much as the mere genealogy and historical logic in the development of capital. If all of them are kleptocrats, between the famous oligarchs and the plutocratic cryptarchy there are more than just quantitative differences.

Why does the cause of Zion serve as a binder between the richest of the Jews and the Goyim? Because the Old Testament provides the oldest and most battle-tested supremacist narrative and can justify everything in the name of the Mission. And given that Zionism was gestated in Puritan England, the term "Anglo-Zionism" is anything but an extravagance.

The Chosen cannot see themselves as a kleptocracy, but as the true creators of wealth. For them, the law of 80/20 is that of "the few indispensable and the many trivial" or dispensable, but there are always some more indispensable than others.

The way in which wealth is measured leads to think that those who have more talent to exploit are the same who create more wealth. But in order to exploit more and better, it must be ensured, before extracting surplus value, that the employees have to work for their subsistence many times more than they would have done otherwise. The benefits of the division of labor, the great lure of society, must be combined with an increased cost of living that can be properly exploited with optimized structures of extraction. Money and its system of creation and distribution meet the requirements to play this dual role, for money and specially debt-money is dual by nature.

80 divided by 20 gives 4; 20 divided by 80 gives a quarter (1/4), so we have a difference of 16. This would be quite close to the difference that may be between the real value of labor and the wages received: elsewhere[3] we said that such a difference should be "somewhere in between" 1 and 100, probably close to a factor of 10. The huge, and for many incredible, difference can only be created by the full-scale manipulation of money, credit and prices, and by the aggregate and invisible debt present at all levels. This in turn facilitates large-scale fraud and plunder, leading to further indebtedness. In the whole picture, surplus value was always a minor issue.

If the rates of extraction and extreme inequality of the pyramid of wealth are already outrageous, the massive dose of systematic division and intoxication it requires for its justification or even for its mere subsistence are equally bad or worse. Such numerical imbalance and inequality can only be compensated for by a extreme weakening and confusion of the population that minimizes its capacity to react —and by an organization of the few only comparable to the disorganization of the many.

The Zion/Babylon code is none other than the old "divide and rule", but with its own accent and an infallible instinct to corrupt and poison everything, transferring to others one's own sense of guilt. Once the virus was transmitted with religion and now the media have taken over. Zion is the fortress of civilization that rises the more the rest sinks into dissolution and anomie: again the opposition between the vertical and the horizontal, between the extraction apparatus and the pool.

Of course, this variant of the "spirit", permanently occupied in converting everything that is not itself into a subordinate matter, is nothing but the most accomplished inversion of what in other cultures has been understood as such: the antithesis of any elevation. However, this projects itself in "Babylon": the ferment of chaos is sown and then the effects are attributed to the confusion of the "natural state".

Divide and rule, Operation Chaos and pushing everything towards the order/disorder disjunctive so that "the elite" is presented as the pole of salvation are just different words for the same thing.

*

The Zion/Babylon code inevitably leads to a syndrome, in which a tiny minority justifies any means in order to maintain its increasingly threatened position —no doubt more by its own actions and initiative than by anything else.

Nothing would be more bloodless, in order to deactivate this situation, than to make public the identity of the highest echelons. The mere fact of putting an end to their anonymity would decisively change the balance of power.

On the contrary, all attempts to excite low passions, to seek scapegoats and bloodshed in the name of the "revolution" reeks of subversion organized by the most powerful to appease the anger of the populace and divert attention from the real culprits. We have seen it too many times already, and one has only to look at who is running "activism" in Western countries and in the United States in particular to know what to expect from them. On the other hand, if hardly anything is known about opposition to lockdown measures, massive pseudo-vaccination campaigns, or insane, brainless and disgusting genetic engineering, it is because none of this is sponsored.

The large-scale war against Nature that has been waged from above absolutely coincides with the war against our own nature; it was already operating at all levels and is now being fulfilled in the most concrete and literal way with genetic manipulation. The whole Nature is "Babylon" for those who pretend to give it shape and rules.

It is necessary to know the full names of those at the top, because after all, even if the responsibility is very organically distributed between the 1 percent that decides or the 10 percent that serves to administer the Siphon, the structure of this Organization points clearly to the top. And it is essential because the very anonymity necessary for the impunity of capital is the most fragile point of this steep house of cards. If everything has been rotting from the top down, if the initiative of this war has come from the pinnacle, it is only possible to proceed with total coldness and surgical precision. If the highest pieces fall we will not fail to see the domino effect of this inverted pyramid leveraged to the extreme.

*

In Bruce Bogoshian's numerical simulation of the kinetics of wealth with a Pareto law, the system tends towards a singularity in which everything ends up in the hands of a single owner while the rest of the population is completely dispossessed. But the real world avoids mathematical singularities; reality is more "twisted and interesting", and what we are witnessing is, by virtue of debt mechanisms that are not even mentioned in his study, a transmutation of wealth into top-down influence and subjection. The global effect of the Great Siphon is far more profound and denaturing than what would result from leaving us all with nothing.

In truth, the combination of Number and Verb, of the mathematical pattern of the 80/20 rule with its temporal evolution under "conditions mimicking the real world", has everything to develop the most absorbing of strategy games: an idea that I propose from now on, and that I hope will spread like wildfire and serve to acquire awareness of our situation and real problem instead of distracting us with the pseudo-problems promoted expressly to keep us away from the issue. Teaching by playing? I think it is more like learning by disgust, as it seems one has be much more disgusted by what is going on before taking action. This game and its metagame will help; at least we will talk and think about it.

Gamers and game developers, you can take note. You have here a whole world of possibilities to emulate the existing sordidness while achieving an incomparable perspective that no business school will give you. Have you ever thought about how many times the 80/20 iterates from the bottom to the top? I leave to the reader the calculus, which will help you to know where you are in the hierarchy, how many "echelons" you have below or above, what are your prospects of climbing or sinking, and so on.

Of course the Pareto distribution is continuous in nature and has no other "echelons" than its successive powers; however, fortunes are concrete and discrete in nature, like any number of individuals. From this circumstance derives a whole cascade of possibilities for the game, which the developers will know how to use with sagacity.

There can be many variants, modalities and levels in the 80/20 game and its Zion/Babylon code, some more focused on the historical inquiry of facts and identities, others on the systematic tactics of division, diversion and disinformation (3D), or on the Assault on the Fortress, the methods of counter-organization against the Organization, the mechanisms of creation of debt-money and its alternatives, an so on. All can be extraordinarily instructive, and the common denominator is the iterated structure 80/20 and the exhaustive exploration of its implications at all levels. This structure is literally the quintessence of the system.

In anticipation of the overwhelming success foreseeable for this "edutainment", we must count on substitutes, biased versions and the most blatant adulterations. Banks, intelligence agencies and even the Pentagon already have lots of developers at their service, and soon we will see improved versions that will lead us to think that extreme inequality protects us from disorder, or that it is due to patriarchy and insufficient vaccination, or that the suffering of the Jews is as incomparable as the position of privilege they hold.

Nor does it seem advisable to play this game online due to the high degree of infiltration and interference; it is preferable to share the same physical space and avoid unwanted intruders. In addition, its dialectic not only teaches us questions of calculus, but also and very especially about people and the instinct that predominates in them: that of the careerist and the born exploiter, or that of the one who tends to swim against the tide. Between the two lies the majority, who seek to adapt while saving efforts.

Self-consciousness is the name of the game; social self-consciousness, of course, because the other one, much larger and neglected, does not know about structures or calculus. But from a structural point of view, the name of the game is asymmetry. The oppressors exploit the asymmetry with an overwhelming, hydraulic logic; those who want change have to use their mobility for the asymmetric struggle and take full advantage of the profound incapacity for real change of the establishment, whose barrage of "innovations" always ends up flowing in the same direction.

*

The staged pandemic has served to show us to what extent what we call "Science" is a massive weapon and a defensive wall for the powers that be. Even for those of us who already knew decades ago what can be expected of Big Science or the media in these times, it is hard to believe the levels of propaganda and suppression that have been reached in the last two years.

The different sciences today are mere prostitutes at the service of very exclusive interests, and this does not only happen in medicine and biology, economics or sociology. It is absolutely general, and it affects even mathematics, pure and applied.

A good example of this is the Pareto law itself. The uninformed reader may believe that this distribution is still something episodic, yet another of the countless curiosities of math. But, far from being something episodic, this principle is one of the main issues of big data intensively filtered by tech giants and large government agencies. It is everywhere, from Google rankings to data traffic, logistics, management and administration, marketing, debts, defaults, etc., etc. —as it is *key in the efficiency of the exploitation of resources*.

Even setting aside for a moment its master role in the physiology of the so-called social body, it still has all sorts of crucial applications in detail for statistics and the selection of data, customers, and problems. And yet the scientific articles available on the subject on the internet are neither very numerous, nor very deep, nor very revealing.

As in many other disciplines, this would be sensitive information and each institution or corporation keeps its findings private to the best of its ability. If data are "the new black gold", the ways to extract and refine them must be even more closely guarded.

Private interest, once again, prevails over the aspiration to general, open knowledge; that old ideal of science that today more than ever seems a chimera. And yet here too we have a great opportunity, in the most opposite sense to opportunism that one can imagine.

Indeed, we can create some kind of open platform for discussion of the mathematical nature of this distribution and its restrictions in time and any kind of variables, physical, economic or whatever. No prizes or financial endowments are required, because there will be people with plenty of motivation to make contributions, either with their names or with pseudonyms. The problem may seem too specious and unspecific, but there are a few main guidelines that should not be lost sight of.

This discussion should be as completely detached from the institutions and its minions as possible. In passing, it must be said that to get new insights on the subject there is no need of the massive data and computing available for agencies and corporations to get new, valuable insights on the subject, but an independent approach on fundamentals they will never have; which will give us another opportunity to demonstrate the insignificance of their crushing means and the ineffectiveness of their criteria.

On the other hand, many mathematicians will have the best chance to redeem themselves for having to sell their talent for unworthy, sterile toils not only meaningless but radically opposed to any ideal.

The power law distribution seems at first a trivial matter but in reality it has a transcendental value for civilization and culture, since it not only raises the question of the continuity between Society and Nature, but also how far we step away from it or how one kind of optimization develops at the expense of any other.

Or to what extent and degree hierarchies, an inherent aspect of civilization, are inevitable, necessary, or pathological and degenerate processes. It may also have deep relationships with population dynamics, general stability, and many other areas that concern us all and that should not be subject of secrecy but should be exposed to the most frank discussion in the public domain.

In today's science there is no solid connection between Nature and human society. Darwinism and neo-Darwinism owe their success precisely to being a narrative halfway between the natural and the social, but the fact that it does not have the slightest mathematical signature evidences its little or no entity for knowledge: natural selection can do it all and everything in between. Here instead we have a mathematical structure yet to be explained and that is crucial in population ecology.

By the way, the so-called elites have never really believed in either Darwinism or competition, ideologies expressly created for the masses. They always thought in terms of exploiting niches and ecosystems, and modern marketing or data mining is just an update of that mindset: this is the thread that connects medieval feudalism with post-industrial society and digital neofeudalism.

The expression "big fish eats little fish" has very different meanings depending on the coordinate axis in which we move. The new brand “horizontal ecology” does us a disservice if it ignores the vertical ecology of elitism settled long before demarcating territories. Zion/Babylon code in action: the vertical axis manages resources and creates artificial scarcity, while in the horizontal axis people fight over them.

*

The discrete, and therefore more concrete, form of the Pareto distribution is the Zipf's law or truncated zeta distribution, so widely used in data traffic and many other instances. It has been said that if the Riemann hypothesis on the zeta function zeros were solved, all the keys to cryptography and cybersecurity could be broken, which is still pure speculation, since no one has specified how that could lead to faster factorization methods. However, Zipf's law might be used to trace financial movements, from the anonymous flows of markets and funds to the fortunes of origin.

Needless to say, large corporations and the states themselves, the same seeking for absolute control of each of the movements of the commoner, will do everything they can to obstruct access to data, always being accomplices of the most powerful both passively and actively. We should never accept any kind of digital money with such built-in asymmetries.

To explore it in depth, the 80/20 distribution must be connected in the most natural way possible with change and the time domain. This would lead us to deal with the frequency domain and 1/f noise, but also with fundamentals of calculus and entropy that are outside the scope of this article, but that have a great potential to change our idea of Nature and even information.

For example, this distribution has been associated with the principle of maximum entropy here and there, which is plausible but too vague and general. If even with energy it is impossible to unequivocally determine causes, much less can be done with entropy. In addition, the vulgar idea, originated in Boltzmann, that entropy is disorder still persists, when it is order that produces more entropy, which is in accordance with the maximum entropy principle.

We have recalled on other occasions, citing Eric Chaisson, that the energy density rate is a much more unambiguous quantity for complexity metric than entropy and can be applied to the temporal evolution of all kinds of systems at many scales, from living beings to stars. We also indicated[4] that the loop created between energy density flow, minimum action and maximum size generates power laws; simulations of this loop can be done with a single CPU.

It can be argued that the Pareto rule does not have a definite cause, that "it just happens". Money attracts money, in the same way that the masses of hydrogen molecules attract other molecules to form stars. But, to begin with, hydrogen molecules cannot clump together to make a star without additional conditions; and as for money, there is no such thing as "money in itself", as it is already pure contrivance and dispositive. Surely one can not get unequivocal causes for statistical distributions, but this is not to say that we cannot get a much better understanding.

Applied mathematics is still very young, but that does not mean either that everything will become more convoluted and unintelligible. On the contrary, there are only real advances if things are made simpler —but that is not the current trend of big data miners, who protected by ever-increasing technicalities and depending on particular interests tend, like everything else, to an exponential opacity. These are just some of the reasons for the present tower of Babel of sciences.

In the new science of data analytics, another word for massive statistics, "correlation replaces causation", but statistics and probability have evolved in unison with calculus. In distributions such as the Pareto power law, one can suspect of the quintessential role of grades in Nature, but in the fundamental laws of physics grades lack any specific entity. This is a huge hole in our understanding that can only be addressed satisfactorily with alternative forms of calculus that give a different meaning to the relationship between the continuous and the discrete in change or motion. Although any of the possible approaches to calculus has its advantages and disadvantages, both standard and non-standard analysis are not suited to the task[5].

*

For Lao-tzu, governing a great nation is no different from frying a small fish; surely he was alluding here to the relevance of grade in measures, however secondary with respect to Virtue, which consists in interfering as little as possible. For Heraclitus, all things come into being by conflict, but that tension is the same principle of dynamic equilibrium seen from the other side.

The very increasing inefficiency of Big Science, its ever decreasing returns, can be assessed with the Pareto law and its powers; although it is clear that, seen from the other side, what macroprojects tend to optimize is the extraction of funds. If something can be done with a billion, it would be foolish to do it with ten or one hundred millions; moreover, what is cheap does not even count. The point, however, is not that there are diminishing returns, but that objectivity and debate cease to make sense in this context. Money and quality of research tend to be inversely proportional.

But if "cheap science" does not even count in the current picture, it is even more important for those of us who do not want to have anything to do with that picture either. Today experts tend to complicate even the simplest things for their own interest; while others are only interested in complexity because it may conceal simpler ideas. It remains to be seen who has the last word.

Modern science is of such a nature that it cannot use knowledge without an accumulated pile of mediations and complexity, making knowledge and simplicity poles apart. This is so because knowledge is at the service of the government and control of external things, while virtue resides in the government of one's own.

Biofeedback, the interaction with one's own biological signals, seems a fairly innocuous subject, and yet it gives us the reverse or antithesis of control theory, since the range of possible control does not pass through voluntary action. Rather, one could speak of a capacity for attunement or receptivity.

Our own body generates many signals with a 1/f spectrum, the frequency or time expression of power laws. And the spontaneous activity of the brain is an ideal source[6] of this type of "noise", which also appears in the pitch fluctuations of most kinds of music. It is so curious that a pattern that appears spontaneously in our brain and continually in music seems "counterintuitive" to mathematicians. On the other hand, stochastic resonance, which can be adaptive, tells us that the optimal noise level is not zero and that noise itself is part of the signal. 1/f noise seems to respond to the principle of maximum entropy of information, while with stochastic resonance entropy would tend to the minimum.

Biofeedback is a modest but significant indication of how one can "do without doing" and "know without knowing"; if we properly connect this kind of "doing" and "knowing" with the formal, external knowledge of physics and mathematics, something extraordinary will have been achieved.

*

Having many possessions makes it impossible to enjoy them, while everything comes to the one who has true receptivity, even without wanting. "By 2030 you will own nothing and you will be happy", a well-known forum tells us. Just see how the same people rallying for the privatization and plunder of everything worry about alleviate your burden preparing to carry it on their shoulders. Absolutely logical.

But they should not be so altruistic. The burden can be shared much more equitably and we can relieve them of such a responsibility. After all, they are the ones who need to learn to live more than anyone.

The 80/20 variant that governs our system is obviously a degenerate case, as are degenerate the econometric criteria used to conceal it. But degenerate with respect to what? The monetary system that has been optimized for grabbing must end and be replaced by a completely different one, something that does not present technical difficulties per se. They are the only difficulty.

If today the economy is so important it is only because of how it has created artificial scarcity, and because in symbiosis with the digital world it has become the most "effective" form of control. However, if they have to intervene to such an extent in everything and cause all sorts of crises, there is no efficacy at all, but only the most diffuse albeit all-embracing form of oppression.

What is optimized only for siphoning is surely pessimized for everything else; so they intervine in everything to the extreme because it cost them nothing when it is not profitable, while they can still dissociate and further disrupt the herd.

If the pendulum has been pushed to the extreme, the direction can only reverse —if it is true that the way of Tao is reversal. Operating both at an "internal" and "external" level, it seems that the "50/50 law" has more practical meaning than that indicated, and one in particular in which dialectics, Nature and calculus coincide. Modern science goes to great lengths to prevent us from seeing it.

Notes

1. https://www.hurqualya.net/caos-y-transfiguracion/
2. https://www.hurqualya.net/arte-y-teoria-de-la-reversibilidad/
3. https://www.hurqualya.net/el-pacto-de-los-cacahuetes/
4. https://www.hurqualya.net/polo-de-inspiracion-xiii/
5. http://milesmathis.com/are.html
6. https://www.researchgate.net/publication/43020262_Spontaneous_brain_activity_as_a_source_of_ideal_1_f_noise

ZERO BALANCE. CONSCIOUSNESS, SYNCHRONIZATION, PHASE PRECESSION AND GEOMETRIC PHASE

2 August, 2021

Consciousness as the last stronghold

Some days ago I read a short passage by Geidar Dzhemal on the difference between knowledge and information. In it, the Russian author invoked the inviolability that the consciousness of the workers still had in Dickens' time, when they "went home to sleep and in the night found the way back to their hearts".

For Dzhemal, as for so many others, the information society, which tends to abolish the difference between the internal and the external, destroys the last defensive bulwarks of our "inner fortress". It is not easy to dismiss this type of lament, which is becoming increasingly common, since the invasion of consciousness by the new technologies, with their unstoppable deluge of triviality, is not a minor aspect, but perhaps the most notorious of our times.

However, one can be sure that, as long as we reach the state of deep sleep every night, we have not yet forgotten the way back home; only waking tends to move away from that fundamental state, and that already by its very nature, regardless of how invasive our technologies are.

And what is most characteristic of the deep sleep state, as essential for life as taken for granted? The balance or indistinction between the internal and the external, precisely. We have then a double paradox, if that is what we want to call it: the waking and dreaming states exist because of a sort of imbalance or dynamic equilibrium, and the fact that this dynamic equilibrium is not consciously perceived while waking makes the return to unconsciousness all the more necessary as the only way to experience a certain state of rest.

Dzhemal opposes knowledge as "understanding of the I" to information as recognition, just as he opposes an "authentic" inner truth to a merely operative external truth; and this also seems admissible and accesible. However, if we do not want that not only our idea of truth, but even the very sensation of reality to be beaten in retreat, it is our recognition that must advance: recognition of our projection in the world, and recognition of the action of the world in oneself prior to that understanding of the self that authors like Dzhemal claim.

Conscience understood as the last stronghold always has to lose, it is made for heroic defeats. This may give it its charm in a world where unthinking facts prevail, but that is not enough; the prospects of success or defeat should be completely secondary to our appreciation of reality.

Consciousness as a navigation system

Few things more curious than the attempt of the analytical sciences to delimit consciousness, or, faced with the prohibitive nature of the task, the neural correlates of the various activities with a "cognitive" component. The problem is that all sorts of neural correlates can be found that fit a cognitive model without the activity under study requiring cognition to take place at all, except in the most trivial way: we catch a fly ball not by some kind of prediction or trajectory calculation, but by simply moving in such a way that the same angle is maintained —without even thinking about it. Not surprisingly, the simplest explanation is the one that always has the most implications, even for the very nature of differential calculus.

One of the recent revelations in neurocognitive sciences is the phenomenon known as *phase precession*: a progressive advance in the firing of the hippocampal neurons related to spatial memory. This phenomenon was first seen in laboratory rats and bats, and only recently has been verified in humans. Today there is speculation as to whether it may be the key to a "universal code" of the human brain beyond specific tasks, since it has also been detected in other brain areas and in relation not only to self-localization in space but also to the processing of sounds and smells, learning and the organization of long-term memory —entering fully into what are considered higher functions of the brain.

In a recent article[1], these processes are referred to as a "human negotiation of their environment". Pending confirmation of the universality of the phenomenon, Josh Jacobs tries to figure out applications: "Then we can begin to better understand how this neuronal coding mechanism can be used for brain-machine interfaces, and manipulated by therapeutic brain stimulation".

Nothing less could be expected. Looking for some theoretical depth, it is worth noting that if phase precession happens to be of great relevance, it is not so much because it might provide us with some specific "universal code", but because it is one more expression of a very general but insufficiently appreciated problem: the need for local asynchrony in information processing.

What the phase precession precisely indicates to us is the importance of synchronization in the brain's activities: but were not this synchronization a truly active matter, there would be no activity as such. For example, in ordinary CPUs, operations are subject to a clock cycle that already implies a global synchronization by design, an organizing principle that governs the results of the parts; this is passive synchronization, designed from above.

There is no "global synchronizer" in the brain, nor in molecules, nor in Nature in general. However, as Koichiro Matsuno[2] noted long ago, the laws of physics do imply a global synchronization, by virtue of Newton's third principle of motion, which takes for granted the simultaneity of action and reaction.

Thus the transmission of signals, and therefore of information, loses all connection with local time and its contribution to an internal global time.

In other words, Newtonian absolute time is external and metaphysical, but neither relativity, nor quantum mechanics, nor entropy as interpreted in statistical mechanics and information theory, modify the timeless concept of interaction. Thus the active role that local asynchrony may have in constructing a global internal time evaporates, and can never be properly recovered if it does not find a place at the level of principles.

Of course, one cannot expect very recent and derivative disciplines such as neurosciences to correct the situation of other much more general and older branches of knowledge such as physics, as that is not how things work. And yet the elementary fact remains: there can only be synchronization if there is local asynchrony in the first place.

Derived disciplines are content to speak of critical action or stimulus times; for example, for Arthur Winfree, the great pioneer in the study of resonance and synchronization in biology, a "phase singularity" was the stimulus time during which it is not possible to assign phase to a process.

To find another perspective we should fit the critical times of the descriptive sciences into the fundamental principles of physics. In *Quaternary Spirit*[3] we already discussed the subject, but phase precession allows us to look at it from a different angle. Of course, we will never find a way to give physical "substance" to the passage of time because it cannot be dissociated from the subjective sensation internal to consciousness.

Indeed, if phase precession seems so important in memory sequencing, it may well be because it itself contributes in a key way to creating our notions of space, time and causality —causality being synonymous with spatiotemporal ordering. This is not merely to say that our ideas in this regard are reduced to particular neurological processes, for these ideas are already the more specific internal manifestation of a temporal profile explicitly and implicitly posited as external, but at the level of measurement or mediation.

But from the most immediately external point of view, if we aspire to a more rounded perspective, we should never lose sight of James Gibson's post-cognitive, environmental approach: it makes no sense to speculate about what is in our head without pondering where our head is in. But although Gibson was a radical empiricist, he did not fail to insist that the most directly relevant aspect from the environment is not shapes or colors but invariants, which do not operate at an abstract mathematical level but with irreducible effectiveness. The aforementioned angle invariance or "optical cancellation" proposed by McBeath[4] to determine where to run to catch fly balls, that sort of "calculus without calculus" that we all have without knowing it, is clearly inspired by Gibson's post-cognitive psychology.

This example shows that even radical empiricism can connect directly with mathematics, and what is even more important, it does so in a way opposed to the computational approach that tries to reduce any process to calculus within a metric and its algebra. This allows us to "invert" the perspective without actually forcing anything, since it is human calculus that has forced the data into its own Procrustean bed.

It seems evident that a phase precession such as that produced in neuronal firing in the hippocampus or other areas has nothing to do with the so-called "geometric phase" that occurs in many physical systems, except for the mere fact of affecting phases and potentials; moreover, the former shows an anticipation, while the latter usually implies a delayed potential. However, both phenomena, still of problematic status, share as a common denominator the problem of local asynchrony prior to the external requirements of global synchronous time.

Besides that, the geometric phase is not limited to giving a "supplementary curvature" to cyclic, adiabatic and conservative systems such as those of optics, the area in which it was discovered, but can also occur in non-cyclic and dissipative systems, such as biological ones, and in animal locomotion itself —its universality has made it an important aspect of robotics and the more recent control theory. And although physicists themselves erroneously present it as exclusive to the interference phenomena of quantum mechanics, it is clear that it can occur at any scale being an element of continuity in the transition between different scales and physical domains.

Although it does not seem to have been the subject of specific studies, phase memory should be studied in perception-action cycles in coordinated rhythmic movements, such as the famous Haken-Kelso-Bunz model[5] or other similar models specifically adapted for this purpose. The geometric phase is so called because it reflects the geometry of the environment and the incidence of non-dynamic aspects (often described as "information") in the global dynamic evolution —"global change without local change". Needless to recall that dynamics refers to motion and forces, while potentials refer to position.

But what the geometric phase exhibits, already since the first cases with polarized particles, is the global or "internal" *orientation* of the system, which however has been left out of its "external", necessarily closed, dynamic description. And the fact that this question arises at any scale and in all types of systems only adds to its interest.

The geometric phase is today conceived as a supplement to the dynamics, but in reality it also makes it possible to modify the very laws of motion and their meaning; and so we have spoken of a "fourth principle" and even of "three and a half principles", although three consistently articulated principles suffice.

Mechanics have cut out the natural substratum basically through the first and third principles. The principle of inertia, which in order not to be purely ideal can only be conceived as a rolling ball, demands "an isolated system with the property of not being isolated"; and the principle of action and reaction, a simultaneity of interactions in dynamics that is at best metaphysical. In a relational mechanics, such as that introduced by Wilhelm Weber or the one proposed by Assis, the principle of dynamic equilibrium replaces the principle of inertia, while the third principle acquires a completely different meaning by depending on potential.

The principle of dynamic equilibrium as formulated by Assis says that *the sum of all forces of any nature acting on any body is always zero in all frames of reference*. On the other hand, the fulfillment of the third principle in relational mechanics implies the so-called "retarded potentials" similar to a geometric phase —but it is understood that they are only retarded with respect to the global synchronization implicit in Newtonian mechanics and its heirs. It is perfectly licit to think that it would have to be the other way around: if global synchronization exists, it should be alien to dynamics, since all interaction requires time. Geometric phase or retarded potential are thought to be passive with respect to forces, but a simultaneous correlation can never be reactive with respect to something that takes time to change, quite the contrary.

This has a myriad of implications that we have not even begun to draw out. To begin with, there are no blind forces in Nature, because there are not even forces of a constant nature: feedback is built in "by default" even in the orbits of electrons, in any extension of the old Kepler problem. To say that Newton explained the elliptical trajectories of the planets is a didactic device at best.

Retarded potentials are never a "trivial loop", even if they do not seem to add anything to the known solution of a differential equation as in Kepler's problem. In reality these equations, from a purely descriptive point of view, are not really differential nor do they strictly define local conservation, but always discount it from the given global conservation, so that the local passage of time follows from the general condition and not the other way around as is assumed. We have already spoken repeatedly of this radical inversion that the standard calculus involves[6] and of some of the possible alternatives. But calculus itself, no less than the principles of motion, veils the problem of synchronization and the relation between global and local time, from which we have derived our present idea of universality.

Returning to the three principles of motion, without this modification of the first and third principles, and indirectly of the second principle of motion, any synchronization process rotates like a tractionless wheel in a vacuum. This is precisely what physics has tried to reduce Nature to. It is now noted that the brain has a much more "sophisticated" "navigation system" than those created by man, but what does this "sophistication" consist of? Is there not something more fundamental than the complexity of neuronal connections?

Current state of the sciences

Talking about the relationship between science and power is in bad taste, a bit like talking about money or politics at the dinner table; but in today's ultra-bureaucratized science there are no other relations than those. What sounds ridiculous today is to raise the relation between technoscience and truth.

Another issue that is hardly talked about is secrecy. Even pure mathematics suffers deeply from it, in order to prevent giving any advantage to "colleagues" who in fact are seen as competitors. It is true that this is not a recent development, since back in the days of the "invisible colleges", and even much earlier, people played riddles and information was revealed in dribs and drabs; but now, giving it all for the applications, everything has taken on a different dimension.

Probably even a plausible demonstration of the Riemann hypothesis would today encounter obstacles to its dissemination for fear that it might serve to break security codes, even if no one has shown how it might lead to faster factorization methods. And while there is not much to worry about that possibility, the fact remains that today the consequences of knowledge matter much more than knowledge itself. This, in turn, has more consequences than the very consequences of knowledge. However, the more ominous the consequences may be, as in biotechnology for example, the less the research is publicly questioned.

Evidently, if concealment pervades even pure mathematics, one need not imagine the levels of opacity in sciences with much more defined interests, from physics to biology, economics, sociology, applied mathematics, statistics and data analysis, behavioral sciences, and so on and so forth.

One can only laugh in the face of those who still present science as an independent and daring exploration of the world; and yet it is true that the West seeks to exhaust its possibilities in a very specific sense, to the exclusion of all others. In this it is undoubtedly coherent.

This sense is the project of domination of Nature and its complete substitution or elimination. And it is in such a sense that it cannot be allowed to triumph, for its triumph entails the destruction of us all. Which does not mean that "Science" seeks to destroy us, for the greatest triumph of power and the greatest defeat of thought is to believe that there can be no other science than the present one.

The present sciences, and the point made about phase precession is merely an example, cannot fail to combine a high degree of technicality and formal sophistication with an empiricism which in reality is only opportunism, and where understanding is spoken of in a purely instrumental sense. Their presumed technical superiority of means only hides an unavoidable theoretical incompetence, since the meaning of theory is limited to its predictive capacity.

It has been discussed, for example, whether "the problem of consciousness" might involve exotic aspects like quantum coherence. Most physicists and neuroscientists consider this a remote possibility, if not pure charlatanism. Of course, quantum mechanics is not universal, but neither is classical mechanics. On the other hand, the geometric phase *is* universal, and is present at all scales, since all potentials are entangled regardless of the kind of mechanics in which they are embedded.

Western logos still does not get out of the impasse created between Descartes and Newton, and it does not get out of it, first of all, because it does not want to, because the disposition that drives it is the radical opposition between the infinite inertia of the material world and a consciousness as an independent hole or singularity. Without this opposition it would never have gotten this far in the first place, so it will not stop claiming its rights.

The modification of the principles of mechanics that we have pointed out ends radically with the separation of Nature, no less than of consciousness. The sum of forces at any point is always zero, but in addition the sum of any number of zeros is also zero. It makes no sense to think that consciousness can be localized: what is unreachable inside oneself, cannot be reached outside; and yet relational mechanics allows us to make the same or similar predictions that modern physics arrived at from a very different start point.

It is very likely that a physics based on dynamic equilibrium would never give rise to the hypertrophy of today's technoscience, which has come so far only by dint of great tensions and imbalances. But at this point, it is still true that it is capable of transferring the whole mass of accumulated knowledge to a new language and a new disposition, without even having to make abrupt transformations that never make sense in science.

Fish or not fish

In a very important sense, the antithesis of modern control theory is biofeedback, since it applies to internal rather than external government and involves a range of control that is not instrumentally voluntary.

Several questions of extreme interest and intimately related appear, which can only be outlined here. It remains for researchers to properly illuminate the relationship between phase shifts, positive and negative, and brain synchronization or motor coordination. On the other hand, possible biological phase memories, which may also affect the brain to some extent, such as that of the bilateral nasal cycle, and their connection with respiratory volumetry or sleep stages, have yet to be clearly identified. In addition, the connection between geometric phases in biological systems, their relevant signals, and biofeedback should be studied.

But first of all, it would be necessary to design experiments in which the geometric phase acquires prominence within perception-action cycles. It is a matter of studying whether the presence of this phase memory can emerge in consciousness itself, and the conditions for consciousness to detect it.

Dynamic equilibrium, as a zero balance, is everywhere and nowhere. Therefore, even if its presence can be admitted in any motion, there is nothing specific that allows us to identify or recognize it directly. Is the geometric phase recognizable by consciousness, if it takes place in the movements of our body? Does it evidence any property, beyond quantitative relations? These are questions that may seem very strange, if one forgets the irreducible ambiguity that has accompanied this phenomenon in physics since its detection.

Dynamic equilibrium is indiscernible by definition, but such a well-studied exponent of the geometric phase as the Aharonov-Bohm effect clearly shows us how a charged particle "feels" or reflects the potential even where the electric and magnetic fields are zero. Can an electron accurately perceive this circumstance and yet be inaccessible to my consciousness? Of course, we are not talking about microscopic scales, but of the parallel transport itself.

Depending on the physical case, its description and interpretation, a geometric phase can be seen as a parallel transport, as a self-induction, as a curvature or flow of the symplectic form, as a conical intersection of potential energy surfaces, as a transition between dimensions, as a torsion or a change in density, as a phase transition, as a point of degeneracy, as a retarded potential, as a resonance, as a loop, as an enslaving principle, as a hole or singularity of the topology of motion, or other ways that need not be mutually exclusive. The better the balance between prediction and description, without one taking precedence over the other, the more credit the interpretation of this or any other phenomenon will deserve.

The geometric phase usually has a smaller contribution to the global phase that than that of the dynamic phase. This, together with the fact of its late recognition, has almost inevitably led to its subordination to already built mechanical ideas. However, we have seen some cases, even in the realm of living beings, in which the non-dynamic contribution —at least in the usual sense- can be many times greater than the standard dynamic contribution. It would be interesting to see whether there is also room for this large amplifications in the human organism, and if the answer is positive, what may be concluded from them.

Is there something non-mechanical within mechanics? But what could it mean that something is non-mechanical? This may have many answers that have nothing to do with the usual discussions about determinism and indeterminism. If we can eliminate the principle of inertia and replace it with the principle of dynamic equilibrium, a good part of what we understand by "mechanics" disappears at a stroke, and it is even possible to interpret that bodies move by their own impulse without incurring in contradiction. Let us not forget that this last principle is much more economical and frees us once and for all from the scholastic arbitrations of inertial reference frames and related conventions.

However, the essence of mechanics is not inertia, but the constitution of a closed system or circuit; and there are no closed systems without the third principle. In the fields what is fulfilled is not the third principle but the conservation of momentum or quantity of motion.

In a relational approach such as Weber's mechanics the third principle is automatically fulfilled by definition. Weber's law does not describe a field but is easily transformed into a field by integrating over the volume. It is obvious that in the ellipses of electrons or planets the third principle cannot be verified, so the question always remains why stable and closed orbits are observed. The answer, as Nikolay Noskov saw it[7], would be in resonance, but resonance does not exclude interaction —emission and absorption- but rather defines its conditions. Correlation is the mother of interaction, and resonance coupling is the effective principle of synchronization in the absence of global synchronization imposed from above.

The analogy between a physical field and the field of consciousness is not entirely gratuitous. But by physical field we mean a portion of space in which a certain evolution in time takes place; whereas consciousness, as we have understood it in our writings, implies a contrast with the undifferentiated background —the homogeneous medium, with unit density, from which all momentary modifications result. This could lead us to other different aspects of equilibrium and its formulation.

In any case, and in the present state of things, it is not necessary to carry the analogy very far nor to try to close it, since, for consciousness, the whole point is whether it is capable of perceiving phase memories, directly, indirectly or whatever. When we have a certain answer to that, other questions may or may not be raised. The subject may question our ideas about causality and its representation. It is also a challenge to find its relation to sound and music.

There are countless forms of equilibrium and to summarize we have been referring to the zero-sum equilibrium, to a unit product equilibrium, and to the equilibrium that can exist between the minimum variation of energy and the maximum production of entropy. The first concerns the motion itself, the second the density with respect to the homogeneous medium, the third an alternative mode to the Lagrangian for the general equations of mechanics. The connections between these three types of equilibrium may be infinite but finding their common axis does not depend so much on algebraic equalities or equivalences as on the intangible methodological balance between description and prediction. Needless to say, none of these four modes of equilibrium receives much consideration in modern physics and mathematics.

Mechanics, as the study of motion in closed systems, is a genuinely mercurial science, that is to say, it supposes a virtual reflection of something that neither moves nor is extensive. Were it not for this, it would not be of inexhaustible interest. And it is mercurial not by a vague analogy, but entirely and by the deepest necessity. Motion is already unnoticed spirit, which as a closed system has separated itself from something prior to measurement: this is the path of descent. And the opposite path of ascent is what was once called the work of the Sun, but in fact this operation embrace one and the other.

The true work of the Sun, the most illuminating and recondite of all, is consciousness itself, and if this is not even suspected it has not been possible to adjust much its ascent and descent.

*

At this point, mean secrecy in science seems almost the least of its problems, which we will not pretend to solve. But, for me at least, science as a whole is worthless if it does not have an intimate, genuine connection with our inner world; nor does prediction alone have any value for what is meant here by knowledge. Yet it is not a question of making the sciences more subjective, since our very idea of mechanics already has in us an excess of subjectivity, of compulsion and civilizing zeal only compensated with an increasing dullness.

It has already been said, to directly contemplate that there is no inertia in the world, to verify that it is basically an overdetermination, is as deep a meditation as it might be, which does not otherwise require any support or instrument, nor any special knowledge.

One may ask what make us feel that we exist if there is always a zero balance of forces. But one may also ask: does one need to be pushed in order to feel that one exists? Evidently not; however, the zero balance exists whether we are pushed or not.

We speak of "Nature" in the singular, but we could just as well speak of infinite natures. Our principles tend to reduce that infinitude to a single plane, just as they try to make that plane a new infinitude "according to the measure of man"; but man already contains within himself all those natures without the need to reduce them. The control of Nature as something external enslaves us and results in a permanent descent with no limit in sight; only its harmonization within us elevates us and gives us our true measure.

The problem outlined should be studied with exquisite care. The fact that the geometric phase is seen in physics as an ancillary phenomenon sets a clear precedent of misunderstanding on the subject. The new is subordinated to the old even when it presents the best opportunity to restate the whole question. The theory of electromagnetism itself shows us in an exemplary way that a fundamental force can be used with the greatest technical virtuosity not only without understanding half of its subject, but without even being interested in what may have been left out.

The same is true for our biology, our engineering or our mechanology. It is clear that modern technoscience does not work for the emancipation of the human, but for its integration with machines within an already defined scheme; and why should it have to do so if those of us who are outside its interests do not care about it either, nor do we succeed in finding the root of the matter.

Both thought and the principle of instrumentation from which technique arises can be seen under a ternary scheme well before the birth of modern mechanics; however, the latter defines a horizon in which the attempt to fill its gaps and contradictions, not even consciously, closes the circle of compulsion in which we move. To dissolve that encirclement today cannot depend more on theory than on practice.

Nor should it be forgotten that simple human behavior and its conscious observance is superior to any technology, which alone does not have the power to undermine it. But in the absence of consciousness, technologies are used both to plug holes in behavior and to enlarge them.

The agent-environment unity is prior to the "interaction" between one and the other conceived as separate aspects. There is a great distance between the I that appears in the child of 4 or 5 years of age and the Cartesian self ready to measure the world, but both mark different stages of an opposition between the I and the other that feeds back on itself. To rediscover the non-I that does not need to assert itself against anything in the middle of a logic that seems designed to eliminate it opens a perspective that only those determined to triumph over this material civilization can appreciate.

Notes

1. https://www.geopolitica.ru/es/article/conocimiento-e-informacion

2. https://www.engineering.columbia.edu/press-release/jacobs-brain-timing-is-everything

3. Miguel Iradier, *Espíritu del cuaternario*. https://www.hurqualya.net/espiritu-del-cuaternario-semiosis-y-cuaternidad/

4. https://www.researchgate.net/publication/15474292_How_baseball_outfielders_determine_where_to_run_to_catch_fly_balls

5. http://avant.edu.pl/wp-content/uploads/SGAW-Gibson-s-ecological-approach.pdf

6. Miguel Iradier, *Polo de inspiración*. https://www.hurqualya.net/polo-de-inspiracion-xvi/
7. http://bourabai.ru/noskov/delay.htm

THE MULTI-SPECIALIST, MECHANOLOGY AND FOURTH-PERSON KNOWLEDGE

30 October, 2021

This is an interview given recently for Ivan Stepanyan's science and technology blog[1]:

Why do you think science is going through a particularly critical time right now?

The different sciences are used to have periodic crises in their models, but now, as a faithful reflection of what is happening in our societies, the accumulated problems have reached an extreme point.

On the one hand, the decline of the unipolar world threatens the hegemony of the "Anglo-Saxon model" of science, which has imposed narratives and ways of doing things that have become so accepted that we no longer even realize it. If we really move to a multipolar world, countries with great scientific potential such as Russia, China, India, Japan and others will rethink their general guidelines of research. However, this alone will not bring about a radical change in priorities, because large institutions have an inertia proportional to their size and, moreover, the basic premise of the current model, which is that the essential aspect of scientific knowledge is prediction, is hardly questioned by those in power.

What is tipping the balance is that the fusion of science and power at the institutional level has reached such a point that debate has been eliminated in everything that matters, which has little to do with what mainstream media reflect. Opacity, consensus and disinformation are manufactured on an industrial scale, causing the most talented people, mainly nonconformists, to flee the institutions and wonder what they can do with all the knowledge they have acquired. But no one wants to talk about this profound discontent.

You have talked about the new role that multi-specialists are called to play in this scenario.

Yes, multi-specialists, people trained in several specialties that don't feel a special loyalty to the rules of any of them, are a new specimen, an unprecedented agent that introduces new dynamics into the ecology of knowledge. We are no longer talking about multidisciplinary or interdisciplinary research, but about a different kind of subject that has another kind of perspective on what is technoscience today. They are both at the apex of discontent and innovative capacity. It is not easy to define their position, and many of them strive for their independence and distance themselves from the institutional logic.

So the panorama we have is one of growing leadership vacuum, as the present model dyes of success because it has exhausted its potential and the ideas are repeated over and over again with predictable twists and turns. And we have a new techno-scientific species that is rethinking many issues but is distanced from the decision centers and is looking for its own place in an environment that is apparently very competitive but tightly controlled from above, from a few institutions whose names we all know.

And on top of all this, we have an organization proclaiming worldwide a Great Reset of the system towards human-machine fusion and the genetic modification of humanity for its own good and to protect its health. An organization without any legal authority but which seems to dictate the script to the UN itself, which uses its official statements to second it. It would seem that instead of talking about science and technology we are slipping into politics, but this is just what mainstream media are campaigning for 24/7. It is just the background that allows us to understand some of the unexpected things that may happen.

This may sound like a futurology exercise.

Futurology is entertaining, but all this is too topical. Now, what is evident from all these statements made from above is that there is nothing new to offer, everything is more of the same and the only thing that changes qualitatively is the increasing degree of control.

There is a total divorce between science and culture because power and inspiration are things that cannot mix, and there is only culture when the sap continues to rise spontaneously from below, not when it is offered from above to those below in the form of consumer products.

Technoscience seems to define the limits of our world. Technology is not only about machines, as techne it was originally more concerned with artistic production. Why do we now want to reduce it to machines and their multiple levels of organization? Obviously it is a reduction, but it seems to be a reduction of a fatal and irreversible nature. It doesn't have to be this way; however, in order to pass «to the other side» without destroying ourselves, we have to create an aperture within mechanics itself. If we don't do it for ourselves, no one will do it for us.

And it is in this context that you address the issue of mechanology. What new turn are you proposing within this field? As far as I know, mechanology is a branch of philosophy that has never reached the status of science.

Mechanology emerged in the middle of the 20th century in France as an attempt to classify machines. Lafitte made a typology of machines according to levels of function, Ruyer according to levels of information, and Simondon according to levels of organization. Their pioneering reflections are very valuable and meritorious, but it is striking that the fundamental premise, namely the very principles of mechanics, which crystallized in Newton, are out of question.

Today we should be compelled to ask, can anything exist that is not mechanical? But the endless fugue of machines, with artificial intelligence and all the rest, is a barely conscious articulation of that question. And yet we should better ask: does anything mechanical exist at all? And naturally, since we have poured our entire worldview into the mold of mechanics, the answer will depend on what we mean by that word.

Newtonian mechanics is summed up in that nothing moves without being moved by something else. But in Germany, in the 19th century, Gauss, Weber and Hertz (not to mention the ideas of Leibniz and Mach) tried to create a relational mechanics crucially different from Newtonian mechanics, although largely compatible with their predictions and with experience. In our time various fundamental theorists such as Kulakov, Aristov, Vladimirov, Assis, Mazilu, Noskov, and many others have tried to develop this relational program in some of its many aspects, but Newtonian concepts still prevail.

Is it not conventional wisdom today that relativity and quantum mechanics have superseded Newtonian mechanics?

They have overcome many of its aspects, but we do not pay enough attention to the unchanged ones, which still are the most important. And there is one fundamental aspect unchanged by each of the three principles: we still adhere to the principle of inertia, to the idea that in Nature there are constant and universal forces that depend only on distance but not on the environment or the speed of bodies, and finally, to the simultaneity of action-reaction, which sets the notion of a closed system and the global synchronization of its parts.

What do we know about physical inertia? Nothing, but it is still the basis of mechanics. The principle of inertia asks us to believe in a closed system that at the same time is not closed. That is, it begs us to believe in something contradictory, which in addition generates all kinds of scholastic distinctions of reference frames. Now, with the principle of dynamic equilibrium, which says that the sum of all forces of any nature acting on any body is always zero in all reference frames, we can dispense with the idea of inertia forever, and we can even interpret that bodies move by their own impulse without incurring in contradiction.

In a relational mechanics such as Weber's, forces may depend on the environment or on factors such as velocity and acceleration of the bodies. This gives rise to the so-called «retarded potential», which actually is only retarded with respect to the «global synchronizer» implicit in the third law. However, by definition, it is assumed that in Weber mechanics the third principle is automatically fulfilled, whereas in field theories such as Maxwell's this is not the case and one has to stick to conservation of momentum. This leads to various confusions and misunderstandings. Actually the retarded potential is telling us about a system's proper time, but being the global synchronizer a supersystem —in fact a metaphysical instance— that becomes irrelevant.

You propose to include the geometric phase within the principles of mechanics. How is that possible?

Sometimes I mention a «fourth principle of mechanics», sometimes «three and a half principles», but it is not so much a question of adding principles as of contemplating the aperture or internal space that the classical principles have within. The Kepler problem still remains the keystone of physics, both at macro and micro level. As we know, the claim that Newton's theory provides an explanation for the shape of ellipses is only a didactic device, but most physicists have been content with that. Hertz lamented that the third principle cannot be verified there and proposed a distinction between material particle, as point of application for a force, and the material point that can have an indefinite volume, but physics conflates both concepts routinely.

For Nikolay Noskov, Weber's retarded potential did allow a satisfactory explanation of the ellipse, with a phase velocity and a longitudinal wave in the moving bodies that coincides with de Broglie's matter wave and implies a resonance. Now, what is the difference between a retarded potential and the geometric phase generalized by Berry for quantum mechanics? The only difference is that in the ellipse of a moving body it is already embedded in the known equations, while in the «anomalous» shift of the geometric phase it is not. And, despite what is so often said, we know that the geometric phase occurs on both micro and macro levels, being truly universal —something that can be said neither of classical nor of quantum mechanics.

The geometric phase is considered a mere appendix of quantum mechanics only because it was identified later and because its contribution to the total phase is usually much smaller than that of the dynamic phase; however, we can see that it already existed long before as a retarded potential, and, on the other hand, there are also systems, including those of living beings, in which its contribution can be far greater than that of the dynamic phase itself. The same pulse wave in the arterial wall would be another of its manifestations, and on the other hand, today it is routinely included in robotics and control theory; only that each case is considered separately.

From this point of view, the geometric phase would be key in the spontaneous synchronization between different systems?

Yes, because if there is synchronization between bodies, it can neither be determined from above in an absolute way, nor can it belong to the domain of dynamics, which is the realm of interaction. Because of the order of development of our concepts, we think that the dynamic phase as related to forces is the active element and the geometric phase the passive element; but what is instantaneous cannot be secondary with respect to that which needs time to operate; what is pure act cannot be passive with respect to that which takes time to actualize. Physics has inverted our perception of many things, including our idea of potency and act. The consequences of all this reach to the present day.

It seems that all this directly affects the foundations. Is it possible today to modify the foundations of such established sciences as physics, or we can only look ahead?

In mainstream physics the foundations are out of question, but in countries like Russia things are different because there is other kind of theoretical concerns. However, the problems remain where they were first met. The real changes can only come from the very foundations, but there is too much built on them.

So other strategies must be devised. Physicists cannot allow themselves to rebuild their discipline bottom-up, however what is at stake is more than a theoretical domain. What is really at stake is the relationship man-machine-nature; but this relationship cannot be clarified without knowing what is non-mechanical. Nature is amechanical, but since physics studies it in the light of the principle of inertia, and of the other two coordinated principles, Nature becomes just another machine.

But Nature is not only what is out there, it is also within us, even in our thoughts. There must be a way out of this anthropological deadlock. And mechanology, when it gets back in touch with the foundations of mechanics, closes a circle that I call the «Tao of Technoscience».

How would you define this technoscientific circle?

Poincaré, in his analysis of Hertz's mechanics, already saw clearly that principles in physics are based on experience but cannot be invalidated by it. Yet in physics, as in almost everything, we have principles, means and ends. The principles are the axioms that determine the elements, the basic method is calculus or mathematical analysis, and the ends are the physical interpretations. Interpretations are not a philosophical luxury for the end of the day, because interpretation, together with description and representation, outline the scope of technical applications.

Theoretical physicists take credit for most practical applications, yet experimentalists and engineers very often make breakthroughs in spite of the fact that the theory says they are not possible; however, in the end theoretical standards confiscate the explanation of these breakthroughs and have the last word.

Even mathematical calculus, as it is conceived today, is a clear instance of reverse engineering known results. But there is another way of going from principles to ends back and forth, which is open-ended and creates another dynamic. And the key is not to think that prediction is the ultimate justification for physics or calculus. The greater the balance between prediction and description, the deeper we go into reality.

And how can this balance between these two elements be found? There are countless interpretations of a physical law, which indicates that they are subjective; that is why physicists consider them secondary.

Right, no doubt they are subjective, but nonetheless they are indispensable for our practical use of science. And if these interpretations can directly affect the very nature of calculus, and the principles themselves, everything changes.

Many researchers still believe that physics must explain the causes, the whys of phenomena; others think that the world is just phenomenon, and that what we need is to describe it well. On the other hand, positivists who think only in terms of predictions and dismiss the whys have little doubts about the causal agency of many of their theories. But even if we could only aspire to truthful and more or less rounded descriptions, the illusion of causality serves often to improve the description itself. In my view, the greater the balance between prediction and description, the more the mirage of efficient causes operating in time disappears. But today, to believe in physical reality is to believe in the reality of efficient causes.

We do not act according to our ideas, on the contrary our ideas are shaped by the way we act and want to act. This is true for everything, including physics. In the 17th century the world became a Great Clock because supposedly it was the best we could do. And what is the best we could do today? To find what is beyond and before the Machine. But, obviously, one can never find that if prediction is the main goal.

The idea of efficient cause is problematic enough in today's science, but should we be content with mere correlations of data? Is there room for something else?

It is well known that final causes were discredited during the scientific revolution and replaced by efficient causes. I for one think that final causes are more important and certain than the efficient causes. However the force that matters in physics is not the measurable force, but the controllable force. Finality is real, but not primarily as design, not as a machine: that is precisely what has created our all-too-human misunderstandings.

We can resume Clausius' entropy, before Boltzmann; entropy is no longer a by-product of mechanics, but a spontaneous tendency towards a maximum. One can reformulate mechanics as Pinheiro has done and replace the Lagrangian by a balance between minimum energy change and maximum entropy production. Then, entropy is already within the "fundamental" laws; given that "order» produces more entropy than "disorder", this completely invert the question and our vulgar clichés about entropy, disorder and finality. It should have a profound impact on our idea of information, equilibrium and evolution.

There is something in man that can only be released if we first liberate what is locked up in *OUR IDEA* of Nature. By unlocking this we will also liberate our own nature, which can never be reduced to an object. Assuming this goal will be an inexhaustible source of inspiration.

In your writings you suggest an elementary relationship between math, physics and consciousness. Could you tell us something about it?

For modern science, consciousness would be the last frontier where all the most sophisticated disciplines converge: the ultimate playground for the multispecialist. But consciousness is not an object, not even a process as they now say.

Anyone who observes his thoughts attentively will realize that they come from the outside, that I do not produce them. But outside of what? They are out of consciousness; it matters little whether they come from the brain or from a screen. Consciousness knows this because it is empty and has no motion. How could it perceive motion were it not still? Yet consciousness is also the indifferent continent where all this takes place.

From this point of view, consciousness seems to us an absolute about which nothing can be said. But the same applies to a primitive homogeneous medium with unit density in which matter and space are not separated, and can equally be full or empty, have zero or infinite dimensions. However, from such a medium one can derive formally different types of equilibrium. The Hilbert space in quantum mechanics also starts with infinite dimensions but it does not include the geometric phase, which is only a reflection of the «geometry of the environment», whatever that means. This additional phase may have many interpretations, such as parallel transport, interference, resonance, transition between scales, dimensions, and so on, but it always implies an aperture with respect to the conservative, closed systems of Hamiltonian mechanics. Not only that, it could also be the loop that closes the contour of a closed system starting from an open background, if the former depends on a resonance. The homogeneous primitive medium is always open and indeterminate, just like consciousness; but the geometric phase represents a peculiar connection between the dynamics of interactions and what is beyond.

You suggest the existence of a «fourth-person knowledge». Can you tell us more about this concept?

We have first-person or subjective knowledge, and we have second-person knowledge, by direct interaction with objects and the world. Third-person knowledge is the generalization of the first two by formal means that gives us the sense of regularity, of the Law. Can there be fourth-person knowledge?

Yes, and in the same sense in which we speak of «a fourth principle of mechanics»; they are two sides of the same coin.

In fact the laws of mechanics are just a particular application of the laws of thought and the semiotic transformation of symbols. The question is, what happens if we totally reverse this direction towards the first person and still keep the formal tools of mathematics and interaction with physical objects? Then we would enter a completely unforeseen territory, which does not belong to any of the «three persons» of our intellectual economy.

But this cannot be done just like that. This is not about accumulation of knowledge, but about its transformation.

What kind of transformation do you mean?

Of a very profound but subtle transformation in the view of principles, interpretations, and the calculus itself. Modern analysis is a dazzling construction, but it has been concerned with the justification of predictions, not with the description of the physical geometry of problems. Often it has not lived up to its name, since, seeking rigor for already known solutions, it has not even made a correct analysis of the physical, not mathematical, dimensions of the problems. We also have the aforementioned confusion between material point and material particle, which prevents us from advancing in the description of real particles, which are bodies with extension. This has all kind of consequences, since it give us the indispensable frame to conceive the non-singular individuation of bodies and processes. And a series of very important equilibria that are hardly considered in modern physics and mathematics. And there is the ubiquity of highly heterogeneous quantities in the fundamental equations of physics, which gives us so many inextricable knots.

Without good analysis you can't have a good synthesis; without directly addressing these questions, among others, knowledge cannot be properly distilled, because it has too many mixtures and impurities. If they are addressed we will find another horizon of understanding; and a convergence that is both formal and informal not only between different disciplines, but between different approaches within any of them.

Fourth-person knowledge already exists virtually within the other three, we just don't pay attention to it as is overshadowed by our assumptions and abusive generalizations. Newton's three laws of mechanics, inspired by Descartes' earlier three laws, seem a great achievement, but they are also a universal leveler. They translate to motion something that still remains different —if physics were only motion, it would be entirely trivial. But why are these three laws not reducible to a single principle? For knowledge in the fourth person unfolds to the same extent that we are able to subsume different laws, even the three laws of motion, into a single principle. And in this very sense, Newton's laws lock an unknown potential that we must learn how to unlock.

My impression is that the current philosophy of technology leaves the technical issues to the engineers and scientists to focus on the more human aspects of use and meaning. Isn't that so?

Of course; and it is further assumed that in technoscience we can only question the future but not the foundations. But assuming this, those who want to think technology are relegated to the position of chroniclers and commentators of a present that is beyond them. On the other hand, in a digitalized world, they tend to believe that information has superseded mechanical and material categories, when it does nothing more than reproduce them at another level. This is all too superficial, but most people think that looking for something new in the three principles of mechanics is like pretending to bring forth water from a rock. And yet this rock has water in abundance.

We do not realize the double meaning of all this. We have predictive sciences, such as physics, and we have descriptive or narrative sciences, such as cosmology or the current theory of evolution, without the slightest predictive capacity, but which try to fill the huge gap of the predictive sciences when it comes to explaining the appearances and forms we observe. Between one and the other there is wild gap, a divide that only statistical and probabilistic means attempt to bridge; but this divide works in both directions. Yet something "as simple" as the principles of motion are a precondition for both the consistency of prediction and description; not to mention that they summarily contain the three dimensions of discourse.

There is nothing more powerful than the tacit concept of the «global synchronizer» established by the Third Principle. And yet it has feet of clay, it is only a metaphysical ghost that is hiding from us the true synchronization between multiple agents, the natural feedback and the amechanical relationship that exists between the global and the local. But all technology since language belongs to the tertiary order. If we admit the existence of another type of tertiary relationship in Nature itself, its impact on human culture may be incalculable. Of course, this cannot occur in isolation, but within a much broader context.

However, many do think that information technologies have definitively overcome mechanicism. Is a Turing machine really mechanistic?

Still there are discussions on this. Turing was asked "what is a mechanical process" and his answer was "something that can be done by a machine"; so for him his machine was clearly mechanical, though not in the mechanistic, but in the functional sense of the word. On the other hand, the same Turing criterion is irrelevant to the real work of a modern computer, being just a meta-analytical artifact. Anyway, the much more basic question we should ask is: are the principles of mechanics mechanistic? For the three principles of mechanics are already a universal symbolic machine. And the answer is no, for their application is not a mechanical matter of fixed rules, but something extremely delicate that often requires great discernment; in fact the new theories have arisen from careful modifications in their many possible strands.

There is nothing mechanical without intention. But it happens that machines are successful externalizations of that intention, portions of spirit frozen in matter. In the middle of the triangle of mechanics there is always a great Eye assigning the articulation of its moments; that is the little secret that the intervenor, nothing divine, would rather we did not see.

Someone said that «software ate the world and now has indigestion»; well, mechanics already ate it before, but we did not even realize it because the predictive sciences find their expansion and their alibi in the narrative sciences that supplement them and in which there is such a wild margin for speculation. But all these narratives also derive from the logistics of mechanics, in which, so to speak, software and hardware merge.

Finally, are there practical applications for these ideas?

Sure, but I am not interested if we don't change the very idea of «application» to begin with. Constant differential calculus, for example, profoundly changes the idea of what applied mathematics is, and with it, physics itself, measure theory, dimensional analysis, probability, and so on. But it is still entirely undeveloped. As our idea of applied mathematics changes, so does the idea of physical reality, of technical application and even of pure mathematics, because our standards for describing mathematical objects change. It is impossible to do really new things without touching the foundations; everything that is built on the known foundations, by sheer historical inertia, is already headed in the same direction.

The priority is to shed new light on mechanics, our concept of "machine» and our relationship to it. We can look for very simple concrete designs that require hardly any means and are more eloquent than heaps of theories. It is not a matter of «falsifying» the old principles, for that is not possible, but of creating «machines» and «functions» that clearly and lucidly illustrate this new amechanical perspective. Once this makes its way into our minds, we will not want to go back to business as usual. Why? Because we want to overcome mechanics, and mechanics is just a successful language created by our minds. From a certain point of view, you can reduce all music to mechanics, but when it comes through us and we pay attention to it, it is something alive and very different. Surely there must be an important relationship between the geometric phase and music that we have so far neglected, just as we neglect its presence in biomechanics and living beings. Feedback and biofeedback acquire a completely different interest if the displacement of the potential defines the contour of a loop. Just as the geometric phase is now used in robotics for control, we can use it for tuning, decisively changing its meaning.

So where does all this lead us? Let those with imagination try to imagine it. Disconstent with the present orientation of technoscience is on the rise, and maverick researchers are looking for a different direction and outlet for their efforts. The time has come for multi-specialists to join forces and spirit to bring something new to the world.

Are you a multi-specialist yourself?

No, I don't have any special knowledge; I am interested in redirecting different aspects of knowledge to a certain perspective.

Since Newton's *PRINCIPIA*, knowledge in all branches has multiplied by about four million; and yet physicists themselves have misunderstood several crucial passages of that obscure book, precisely by over-specializing and restricting its meaning while at the same time generalizing its scope without measure. Multi-specialists have the freedom of perspective and the technical competence to do things, and a good number of them longs for meaning and purpose.

Notes

1. https://all-andorra.com/category/blog/modern-science/

BEYOND AND BEFORE THE MACHINE

7 December, 2021

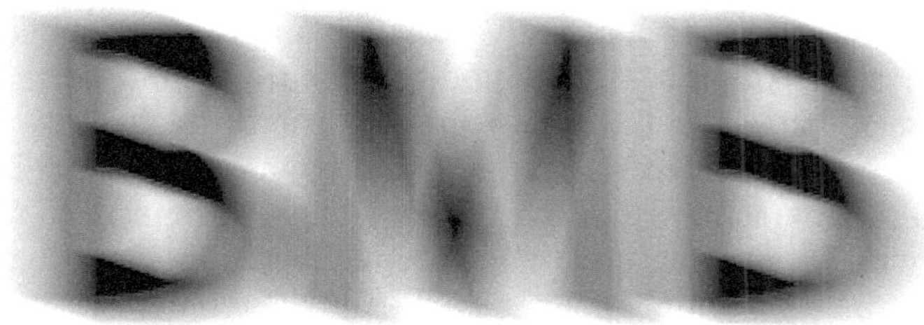

There is something in man that can only be freed if we first liberate what is locked up in our idea of Nature. By unlocking this we will also liberate our own nature, which can never be reduced to an object. Assuming this goal will be an inexhaustible source of inspiration.

Can machines be used to break the encirclement of machines? Certainly, today there is very little that the naked human being can do against them, or so it seems. But neither can machines do anything on their own, but only as part of a vast technological web in which they interact with humans.

Everything historical is ultimately contingent. Today's prevailing system, as well as the dominant scientific theories, are only one of many possible versions of developments that are neither universal nor necessary. The same can be said of the current technology deployment. One and the same body of knowledge can give rise to the most diverse technological applications, but on the other hand, different theories can also achieve applications nearly indistinguishable.

In technoscience, as in power, we can recognize at least three basic levels: the symbolic, the structural and the functional or instrumental. Far from dealing exhaustively with any of them here, we will be content to indicate, in the most schematic way, widely ignored aspects of these levels which are of special interest and which are contained within a different level which is not technical at all but remains the indispensable background.

*

When we speak of machines we no longer think of something as basic as Newton's three laws of mechanics, and yet in them the instrumental principle that still guides this civilization is already fully realized. Machines long predate the birth of mathematical physics, but it is only with mathematical physics that we begun to see the entire cosmos as a closed system with the clock as its model. Less than a century later, the West had already invented «Nature» as a nostalgia for that without separation.

The maximum of intelligence is put into the design of machines so that their use requires the minimum of wit; this gap between the side of creation and the side of the user is constantly growing, and the mechanisms of power do their utmost to exploit it. Before modern physics, nothing seemed more stupid than a machine. After the pendulum clock and calculus, people began to doubt whether man imitated Nature or Nature imitated man. And later, when it became clear that machines can interact with the environment to an increasing extent and without definite limit, many are inclined to believe that there is nothing in Nature that machines cannot eventually overcome.

Artificial intelligence gradually expands the complexity of its feedback cycles with the environment; in this sense, not only it does not create closed systems, but their range of operation is permanently broadening. However, we still maintain the idea that the physical ground that makes it possible, with its atoms and particles within its fields, is a closed system by definition —it is a mechanical system.

Much more attention has been paid to the major differences between today's physics and Newton's physics than to what they have in common, which, as befits the least noticed, is usually the most important. On the other hand, there is a tendency to believe, without the slightest justification, that a predictable system is a closed system; and since it has also been inculcated that the essence of intelligence is the ability to predict, this combination of prejudices, interests and misunderstandings directs the entire flight of machines, whatever their level of organization.

Everyone knows that Newton could not explain our solar system as a mechanism, nor has anyone since; and yet we still judge the motion of celestial bodies by his three laws of mechanics. These laws remain the irreducible scheme of any closed system, and prefigure all the conservation laws which give their fundamental character to all later theories.

Newton's three laws, the very foundation stone of modern technoscience, already involve the three levels —symbolic, functional and structural with respect to the most recent theories- that we have mentioned, and without cleaving this rock and bringing forth its water we will never get out of its charmed circle.

*

There can be no proper mechanology without due consideration of the three principles of mechanics, but since they are universally taken for granted, it is necessary to recapitulate what we have already said on several occasions.

Subsequent physics has introduced dramatic changes one after another, but one fundamental aspect remains unchanged by each of the three principles: we still adhere to the principle of inertia and inertial reference systems, to the idea that in Nature there are constant and universal forces which depend only on distance but not on the environment or the velocity of bodies, and finally, to the simultaneity of action and reaction, which defines both what a closed system is and the global synchronization of its parts.

So here there is a principle of reference, a principle of action and a principle of regulation. These are three completely different levels, which nevertheless lie on the same plane and which can be applied in both the continuous and the discrete domain. It soon became clear, for example, that Newtonian absolute time was just a metaphysical principle; but a later theory such as relativity does not really change this state of affairs, since it is explicitly based on the simultaneity of events and thus on their global synchronization.

Much has been said about the dispositive as the essence of technology, but the disposition of the modern mechanics underlying it still is waiting to be properly contrasted. On the basis of the relational criterion that physical time is nothing but the measure of motion, it has often been argued that physics would not even be possible without assuming the constancy of forces or the equality of background and coordinate systems; however, it is from purely relational criteria that it follows from Weber's law that both the principle of inertia and the constancy of forces, besides universal synchronization, can be dispensed with, if the concept of retarded potential is correctly interpreted.

Such a relational mechanics is in agreement with known observations and predictions, even if it also allows for important divergences. It can also be argued that its internal logical consistency is greater, since the principle of inertia forces us to consider a «closed system that is not closed», which is obviously contradictory. On the other hand, the principle of dynamic equilibrium, which makes it unnecessary, only states that the sum of all forces of any nature acting on any body is always zero in all reference frames. Newton's laws can be summarized as meaning that nothing moves without being moved by something else; but in relational mechanics, bodies can be interpreted as moving by their own impulse without incurring in contradiction.

It might be thought that two systems of mechanics leading to the same predictions only raise a question of formalism, but the fact that they can yield radically opposite meanings even when they are compatible should already strike us. And it reminds us that the principles of mechanics are not univocal even in the most basic cases, let alone the more complicated ones that demand increasing degrees of discernment in their application.

One system tells us that without uniformity of measurements it is not possible to do physics, while another, without uniformity of time, forces or reference systems, allows us to reach the same predictions with less artifice. The problem is almost the same as the one that arose when moving from classical electrodynamics to relativity; however, in relational mechanics it is not necessary to postulate a four-dimensional space-time. It is another way of seeing that there is room within the principles themselves. There is plenty of room at the bottom, but not where most people think.

*

Modern field theories still are in the same middle ground that Newton already occupied between Descartes and Leibniz, that is, between mechanics described in terms of geometry and purely relational mechanics —or between purely mechanical causality and acausal or amechanical description of phenomena. Known physics can be poured into a relational mold, but for most, if a framework does not lead to new predictions, it is basically unnecessary; if, on the other hand, it yields too many new predictions, it becomes undesirable.

However, the mere fact of overcoming the principle of inertia should be for us at least as important as all the predictions, since it eliminates at a stroke, within the theory itself, the artificial separation we have created between us and Nature. Even today, we would probably need another three or four centuries, as it has taken since the introduction of this principle, to understand thoroughly the consequences of this step.

From a relational point of view, it is clear that one can do physics without the principle of inertia, but it is another thing to dwell on what that means. But from the opposite perspective, it is possible to return to Cartesian physics in a consistent way, as some recent researchers have done, accepting that all motion is rotation and involves an acceleration. In this case, what is suppressed is not inertia itself but the inertial reference system, so that all forces are different expressions of internal «inertial forces».

If one observes closely enough the displacement of an automobile, one can verify there is no such thing as a truly uniform speed, and that the forces of traction and friction are never exactly balanced. The construction of clocks poses the same problem. In relational mechanics, whether a body is in motion or at rest, there is always a perfect balance of forces; in the new Cartesian geometric mechanics, there is no motion without imbalance; and yet both dispense with the inertial frame of reference. It should be enlightening to thoroughly contrast these two perspectives and their coincidence.

This and many other possible observations show that we know nothing about inertia. Inertia is the basis of all modern physics, but it must be necessarily fathomless to the very physics that has been built upon it. Physicists do not pretend to know what reality is, but if today it is generally believed that reality has a physical nature, or if we believe in the reality of matter, it is just because we believe in the reality of inertia. And if this universe is believed to have arisen from an explosion 13 billion years ago, it is not so much because of any experimental evidence but because physicists seek a justification for the plain fact that everything is in permanent motion.

The principle of inertia, by its very definition, is the decisive cut-off point between the uncontrolled Nature and the laws that man postulates in its regard. But for the autonomy of human laws to become definitive, it was necessary for the other two principles to be consistent to form a closed circuit, as Newton finally achieved. Organic chemistry is only that part of chemistry which deals with the most complex molecules, a definition which is already indebted to physical considerations. Is an organic physics conceivable? Here we would have to proceed in the opposite way: the known «inorganic physics» would be a mere part of this more general «organic physics». It would, in any case, be a physics in which the principle of inertia is not shielded.

And, to the surprise of many, it can be seen that a renewed Cartesian mechanics, always considered as the epitome of rudimentary mechanism, is at least the most geometrical expression of this organic physics. The best exponent of inertia is a rolling ball; but Newtonian mechanics does not even contemplate the dynamics of an oriented point with rotations. When its variables are included, we have six dimensions for motion within ordinary Euclidean space. A point with freedom of rotation generates vortices as it moves, and vortices are the only natural gear known to nature. They also allow the passage from ideal points without extension to real bodies with extension in space, source of so many misconceptions in physics and in our representation in general.

According to relational mechanics, the geocentric system of Ptolemy and the heliocentric system of Copernicus are dynamically equivalent in all their aspects, including the Coriolis forces or Foucault's pendulum; but, as is well known, this is not the case for Newtonian mechanics. The trained habit inclines one to think that the relational framework must be missing something specifically physical that makes it a mere exercise in kinematics. However, from the point of view of geometric mechanics, the apparent retrograde motion of the planets and Ptolemy's epicycles also have a physical meaning unnoticed in Newtonian mechanics.

Classical mechanics does not even describe rotational motion correctly; but this should not be surprising because classical mechanics has never been much concerned with descriptions, but rather with shielding its predictions. It is clear that modern field theories and particle physics violate the laws of classical mechanics in many different ways, but nevertheless, they try to fit into the framework of our ordinary experience, and we happen to confuse that framework with that of Newton's three laws.

How is it that much more complex laws can be accommodated within the three laws of motion? In the only possible way: by reducing to the same plane things that may be on very different planes. And so these three laws continue to have incomparable power and weight, because they alone allow us to maintain the illusion that there is only one plane of efficient causality. Our ancestors could not believe too much that the Earth was flat because, in general, they did not even bother about it. It is we who try to reduce everything to a single plane of causality, even if we know that this is an impossible task.

Inertia is both the soil and the veil of physics, and it is a veil precisely when it gives us the illusion of being soil. As a principle of reference, it is necessarily reflexive; but the leap from the ideal to the presumption of reality is automatic, passing to deny that there is any viable way of questioning it, despite the fact that it can be disputed from opposite extremes.

*

The three principles of classical mechanics have not only been a foundation for physics but have acted as a general principle of reduction of experience to a single plane; they form a line of descent which can have no end for itself.

This course of mechanics has been assimilating things that were initially alien to it, both in the domain of physics and in technoscience in general. In this sense, it operates as a great universal symbolic machine, translating external processes into its own language. All subsequent symbolic machines end up in its funnel, since only in the principles of mechanics software and hardware merge.

The phenomenological analysis of the temporality between memory and anticipation makes a distinction between a primary retention of perceptions and a secondary retention in the imagination. Later it has been argued that the various technologies, including language, constitute a tertiary sphere of retention that acts on the first two. Undoubtedly, Newton's three principles involve also these three retentions on a much more elaborate level, both formal and operational, which nevertheless feed back on the previous levels.

Just as consciousness does not perceive time, but creates temporality through its own activity, so the laws of mechanics are not a simple generalization of the physical plane, but rather they end up creating «physical reality" proper from a more indeterminate background of experience. Both types of movement are incommensurable even though they are intimately connected. In the movement of consciousness, however conditioned it may be, the inner spontaneity is never extinguished; whereas mechanics, by means of a cut fraught of consequences, definitely establishes itself outside.

There are other formulations of non-inertial dynamics compatible with modern mechanics and physics, and one may wonder whether adopting them would make any difference at this point; but this is not the question now. Modern technology always seeks to release energies in Nature to put it at work; here, however, we seek only to free Nature from inertia and contemplate what manifests itself instead. Although it sounds similar, both things are diametrically opposed.

Reverse engineering on Nature is thought to belong to the sphere of technical applications, not to theory, when the reality is quite the opposite: theory, starting with its primary means, calculus, is reverse engineering on known data from Nature; but the descriptions being subordinate to the results, departs so much from the physical geometry that in general machine designs follow a completely independent logic.

*

Mechanics is the control of forces through other forces, which obliges us to discard other quantities that may be measurable but not controllable. This definitely marks the limits of both mechanics and classical control theory.

In our writings we often speak of the so-called *GEOMETRIC PHASE* not only as a non-standard shift of the potential but as a virtual displacement of the three principles of mechanics from a relational perspective; hence we also speak of a fourth principle or «three and a half principles», without intending to add any more. Holonomy and geometric phases, first recognized in optics and then identified in the most varied contexts including animal locomotion, are now commonplace in robotics and control theory, where they are included for the most instrumental purposes.

The phenomenon of the geometric phase could not be more ambiguous and is open to all kinds of interpretations. Its contribution to the total phase is often smaller than that of the dynamic phase, but it can also be larger. Since it has only lately been noticed, it has been assigned a purely passive and auxiliary role, as always corresponded to potentials in mechanics. But that which is instantaneous cannot be secondary to that which takes time to operate; that which is pure act cannot be passive in relation to that which depends on interactions and takes time to actualize.

Speed is of supreme importance for interactions, but of no importance for what does not depend on them. Nor can time in the modern sense, which depends on the idea of velocity, be of any importance for this. Mechanics and its control interrogate the world through forces, which are reinterpreted by the three principles; these forces can then be diverted to multiple ends, channelled and mediated by all sorts of logical or information-associated categories.

What interests us now, however, are the most basic forms of relationship between a force or kinetic energy and a potential when consciousness is placed in between them. The antithesis of modern control theory is biofeedback, since it applies to internal rather than external regulation and assumes a range of control that does not pass through the voluntary. Biofeedback, the use of machines to tune into internal states and functions of the organism itself, seems very limited and innocuous, yet it is a threshold for exploring this relationship and a turning point for reversing its meaning: using forces in a minimalist way to probe potentials.

Holonomy, global change without local change, is the imprint or signature that the ambient background imprints on a presumably closed system — when we speak of a geometric phase, we mean that geometry of the environment that is not integrable in the evolution of a closed system. However, this factor can also be present in the fundamental interactions themselves: if we apply Weber's mechanics to the Kepler problem, what we have is a variable force and a retarded potential, with a feedback between the length of the interaction and its strength. The difference is that in the first case the factor is not included in the equations and in the second it is masked by them.

It is one thing how fundamental physics wants to see this question, and another how it can be directly perceived by us. Newtonian mechanics is certainly drawn from experience, but that does not mean that it contains it completely, not even at the level of mechanics, let alone that it can be invalidated by it. We start from the idea that the origin of the so-called fundamental laws has always been global evolutions, and that only the predictions have a local character; it is the reverse engineering of the calculus that has made us think otherwise.

We also presuppose that any phenomenon can be seen as an ephemeral variation or dynamic equilibrium within a primitive homogeneous medium of unit density where there is no room for distinctions between space and matter, motion and time, emptiness or fullness, consciousness and object, zero dimensions or an infinite number of them: this homogeneous medium is the support of any appearance. But this does not exclude quantitative relations of density or dynamic equilibrium in more restricted physical frameworks. Since even in field theories the homogeneity of space is more fundamental than forces, although they deal with a secondary or derivative homogeneity so to speak, there must at any time be a connection between this primary medium and any dynamical evolution analogous to the connection of the geometric phase if this can be treated as a torsion or density change.

Similarly, there would also be a direct connection between any state of intentional consciousness, consciousness of a given thing, and undifferentiated consciousness. The question man has always asked himself is what his conditioned consciousness can gain by uniting with this other undifferentiated consciousness. And the answer can only be nothing; on the contrary, in that direction one has only things to lose. However, the reversal of potency and act that physics has carried out between interaction and potential may also be applied here.

*

Some organic functions, such as the blood pulse, seem to have the same kind of feedback that the retarded potentials in fundamental physics. Probably also the bilateral nasal cycle of breathing exhibits a holonomic phase memory. If we step outside the inertial framework and global synchronization, retarded potentials should be ubiquitous in nature and in the human body, and only need to be identified. Retarded potentials may only appear to be so from the viewpoint of the intangible global synchronizer, but beyond that, and since they operate in the open, they must be indices of the environmental geometry and the connection between various systems.

Cybernetics and artificial intelligence reproduce at many levels cycles of perception and action already found in organisms and which in the nervous system correspond to the afferent and efferent impulses of the sensorimotor apparatus. Simplifying, we can see the parallel with the mechanical distinction between interactions and potentials, with a mediating connection that can be an index of a certain temporal density, a specific or proper time. If they all express the influence of the medium at various levels, there should be a continuous scale in these mediations.

The laws of mechanics based on inertia operate as a levelling or reducing principle that by compensation tends to explain everything that escapes it in terms of material complexification: in this sense they behave like a principle of material descent, even if ultimately there is not even such a thing as matter. The question is no longer what matter is, but that everything that comes under our control, either actually or virtually as a function of prediction, is reduced to an object. The principle of instrumentalization is already inscribed to such an extent in the principles, that no application needs to be justified anymore.

It is possible to transmute the very principle of instrumentation and its theoretical shielding, in fact a legal shielding. But here we are also and especially concerned with the symbolic, for the symbolic machine that is mechanics reduces man's symbolic potential to a single plane in a perpetual process of construction.

Seen with some perspective, Newton's principles are a great cosmic trap from which we have not yet managed to escape. The «fourth principle» or zero principle in Newton's laws would already be included in the law of gravitation itself —or by extension in any fundamental law of interaction- which had to be justified by the other three principles, although its application in celestial mechanics could not be more dubious. A «cosmic trap» not because it isolates us from the cosmos, but from its background. And the principle of equivalence — and ambivalence- between gravitational and inertial mass, which was already known to Kepler and Galileo, closes this trap whose door is only opened by eliminating the inertial frame of reference.

We can use the internal and external difference of retarded potentials and geometric phases as a thread to get out of this trap, going in the opposite direction of its current technological exploitation for control or quantum computing attempts. We have two basic modes of intelligence, the one that is continuously creating temporality, and the one that is merely aware; the one that proceeds by identification and the one that unveils identity. The latter is always out of subjective time, but intelligence being one, one cannot have both at the same time as we use to pretend.

There is nothing mechanical without intention. This is the reality that makes it possible for us to externalize our spirit in machines. And we must say spirit because it is not only our intelligence, but also our purpose or will. This coagulated spirit of the machines in turn makes it possible for both understanding and will to become more and more divergent in their users.

In something as simple as biofeedback, action and perception converge and tend to merge, in what may well be called self-action, and also self-perception. However limited its scope may always be, it cannot be denied that this confluence exists; but the holonomy of a geometric phase is also a self-interaction of the system as a whole. In field theories, where conservation of momentum replaces the third law, self-energy and self-interaction cannot be eliminated either, which only seem striking if we insist on separating the particle from the field. Similarly, the feedback that emerges in an orbital motion is only striking because our equations are cut off from the background; there is no real feedback, there is simply unity, but physics would not have got very far by sticking to that concept.

Since it is not a question here of achieving further predictions or applications, it makes perfect sense to apply a retroprogressive method that goes from the relatively complex to the simple. In this way thought-mediated intelligence could aspire to accomplish what it has often only pretended to do, namely to grasp the emptiness of time: a time zero which has nothing to do with the standards of current physics, and which can nevertheless be the ultimate object of fundamental physics, so different and even opposite to the extremely speculative theoretical physics with which is usually confused.

*

The horizon of man-machine fusion that is being imposed from above is based on the belief that there is no distinction between natural life and machines, since everything would be mechanical. We know that even classical mechanics is not strictly mechanical, but the mechanical administration of Nature separates it hopelessly from its basic reality, forcing it to depend on other controlled instances. Clearly it is always one same project of domination.

What is predicated of a supposed external Nature is eventually fulfilled in our own nature. It is really incredible that the idea that Nature is dead shit in motion should have been considered acceptable, let alone exalted; but we are still at it, and we see no anxiety to change the basic idea even when it can be done without losing anything and gaining much in universality, since only by abandoning the principle of inertia can the motion of a body be described in the same way in any frame of reference.

In recent times, nanotechnologies and breakthroughs in the manipulation, modulation and tuning of individual quantum states offered the ideal experimental framework to rewrite atomic physics and shed another light on the controversial transition zone between the classical and the quantum domain, for precisely the distance between manipulation and modulation/tuning already exposes the whole problematic of the relationship between both domains. At last it is possible to move from the obtuse lack of perspective of particle accelerators to the more exquisitely minimalist interrogation of energies from every conceivable angle.

New experimental specialities such as continuous quantum measurement or quantum feedback emerge one after another, and the possibilities, interrelationships and bifurcations are just beginning to be explored; they even talk about self-feedback for cases where a resonator interacts not with a controllable system but with a many-body environment. But instead of taking advantage of these fortunate circumstances to reformulate the theory, what is done is to exhibit results by saying as little as possible about how they were achieved. Thus, technoscience as a whole has taken the traditional opacity about its means to a new dimension of secrecy, of esotericism in the worst sense. The occultists of old rambled on about the occult, while the new watchmakers are adept at concealment —especially when it comes to hiding the truth from themselves.

If knowledge that was supposed to be in the public domain has become increasingly private, in good reciprocity it also becomes more likely that knowledge that was previously irreducibly private will find increasingly open avenues of access.

It is common knowledge that the principles of action have a blatant teleological component which has never been explained. The Lagrangian is only an exact mathematical analogy, and yet it is the basis of modern field theories. There is, however, the possibility of employing exact analogies in the opposite direction to the prediction of states of motion in time. Motion alone is insignificant, and time defined according to motion is also insignificant. It may well be said that everything that moves is only a symbol of something else that does not move, but physics only conceives of explaining that something else through motion and interaction.

If the word «mechanics» is used as a synonym for a consistent and rational system, it is clear that one can dispense with the principle of inertia and retain all the rational connotations of mechanics and then more; but on the other hand, without inertia and the other principles that guarantee its closure, the word «mechanics» loses all the heaviness associated with it, and what we have is pure dynamics. This shows that the transformation of the first principle, in genealogical order, can radically modify both meaning and sense, if the means and ends are also redirected.

To contemplate the absence of inertia is an excellent way of putting the world and our relation to it in abeyance, whether it is done at the most immediate level or within the most meticulous theoretical reflection. It should come as no surprise that it has such reorienting power, when the very principle of inertia, as a frame of reference, delimits the primary interface between the self-positioning of the I and what the I can measure of the world.

Too much has been said about the role of the observer in modern physics, and it has even been claimed that its inclusion would close the gap open with classical mechanics, but since the principle of inertia remains untouched, this is manifestly false. On the contrary, subjective factors have been introduced where they were not needed, besides causing other unnecessary breaks and degradations in the classical continuity of the equations.

*

Neither classical nor quantum mechanics are universal, but the geometric phase is, and this alone makes it an inescapable bridge between both. One might think that the transition zone would be more or less at the molecular level, but the reality is simpler, more elusive and more interesting: since there can be no separation between both domains except in our formalisms, there must be a slipknot between the partial ends of the two theories.

But applied physicists and technicians seek to bridge both domains without dwelling on what they have in common, which is quite a task. It is a matter of exploiting resources while ignoring unity as much as possible, as if they fear it might be a threat to their designs. Something very similar happens with attempts to establish a direct connection between brains and machines, where they try to translate, transmit and reproduce impulses while ignoring the elusive background of the experience that contains it.

If we cannot say that classical mechanics give us the «external» side of mechanics, nor quantum mechanics the «internal» side, it is, among other things, because quantum mechanics also assumes a determinism of local motion; that is, it assumes that global consistency arise from the integration of the local despite the fact that time synchronization is given in advance. For the vast majority of neuroscientists, the challenge of understanding the higher functions of the brain does not involve quantum mechanics at all, but rather questions of complexity and organization at various scales within the classical domain; however, in any case, how synchronization occurs, which is the real crux of the problem, remains unexplained.

We should never have accepted the principle of inertia even if, after the long time of its laborious construction, it seemed the easiest way forward. We should have rejected it from the outset as a matter of principle. But the bait has been swallowed all the way down, and now that we have observed its implications at length, it only remains to apply thoroughly the retrogressive method from interpretations to means, and from means to principles, to recede within the principles themselves from the later ones in historical order to the former.

Certainly, within all the tertiary retention of consciousness that exists in the materialized form of technology, there is a very special one in the principle of global synchronization implicit in the third «law»; just as there is a «secondary retention» in our picture of interaction that corresponds to the second, and a «primary retention» at the most immediate level in relation to the first. Even if it is true that in the civilizing process the tertiary always involves the secondary and the primary, the same is true at the level of principles, and here we have the best chance of asses it.

Global and local are recurring notions of both power and physics, but they are understood in the sense of space, not time. The retarded potentials seem to indicate that the global synchronizer leaves timelines in the shadows. There is no need to go very far in Nature to get a more precise idea of what this means: the model is above our heads, in the sky surrounding the planet.

Those who push us towards a horizon of biodigital convergence will try to use the connection that holonomy involves as a recourse to plug us into the network as directly as they can, ignoring as much as they can its meaning and implications; exactly as it is already done, excused by the same adopted principles, with all kinds of biological manipulations. And in good reciprocity, we will use that same connection to make a deeper contact with that physical reality they want to cancel.

There is, of course, another horizon and it is up to us to broaden it and keep it clear. The descending path of subjection entails incessant complexification, endless technicalities and increasing opacity; the ascending path must seek as much as possible simplicity, universality and the relegated intelligibility. And, of course, there is still something beyond these two paths that must never be forgotten.

So we should investigate, for example, the most basic relations of the geometric phase and the retarded potential with the movements of the human body, organic functions, acoustics, musical creation, and finally with consciousness itself within an amechanical perspective.

How is it that the relationship between the geometric phase and music has not yet been explored? Anyone would think that something present in parallel parking or screwing a light bulb lacks any interest, and yet, properly contemplated and connected, it can take us from the most trivial to the most sublime.

*

To grasp the naked reality, any kind of technical artifice or concept is superfluous. However, we see that there is a way back that can take up again the mediations and complexity of scientific discourse towards simplicity. This retroprogressive path inevitably has its own rules, justified by the fact that neither its practices nor its objectives coincide with those of the present technoscience.

Much has been written about machines and desire but here we have pointed to biofeedback as an exponent of a possible inflection. It hints at a will of opposite sign to that «will to will» governed by compulsion; and it also hints at an adaptive internal perceptual continuum that also plays an active role at a different level. Even if any external signal has a very limited range of action, in certain cases some of them may be sufficient to establish certain conditions of equilibrium and the idea of a general balance that could be far-reaching, either by exact analogy or by analogical transposition beyond the quantitative. The geometric phase of the signal would be the eye of the needle between variable action/perception cycles.

In this context, various feedback experiments with biological and physical signals can be carried out and certain «applications» hitherto unthinkable can be conceived. In the face of technologies of unlimited possibilities, powered by the digital plasticity, there are other technologies of real limits that do not lend themselves to escapism or the flights of dataism, even if their signals are digitally mediated. Analog or digital, the important thing is not to catch mice, the important thing is where it looks. There is in the digital and the analogical another life and another impulse.

Biofeedback is simply a detour that allows us to use machines to return to the reality of our physical organism, just as it allows us to use graphic signals that are nothing more than representations to tune functions that exist independently of our mental pictures. Our very organism, at the most basic level, shows signs of something that does not allow itself to be reduced to the afferent and efferent structures of the cycles of perception and action, and the liberation of structures is precisely the way forward, not only in technologies but in general.

The retroprogressive approach makes it possible to connect formal and informal knowledge in the fourth person, which for logical thinking can only be conceived in the same sense that one can speak of a «fourth principle of mechanics», or to the same extent that its conventional «three laws» merge into a single Principle without seams nor closures.

Since the third principle of action and reaction already implies equilibrium just as the first principle of dynamic equilibrium that replaces the "law of inertia", there should be nothing to prevent their more direct and fluid connection except the way of describing the force or the action principle. And on the other hand, we have already seen that forces can also be defined by an imbalance between rotational and translational motions, and that this also has implications for time and for the space occupied by matter.

The Lagrangian action principle can also be defined as a balance between minimum energy change and maximum entropy production. Thus, entropy is already within the «fundamental» laws themselves, and since «order» produces more entropy than «disorder» and entropy tends spontaneously to the maximum, the question and the vulgar clichés about disorder, entropy and finality are completely reversed, and we can have another idea of equilibrium, «information», «design» and evolution of natural systems.

We have spoken of two great modes of intelligence, and to study the purposeless intelligence of Nature leads our utilitarian intellect back to that timeless intellect; physics itself has pretended this, trying to hide its enormous utilitarian bias. But it is not possible to deceive oneself in this respect: where there are «universal laws» there is no room left for natural intelligence. Inevitably, some of us start from a certain a priori knowledge that has little interest in the results of experience, while for others results are everything while knowledge is rather a presumption.

Fourth person knowledge, as continuity, exists always and we are not going to invent it; the only thing we can modify is the connection between this consciousness and the other three moments of knowledge, in the first person or subjective, in the second person by interaction with objects and the world, and in the third person by language and generalization. The thinking mind, the consciousness of objects, is discrete by its very nature; therefore any continuity that exists does not depend on it. For thought, «consciousness itself» is both an empty set and the ultimate possible synthesis, exactly like the primary homogeneous undifferentiated medium that underlies everything and cannot be produced. Both are the Alpha and the Omega whose supreme identity is to be found.

Today there is much talk of the discrete digital infinity, inaugurated with language and that with code we want to take to the ultimate consequences; but there are never ultimate consequences for what has already established an arbitrary cut with reality, however much it may want to swallow and assimilate it, since it is the cut itself that provokes its insatiable hunger.

Between the continuous and the discrete we have the finite, but the differential calculus, to give an example, even having the limit as an a posteriori foundation is still basically an infinitesimal method, not a finite method. And yet finite difference analysis is much more in tune with the relational ideal of homogeneity in the quantities of the equations and allows us to approach it if it is substantiated and applied systematically in dimensional analysis and measure theory. We said that there is no better way to delve into pure mathematics than to refine the method of its physical applications, and here we would have a good example.

In the continuous→finite→discrete→ sequence of calculus and data processing, and its discrete→finite→continuous counterpart, the finite has never had a substantive role, but a purely contingent one. And yet it is only by giving it back that substantive character within the calculus itself that we can better see, not only the mathematical continuum, but the homogeneity that lies beyond the physical continuum as its most undifferentiated nature. Apart from the fact that the physical continuum may have properties that physics itself does not use to consider, such as non-differentiability, relativity of scales for forces or fractional number of dimensions. The homogeneous, the most undifferentiated, always contains any order of complexity, and thus one can conceive a homogeneous←continuous←finite←discrete←homogeneous sequence.

Mathematically, homogeneity means nothing. But from the physical point of view, physical homogeneity would be prior and posterior to any conceivable order of complexity in the physical continuum. On the other hand, the very idea of law to which thought clings is a permanent search for the continuity that it lacks in its very nature; a search that also by its very nature cannot be completed. But the sense of this incompletion also changes radically if we reverse the object of the search. In the same way, basic logical-mathematical concepts such as identity, equality and equivalence are completely transformed when they are applied to the multiple possible physical equilibria as these in turn also tend to the maximum of homogeneity. If there is a retroprogressive method in the physics-mathematics continuum there must also be, in terms of code, a retroprogramming that has nothing to do with the use of obsolete software.

Measurement is a synthesis of quality and quantity, but quality takes precedence over quantity. The increase of homogeneity in the physical quantities of the equations facilitates the transparency of the relations, and with it, the emergence of proportions, which are relations of relations, returning in a certain way the quantitative to the qualitative. All this should be seen under the prism of the finite, constant differential calculus, and, beyond it, of homogeneity, for it is in homogeneity without qualities that qualities unfold without obstructions.

We also assume that there is always a non-intuitive immediate knowledge that is identical to fourth-person knowledge. The close correspondence between the method of constant differential calculus and the execution of tasks without knowing how they are performed, such as catching a fly ball, show us that this immediate non-intuitive knowledge is something always underlying and not some kind of unattainable ideal.

The difficulty then is not in attaining «intuition,» for we know that this is but an extension of our habits beyond their ordinary reach; but in connecting our intuitions with this other constant knowledge. The way to this cannot be axiomatics, with its obstructive idea of foundations; it is not a matter of deriving reasonings from the principle but of leading them back to it. True intelligence lies in seeing in the undifferentiated, the differentiated can already be seen by anyone.

We can see Nature as an endless dance around a point outside any movement, and also around a point of dynamic equilibrium that would be everywhere. This allows us to establish a double analogy: an exact analogy, of the same type as those used by physics for its predictions, and an analogical transposition beyond any measure, in which the medium-consciousness uses physical descriptions and representations as a symbol of itself and for itself. This does not require complex models, but rather the elimination of certain presuppositions about the nature of physical reality.

In the present state of technoscience, an extremely complex and tortuous theory demands, by definition, apparatuous applications, that is, applications that have to be realized through equally sophisticated machines and with as many layers stacked on top of each other as the theories themselves. But if our idea of retroprogress has some depth, between knowledge and power there must be a form of practice with objectives, angles and an intention incommensurable with the ideas of current technology. Even today the idea of practice in the arts still is a model, and the same could be said of many old «techniques», without ignoring the fact that now we proceed from different conditions. It is essential to create new examples, but practice, in the higher sense of the term, must always guide both technique and theory.

Some may take the path back to the simpler with a wealth of formal knowledge and others may immediately hit the target with their more undivided attention; but while some pull and others push, it is excellent that both can share one same orientation. Bridging formal and informal knowledge has been impossible since the beginning of the scientific revolution.

*

A large part of the scientific ideas that we indicate come from the works of André Assis, Gennady Shipov, Mario Pinheiro, Miles Mathis, Nicolae Mazilu, Nikolay Noskov, Peter Alexander Venis and other researchers of whom I have already given references in previous writings. These works point, from different angles, to a transformation of our vision of mathematical physics from its foundations, which apparently has no place in a discipline that only allows itself to grow forward, towards the speculative sector of theoretical physics.

And, naturally, the ultimate horizon of theoretical physics can only be the unification of the forces of Nature, at the level of its laws and their evolution in time. But any attempt at unification without due attention to the foundation is like wanting to produce unity without having to go through it — just as one would want to unravel «the mystery of consciousness» without having to go through self-consciousness, and vice versa. However, the always unappreciated unity of life is infinitely more important than any of our theories, and the mere fact that these poor theories of ours seek attunement with it already enhances their value.

We hope to open soon a new website with a collaborative spirit and the intention of distilling that fourth-person consciousness contained in, between and beyond the other three. To begin with it will be a virtual space, but the goal is to create a communication of a different order than the one that prevails today in the scientific community or the web itself. It is rather a matter of finding a communion in aspiration and knowledge, not by adhesion but by a common orientation that counts on the goal itself as an effective means.

The Principle is the goal, and only the Principle itself has the capacity to align otherwise highly divergent efforts. And it is good, and it is inevitable, that we have to look again to the Principle from the complexity of multiple perspectives. We have noted the role that in these circumstances the multispecialist, the antipode of the current «expert», is called upon to play as a knower of various specialties who is not compromised by the corporate interests of any of them. He alone today has the freedom of perspective and the technical competence to create a new course in a totally controlled and bureaucratized techno-scientific landscape.

Mathematicians also have an important role to play. They too have technical competence and a freedom of perspective regularly frustrated by having to offer their services for purposes too often unworthy of an uncompromising intellect, which is precisely what characterizes mathematics. Somehow they will have to get even for the now clearly perverse misappropriation of their talents. Or the programmers of the world, for example, including graphics programmers. But the need to direct our talents in another direction is general.

Let us also keep in mind that the relationship between pure and applied mathematics today is mediated by a tool par excellence, calculus, which has reversed the natural relationship between predictions and the physical geometry of problems, which is why it is of enormous importance to recover a less instrumentalized point of view. The balance between prediction and description is critical to the depth of our vision, and this balance was abruptly broken with the establishment of differential and integral calculus. However, this balance can be recovered and there are alternatives that yield solutions.

For starters, we will propose a series of ideas, themes, and argumentative threads already pointed out in previous writings so that they can be elaborated and developed by those with interest in them. We will try to pose problems that make sense and with a needed against the current orientation, problems in which an advance is possible without the need for special means. Some of the topics of interest will be morphology and natural philosophy, retroprogress, fourth-person knowledge, wave mechanics, the connection between waves and vortices, the connection between light and sound, the hypercontinuum of dimensions within a homogeneous medium, the individuation process, indeterminacy relations and dimensional analysis, the ignored aspects of differential calculus, the «anholonomic music», the link between the retarded potential and the geometric phase, prime numbers and the Riemann zeta function, consciousness and number, power laws, retroprogramming, diagramatology, the relationship between thermomechanics and thermodynamics, equilibrium, new links between reversibility and irreversibility, the reversal of reverse engineering in practice and theory, experimentation with biofeedback, or a general theory of health and aging that only for powerful reasons still does not exist.

Everything can find here another value, both the contributions with a high technical level, as well as the connections between multiple subjects, the emergence of a new idea of practical application and functionality or the never sufficient appreciation of the various symbolic levels of any logic or object of knowledge. Precisely with modern mechanics begins the reduction of our idea of becoming to efficient causes that are no less imaginary than others but that polish the tabula rasa of the permanent process of desymbolization.

*

We know that art and wonder are at the other extreme of modern technology and theoretical science. Let us think what music and the plastic arts have managed to do with two of our senses, vision and hearing, and let us think of the present physical theory of light and sound. According to this theory, light and sound waves are more dissimilar than equal, and no real connection can be established between them. And yet for this same theory it was not hard to reveal a fundamental continuity between both types of waves that has sunk into the shadows, waiting for someone to rescue it. A continuity that can also be established at various levels without the slightest detriment to theoretical rigor.

To speak of the relationship between light and sound is to a large extent to speak of the relation between space and matter, those other two great unknowns in our representation of the world. But we have already seen that even today the most basic relationships of combination between the motion of rotation and translation, or between rotors and oscillators, are still ignored, though they are at the very basis of all our technology. But beyond and before that, this combination of waves and vortices, in the broadest sense of the term, generates the infinity of organic forms that we observe in Nature. It has been precisely the excessive utilitarianism of theory, its opportunism and short-sightedness, that has prevented us from opening up to the contemplation of the six dimensions of space in our ordinary experience and the extraordinary prodigies that their equilibrium unfolds.

In 1820, Oersted discovered in a very simple way the relationship between electricity and magnetism, and we have already seen what is still coming out of it. Recovering the links between light and sound that our own theories have buried, and following them to their source with the proper disposition could have equal or greater consequences, although in almost the opposite direction. Let us not forget, moreover, that even of electromagnetism, the natural force we think we know best, we have just an utilitarian concept.

When in technoscience we speak of principles, means and ends, we should keep in mind that the end of the theory, the interpretation, is the starting point of experiment and application. Interpretation, far from being a philosophical luxury that the physicist allows himself at the end of the day, outlines with its description and representation the field of technical applications. In 1956 Bohr and von Neumann arrived at Columbia to tell Charles Townes that his idea of a laser, which required the perfect alignment in phase of a large number of light waves, was impossible because it violated Heisenberg's inviolable Uncertainty Principle. The rest is history. But now the tale is that quantum mechanics predicts and makes lasers possible.

This is more the rule than the exception. Theorists are specialists in confiscating experimental achievements and having the last word on them; for a way is always found for the theory to finally «predict» the results that are in plain sight. Of course, a theory that can come to predict anything a posteriori is not a great theory, but a great rationalization, and this is true for any kind of theoretical standard, whose main virtue consists precisely in providing a standard for calculations. Think, for example, of a theory as «restrictive» as quantum electrodynamics, which subtracts infinity from infinity in recurrent cycles of calculation until the expected result is reached. Today everything is justified in the name of prediction, but Ptolemy's epicycles also had a predictive capacity unsurpassed for their time.

It could be argued that, at worst, the theory is not obstructing the pursuit of new technical applications. This may be true to some extent; but let us keep in mind that the divergence between Bohr and Townes, between theory and the average innovative application, is minimal compared to that which might occur if the technician were to start from the beginning from a different interpretation, different means and different principles —since other goals and uses could also be conceived that are hardly commensurable with the present ones. From principles to ends, passing through the means, there is a round trip, a double direction that creates the continuum between science and technology, but the amplitude of the circle of technoscience possibly passes through the degree of reversibility, or unobstructedness, between principles and ends. On the other hand, if we could see without obstructions from interpretations to principles through means, we would probably not miss formal knowledge nor technical mediations.

Overcoming the principle of inertia within mechanics is equivalent to taking the first and most important step in this process of unobstruction. The same division between principles, means and ends within a continuum already exists at the very heart of the three principles of mechanics, and it is no small thing to understand that the third principle, with its systematic closure and its global synchronization, is the whole horizon to which classical mechanics can aspire in its symbolic economy, the limit of its interpretation; for if one were to see this beforehand, one would perhaps seek to understand things differently. But there is also a correspondence between these three principles and the three moments of temporality or the three aspects, structural, functional and symbolic, that we have distinguished in technique itself.

All this brings us back to fundamental symbolic and semiotic issues. The same ternary division of man into body, soul and spirit, and other analogous divisions that applied to cosmology in the cultures that preceded us, are still in correspondence with modern principles even if their intention and range of application could not be further apart. Discounting some eccentric logician like Peirce, perhaps the last one with some awareness of this correspondence was the one who did most to establish their definitive separation, Isaac Newton himself.

This recurrence is not the product of logical recursion but the effect of a «symbolic resonance» that is not inscribed in language but rather circumscribes it. And, needless to say, its consideration is totally irrelevant, if not counterproductive, for the modes of reasoning of modern science. In a context of efficient causes, dwelling on them could only create «destructive interference,» to continue with the wave simile. And yet, beyond this limited context, they allow us to contemplate «constructive interferences» between various levels that can also have their own meaning.

Thinking by correspondences is prior to the development of logic and in such a sense still involves us without our realizing it; which does not detract from the fact that the same correspondence is one of the most basic and fruitful ideas of logic itself, mathematics, physics or the theory of truth. On the other hand, various theories of the emergence of consciousness in the brain are based on neuronal resonance and its problematic synchronization. However, the same three principles of relational dynamics contain a local temporal adjustment of potentials in which various levels are already environmentally connected. We leave this paralogy open for those who wish to delve deeper into it.

Fourth-person knowledge is not to be confused with the collective intelligence that emerges from discussion and mediation, in technoscience for example, since such collective intelligence ends and is always exhausted in the same tertiary order. If we assume that it exists always and at all times, the only thing we can do is to try to connect our tertiary representations with it, by more or less rigorous analogies; or to connect our always incomplete intuitions with a knowledge that we do not intuit but is already immediately given. Since there is harmony between this knowledge and the «fourth principle of mechanics,» the analysis and synthesis of the operation of the latter serves to get an idea of the former.

Elsewhere we have seen how the evolution in time of a spherical wave in the six dimensions of space serves as a very simplified model of individuation of an entity; and we have also seen that considering six dimensions is not an obstacle to the intuition of organic forms but rather the opposite. Similarly, analogous models can be presented for knowledge, whether at the most reductive level of information, or at any other broader level; if on the one hand a wave or oscillator allows the emission, transmission and reception of a message, on the other hand we can directly apply to it the interference or parallel transport that characterizes the very same geometric phase that they try to exploit for things like the so-called «quantum computation».

The difference is that while the technologist tries to obtain a performance from operations whose physical representation is pointless, from the point of view of this other knowledge the correspondence between the representation of the evolution and the environment surrounding it has a critical value for the extent of its resonance in the field of the knower itself. Needless to say, we use here the word «field» by analogy, in a much simpler, more general and naturalistic sense than the present field theories. In this way a radically different horizon of knowledge from the present one can be created, capable of distilling its own quality of convergence.

Although the wave-vortex model is absolutely general and does not pretend to deal with the complex questions of electrodynamics, it is not superfluous to recall that the orthogonality of electromagnetic waves is not a geometric aspect but a statistical average between space and matter; nor is it superfluous to recall Hertz's distinction between material particle, as a point of application for forces, and material point, as a volume that can contain any number of material particles —and that both already comprise the wave-particle duality that would later come to the forefront in quantum mechanics. Or remember that de Broglie's matter waves imply a vibration internal to the bodies, which the theory cannot fit properly because material particles are not considered to have volume either.

Bearing these things in mind, among others, one can come closer to understanding how it can be that the interactions of matter and light, in the laboratory no less than in our common experience, take place within a homogeneous and otherwise unchanging medium, and that if this is understandable it is because one oneself is that medium which is nowhere and nowhen.

Venis' sequence connecting waves and vortices also evolves in six dimensions combining oscillation and rotation, but in this case it is not a mechanical model but a phenomenological morphology independent of any metric or measuring process. In its projective space of continuous dimensions one can contemplate not only the infinite becoming of forms but also their permanent connection with the surrounding environment.

It is extraordinary that we are unable to have a complete idea of the evolution of a spherical wave in three dimensions, which implies a continuous deformation at all points as in Huygens' principle of propagation of light, and yet we can «intuit» retrospectively, with the invaluable help of logic, its evolution in six. But what is also intuited in all this representation, without ever finishing to encompass it, is the pure transience and the necessarily superficial character of its manifestation.

The original «holographic principle", that of Gabor, already unknowingly included the geometric phase loop. However, it is a supreme irony that the ultra-speculative theoretical physics has only granted it the status of a universal principle after deducing it from the ultimate horizon of a supposed gravitational singularity, aka «black hole». So the «holographic principle» of the speculative horizon of physics says that any possible event or information of the physical world is contained in a two-dimensional surface, something that has not ceased to sow all kinds of perplexities, even if we always had to know that all our knowledge comes through the agency of light.

Of course, in a relational mechanics in which retarded potentials are ubiquitous and therefore also present in gravity, such a singularity is impossible because as the velocity of the masses increases the force binding them decreases. One has to go to a theoretical extreme such as a black hole to see clearly how general relativity continues to pay tribute to absolute time and absolute forces independent of the environment. But if that which for the theory only arises in the extreme is already apparent at any time and under any conditions, there should be no internal need to bring it out. Here the irreversible is in the evolution of the theoretical horizon itself, not in Nature. But this contains something even more problematic, when physicists like to think that the fundamental laws are reversible, whereas in an open medium reversibility does not seem to makes sense.

Any closed system possessed by an absolute dynamism must lead to a singularity: it cannot have any other theoretical horizon. And here the counterpoint for human creations, in the form of a «technological singularity", cannot be missing. However, a geometric phase can also be seen as a gap or singularity in the topology of motion, not in the pathological sense, but in the much more elementary sense of not being integrable in the contour of the closed system.

And this non-closed contour concurs with the «principle of individuation» of forms so wonderfully reflected in Venis' sequence. And although one would not have to say that the individuation process is prior to the individual, it still becomes necessary in a time that has pretended that the individual is the only principle of everything; an individual that would be as closed and atomic as the laws he assumed and applied with utmost generality. Let us say, for the rest, that any particle must be an ephemeral configuration independently of its duration or stability, by the mere fact that it does not become apparent without interaction.

At any level, the geometric phase is the interface par excellence between systems that are otherwise defined as closed: this is its irreplaceable strategic character, which can only become apparent when, firstly, everything «fundamental» is considered closed, and, secondly, «connectivity» between more and more «closed» or «atomic systems» begins to proliferate. But since it is excluded from our principles, it remains just one more contingency.

In a rather opposite but not exclusive sense, the geometric phase can also be interpreted as a transition of scales, as well as a transition between dimensions. Regardless of the fact that the Venis sequence is only a projective unfoldment still out of touch with the quantitative domain of physics, it offers us an invaluable and still entirely undeveloped link between modern physics as a domain of nomos and the becoming of physis eclipsed by the Law. Even in the domain of differential geometry and calculus it raises a number of questions that have not yet known the contrast between our intuition and our reason. In any case, the understanding of the evolution of an individual entity as such seems to me infinitely more interesting than forced speculations and blind applications.

Depending on the physical case, its description and interpretation, a geometric phase can be seen as a parallel transport, as a self-induction, as a curvature or flow of the symplectic form, as a conical intersection of potential energy surfaces, as a transition between dimensions, as a torsion or a change in density, as a phase transition, as a point of degeneracy, as a retarded potential, as the gap in the Lagrangian, as a resonance, as an holographic interference, as a loop, as an enslaving principle, as a hole or singularity of the topology of motion, as a proper time or timeline, as a memory, as an interface or other ways that need not be mutually exclusive.

Why does a geometric phase admit so many interpretations? Poincaré already noted that any «fundamental law» expressed by an action principle admits an unlimited number of possible causes, which would therefore become irrelevant. The geometric phase, on the contrary, admits an unlimited number of global ways of seeing the gap of local causality.

Since global synchronization, by definition, cannot be within dynamics understood as the domain of interaction, with respect to the potential it is dynamics itself that reacts, and it is in dynamics that one should speak of a proper time between actions and reactions: the ideas of locality and causality are due precisely to this, not the other way around, as is currently thought.

The assembly and overall design of human machines yields a fairly unambiguous causality and sense, even though the natural forces on which they are based, such as gravity or electricity, can never have them. Thus, our idea of causality is based on the synchronization of parts (end), the application of forces (means) and materials (principle).

From this point of view, it is clear that the creation of the mechanical clock in the Middle Ages is by far the main development of modern technology, the one that most defines its spirit. What the industrial revolution adds is the exploitation of forces, and the computer-driven digital revolution is inscribing this spirit more and more deeply into matter through an increasingly minimalist control of forces. And the ultimate limit of this tendency would be the exploitation of quantum potentials; but this obfuscation with things that can only offer diminishing returns prevents us from seeing the loose end of the matter.

If we ignore the efficiency of the physical medium, the very disposition of mechanics obliges us to believe in efficient causes even if we refrain from identifying them, and it is up to forces of whatever kind to fill this gap; insofar as we have it, whether we call it homogeneous medium, vacuum or whatever we like, we can contemplate physical processes beyond causality and see at the same time how we insert our human ideas of causes in between. This coincidence of the causal and the acausal, which can certainly be understood at many different levels, points itself to fourth-person knowledge.

*

Finally we need to assess the meaning of the retroprogressive movement in general and its virtue in the present juncture. On the one hand, retroprogress is an integral part of the permanent turning back on itself of reflective consciousness; on the other hand, from a natural point of view, it is part of the animal need to re-appropriate the possibilities offered by the environment. But today both movements are systemically and systematically obstructed in this terminal phase of a certain civilizing process, and the escapes available for this increasingly violent retention are just pathetically gross substitutes.

Today neither the theoretical sciences can afford the luxury of freely evaluating their own foundations, nor can technologies revise the standards on which they are built. A typical example of a standard that has been shown to be irreversible is the arrangement of letters on the keyboard, expressly conceived to keep manual typewriters from frequent jamming; today this limitation makes no sense on the computer, yet no one has been able to change it. But the theories also have defined themselves as standards, and it is not by chance that today we speak of the standard model of particle physics or the standard cosmological model. There is also a whole historical sedimentation of standards in multiple layers at both the theoretical and technical levels. This sedimentation or accumulation, this systemic obstruction of the possibility of return, is the «living» image of the inertia of civilization, or to put it better, the irreversible in the accretion of its structure.

And everything suggests that we are at the end of a long cycle of accumulation that is already more than half a millennium old; although this period would be very little had it not remade in its own image all that preceded it, with its more vast, slower and surely also more powerful cycles. Knowledge and power were the two feet on which the fatuous man of the Renaissance believed himself to stand, but in the first phase of its unfolding the very process of the expansion of knowledge prevented to asses the movement of power in the opposite direction; now, at the end of this process, there is no knowledge that is not felt as an exercise of power, and a deliberate power in fact. The first liberalism also presented itself as a suppression of frontiers; but in reality what it did was to destroy the existing structures in order to impose its own, which now, having colonized the whole territory, are only perfected in their control and closure. That today's so-called liberals do just the opposite of what they always preached speaks eloquently of the fulfillment of the cycle.

Knowledge and power compensate each other with the most exquisite equilibrium even where their proportion seems to be most unbalanced, of this we can be sure with the most apodictic of certainties, for there is no better safeguard for the world, for any world. Many will think it is pointless to question the principles of mechanics as we have done, but apart from the fact that here we are dealing with an absolute necessity, it places us on a line of minimum resistance and maximum opposition, which is at the same time a line of minimum opposition and maximum resistance. And the axis of the cycle we are talking about revolves at the confluence between historical inertia, physical inertia and our idea of both.

The last consequences of the idea of physical inertia are to be found in the systematic violation of all inter-species barriers by biotechnologies. Despite the vast mountain of empirical evidence to the contrary, genes are still considered as carriers of a digital code. And ultimately it is thought that they have to be a language, because it is assumed that without it biomolecules would be inert and stupid. But here too, once again, the reality is almost the reverse: to be a reliable vehicle of biological inheritance, DNA must of necessity be an extremely stable, and therefore passive, molecule. What really play an active role are the globular proteins and enzymes, truly the first level of organized life in their own right, which are able to build DNA and all kinds of proteins with a variable response depending on their environment. Life has made DNA, not the reverse.

Biotechnologists cannot claim ignorance of this and many other elementary realities of molecular biology, and yet they speak and act as if such realities did not exist. Much more than a language for the plans of a building, genes are themselves building material, which, yes, may be more in tune with some shapes and functions or others, which in any case come, where else, from the environment surrounding these massive and clumsy molecules. But this cannot be reduced to code, and the idea of code and its program continues to have an irresistible attraction for those willing to combine and recombine everything.

We do not act according to our ideas, but rather our ideas reflect what we do and want to do. This is always the case, but now that technoscience is indistinguishable from politics and we have entered fully into technopolitics and digital techno-feudalism, we can verify firsthand what this means. This does not detract from the fact that such willful ignorance is criminal, especially when we are talking about full-scale manipulation of life.

Globalism is the greatest enemy of the planet, and those who push it are the last people who could talk about saving it. Its fate would not matter to them in the slightest were it not for the fact that they still have to live with its troublesome inhabitants. Globalism sees the planet as a portfolio of assets and all life as a reservoir of potentially useful genetic resources. And the best proof of the extent to which they despise life is to be found in the very biological theory and practices they promote in the most determined way. Any fabricated threat pales in the face of this.

Every power tainted with the meticulous and deliberate adulteration of life is destined to fall, for one cannot confuse the Tree of Life and the Tree of Knowledge with impunity; especially if you are willing to ignore life to such an extent. Such powers attract the worst misfortunes upon mankind and the rest of the species, things more undesirable than extinction And those of us who are witnessing all this cannot remain silent, for the loss of man's dignity has already gone too far, and by keeping silent we prepare ourselves for still far greater indignities.

It is said that we only really fight for our immediate interests but that is not true. One should not only worry about the consequences for oneself of the stupid machinations of those who claim to be masters of life, one should begin by rejecting their whole program and their repugnant means from beginning to end and in the most radical way. For the self-serving premise from which they start, that living beings and our machines are the same, is manifestly false even for a child, and the most that the development of such an approach can achieve is a refining of simulacra; in a word, of lie and deception at all levels.

So their starting point is simply unacceptable; we will not say their principles because it is obvious they have none. Moreover, it is unacceptable not only morally but also from a theoretical and technical point of view. And the first thing needed to reject this power is to reject its assumptions and pseudo-principles, which is already half way to have principles of your own.

Most technicians and scientists do not realize how far the colonization of their minds has gone because they do not care either, as they are already satisfied with being able to work within a framework with established rules and within a community that seems to give meaning to their effort; but the best among them have little to do within the present mediocrity. The sciences have created a whole narrative of mavericks, revolutionaries and misunderstood prophets, but the truth is that today, even with all these so-called «disruptive technologies», they have more inertia and things to hide than the Church that prosecuted Galileo. The tables have turned completely.

One can be sure that the only ones who are bringing science into complete disrepute are not those who criticize it, but the very scientists who have allowed it to reach this state. And, for one thing, even that could be useful to the powers that be, just as the erosion and discrediting of politics is useful to them. They have foreseen scenarios where the bulk of the population turns its back on scientific knowledge so that a small organized caste can work even more freely —and that is almost what is already happening. It is worth asking with what the power will legitimize itself at a given moment, when nobody listens to its priests anymore. But even if they have succeeded in degrading every conceivable angle of knowledge, they will not prevent reason from having the last word, and we certainly will not give up reason.

It has been repeated that what characterizes technocracy is not the pursuit of ends but its emphasis on means and their efficacy; but this is not a very accurate judgment, and even less so in the present circumstances. Already in physics we see that the means, predictions, became ends, from the very beginning and in the most blatant way possible; what can be expected then of everything else. But precisely the fact of elaborating ad hoc theories has made them the straightest way for one thing but inappropriate for everything they ignore.

Our most celebrated physical theories, or calculus itself, often exhibit great inefficiency for tasks that can be solved much more simply by other means. And when they try to connect different theoretical layers together, the combined inefficiency can grow exponentially because each framework has been optimized for one type of problem at the expense of all the others. And the same is often true of technologies and the attempted technical solution to many real-world problems. There we have the great efficiency of our economic model for just one thing, and its enormous destructive effect for almost everything else. The myth of efficiency to which the technocracy clings is already an object of scorn; absorption by ever narrower ends finished it off.

*

If retroprogress does not seem to us to be the natural dialectic it is, more than anything else, because of the very unilateral exaggeration of the idea of modern progress itself. A type of progress that they still want to force with a funnel even if we do not have the slightest interest in it. What is not yet pondered is the extent to which science, in pursuit of its predictions and solutions, has inverted the perception of problems; nor has it been even remotely pondered that there is a simpler way of dealing with them without renouncing reason and simplicity at all, but rather quite the opposite. Retroprogress traces back this line of resignation of reason seeking its restitution.

To maintain this fiction of irresistible and irreversible progress, we are sold the most ridiculous lines of escape, when all we want is a little bit of unadulterated reality. Nor is almost anyone unaware that these drastic and irreversible changes that they want to force are staged precisely not to change what matters, but quite the opposite: they are intended to modify the entire population so that the small group of perpetrators remains intact. In a word, any appeal that progress could have has been definitively buried by its latest promoters, in whom unbridled progress and extreme reaction coincide as never before.

As we have said often, the real Megamachine of globalism, the one that envelops and imposes its meaning on everything else, is the inverted pyramid of wealth distribution with its 80/20 power law reiterated almost to the limit, and, according to some simulations, tending to another «singularity» in which all the wealth would belong to one holder and the rest would have nothing. This Great Siphon extracting and pumping wealth from the physical base of resources to the top of the dome has an elementary mathematical structure in which many things are nevertheless inscribed. Despite the simplicity of this structure, which also plays an ubiquitous role in data mining and data analysis, logistics, marketing, management, administration, and a long etcetera that boils down to optimizing the exploitation of resources, its full-scale study is systematically ignored in all the sermons of economists and sociologists even when they pretend to deal with the «problem of inequality".

Which demonstrates once again to what extent the sciences, whether natural or human, with their «formidable analytical resources», are controlled discourses integrated into the official narratives. And it also demonstrates to what extent this public discourse has little to do with the real analysis of data that is routinely done behind closed doors.

If in discussing technology it is inescapable to contrast Machine and Nature, here we have a case of particular interest, since power laws and its manifestation in the frequency domain are also ubiquitous in all kinds of physical and biological processes, including our own brain, a very complex form of society. Some have wanted to see this as evidence that inequality is simply «natural», but it seems obvious that the «freedom of scale» that capital has had to grow, the condition for its permanent accumulation, is, more than inheritance, the appropriation of the mechanisms of money and credit creation —for it is also obvious that the inverted pyramid of wealth reflects the mountain of aggregated and accumulated debt of our economies.

In the Great Siphon, monetary, digital, media, state and other large institutional mechanisms are integrated in a way that is at once hierarchical and non-hierarchical, as vertical as the slope of inequality and as horizontal and «anonymous» —in an anonymity proportional to the quantity- as the flow of money; and while the distribution itself is continuous, the number of elements among which it is distributed is always a discrete set. It is clear that only a small part of what is being learned about this power law ends in the public dominion, but there is a great key here to the relationship —and the divergence- between Nature and civilization, and between the supposed liquidity of money and the most vertical hierarchies. Independent studies will be needed to unravel this link, but this is not one of those intractable problems. Contrasting the hydraulic system of debt with the limits of natural processes will bring sunlight into this rarefied crypt.

All the monetary engineering, financial engineering, social engineering, media concertation, knowledge engineering, genetic engineering, etc., are directly indebted to the Great Siphon structure and work to protect it, since it not only accumulates at the top almost all the surplus purchasing power but also channels it. Any society worth of that name will have to do everything necessary to ensure that this structure of massive suction and optimization of collapse never happens again, for such a monument to dysfunctionality not only generates grotesque inequality but corrupts everything it touches, molding it in its own image and likeness. This is what must end.

Any important truth can be understood at many different levels, and conversely, if a process is only known at the level of prediction, as so many are in physics or probability, we understand it poorly or we don't understand it at all —no matter how deftly the equations are handled. Since power laws occur equally in Nature and in human processes —from fundamental physical laws to the size of cities, corporations, earthquakes, grains of sand and stars- and a huge mass of statistical information has been accumulated about them, the degree of knowledge that the best experts may have about it is surely halfway between effective understanding and predictive ignorance. One may have almost everything to fully understand the phenomenon, except the necessary disposition to assimilate it.

But there is no need of math to know that in any human group those who take the initiative are a small fraction, and that their proportion and influence extends in a more or less continuous way as far as the number of the population allows. Nor is mathematics necessary to understand that the advantage gained from this initiative can be accumulated and amplified over time through inheritance and the control of the most vital mechanisms for the functioning of the whole.

We understand this without difficulty from our own experience in social interaction as well as from the accumulated experience of history itself. In this case, the critical mathematical elements are the ratio 1/5-4/5 of the «law of the vital few» and the number of times the power is repeated, which must necessarily depend on the scale. Because it is undeniable that this tendency also exists in small communities, and what gives it its monstrous implications is the unrestricted expansion of scale.

This unrestricted expansion in space is opposed by increasing restrictions in time. A very basic aspect of the increase in complexity is the correlative increase of systemic restrictions: this increasing restriction, this fragilization, is something as basic as the aging process we experience in our body, and yet it is generally neglected in the descriptions and analysis of development. Which is not surprising if one notices the implications.

The cycle of expansion is followed by a cycle of contraction: but here we are no longer talking about economic cycles but about vital cycles. Actually, this whole cycle of concentration of wealth that begins with the expansion of Europe until it ceases to be Europe to become "the West» is a cycle of dissipation of energies that had accumulated for at least another five hundred or maybe one thousand years. This first phase grew inward and distilled qualities, while the second phase grew outward expanding the reign of quantity to its ultimate frontiers. Already within the expansive cycle a few awakened enthusiasm and took advantage of the enormous surplus of inner energy of the population, while now to subsist the system takes advantage of the lack of energy, channeling the scarce surplus as best it can.

Evidently, here the collapse will not coincide with any kind of singularity either, since all the surplus wealth accumulated at the top must somehow buy the adhesion of all that remains below; but this massive buying of wills translates mostly into serfdom, generalized corruption, increased restrictions and inefficiency, and the sedation and permanent loss of energy so necessary to be able to take a different course.

That is why now more emphasis than ever is placed on the irreversibility of technological changes: what is irreversible is the end of this civilizational cycle, but those who pretend to guide it prefer to see man disappear than disappear themselves. They do not want to have witness to their end.

There are many levels of understanding for the power law and the principle of development; even the evolution of a spherical wave in six dimensions gives valuable knowledge of the self-dual nature of expansion and contraction. But whatever level of knowledge we acquire, it remains useless, if not counterproductive, for the maintenance needs of this system. So, as with everything else, what should concern us is what value it can have for the communities we build, now still on the margins, someday outside of this system.

Today the structure of domination is fundamentally economic, as in other times it was priestly or military. Although the objective is that no such thing exists anymore, we still do not know to what extent this is possible, because to this day it remains a fact that there is no organization without hierarchy. The only way to minimize this fact is the drastic reduction of scale.

The opposite of death is not life but birth; life itself has no opposite. With this in mind we better understand the omnipresence of the forces of death in motion, which do not pretend to be directly homicidal, but are the very embodiment of a resistance to dying that coincides with the opposition to birth. And our relationship with machines, that coagulated spirit, also concretizes in its own schizoid way the resistance and contraction of that which resists death. Because the truth is that machines continue to have an unbreakable connection with us even though we are being strangled by the umbilical cord.

*

Those who force against nature the irreversibility of the current changes do harm to all those who suffer them but also and especially to themselves. Certainly they still have the traction of the gears of power, but it is the only thing they see of a much longer shaft. By insisting on the irreversibility of the changes, they multiply exponentially the opposition and cut off the possibility of retreat. No matter how many false challenges and false alarms they fabricate to make themselves indispensable, everyone knows that they are not the solution but the problem. Even to «reinvent» themselves in a minimally credible way, they would need radically new things that they are the last ones to be able to create.

The idea of life and knowledge that is held here is diametrically opposed to that of those who believe in the irreversibility of progress, for such irreversibility is nothing but the inert accumulation of its many layers. We all feel that the fullness of life lies in the liberation of structures, but modern technology tends on the contrary to their systematic proliferation and concealment. And what power and what knowledge are those that need so many devices?

If today's mainstream idea of biology is just the last consequence of a way of understanding the physical world, there is in physics, which would be the holistic science par excellence if this were included in its principles, a still unheard potential to describe satisfactorily the most important aspects of biological life, aging, evolution and health; and also to confront at least the question of consciousness. The «unreasonable ineffectiveness of mathematics» for biology is notorious, but this should not have surprised anyone since, to begin with, the very application of mathematics to physics and its «unreasonable effectiveness» has been nothing but a concatenation of reverse mathematical engineering operations for very specific purposes. For this very reason, there is also another inverse path to simplicity here, which is barely hidden, and that we hope to be able to show.

Despite all its shortcomings, mathematical physics still maintains a connection with the noblest lineage of knowledge. Molecular biology, precisely because it ignores the wonders among which it continually moves, in its present state is nothing more than a condemned fruit. This judgment may sound too manichean, and yet it is physics that must redress the grievance. The sum of the knowledge of a hundred thousand experts in molecular biology is ridiculed by what a tiny enzyme can do; if one day they are up to its level, perhaps they will no longer want to do so many experiments. And yet, that enzyme knowledge depends more on what has been ignored in fundamental physics than on the whole biomolecular tinkering.

It is not man that has to be overcome, but science, and even more so technoscience in its current concept; Man will remain an unknown halfway between Heaven and Earth, between what he is capable of grasping from the infinite sphere of knowledge and what he is capable of concretizing with it. But I do not believe that this planet will allow at this point the coexistence of two different civilizations, one dedicated to the manipulation of life as they please and the other condemned to suffer it without being able to do anything about it.

No, that will not happen; the tension between Heaven and Earth will not allow it, nor that between our knowledge and our action. Therefore, either this whole race will come to a swift and violent end which stop the impious experiments, or an opponent will soon emerge capable of balancing this extreme situation. One can see lines of forces of the planet opposed to the forces of globalization; but in the scene of knowledge nothing similar exists yet, and without this guide, any opposition is defeated in advance.

Which does not mean that a new orientation of science, technology and practice requires the support of any power, but quite the contrary. If it is not able to emerge spontaneously and by its own means, it will not be worthwhile either —if it is a question of opposing a simplicity that is at least as pervasive, as all-encompassing, as the complexity with which modern technologies surround us. To come up with it is not something within the reach of money.

More than a few have sought to free themselves from the structures in the domain of thought, but until now the domain of quantity that has given this system all its capacity for control has remained unchallenged. However, the supposed technical superiority of its analysts has much of an imposing facade, and one can verify that not even the physical dimensions of elementary calculus problems have been correctly analyzed. Without good analysis there can be no good synthesis, but that has always been secondary to the growth of the specialties. Today the panorama of knowledge is simply indescribable, and not only because of technical and historical problems, but also because, as power and science increasingly depend on each other, the number of things that experts must ignore to stay in their positions neutralizes the accumulated knowledge they hoard.

Our perception of reality is always very superficial. Earlier we noted the concurrence of the discourses of power and physics in certain concepts, although, on the other hand, no one ignores that there is a decisive difference between force and power. We also noted that motion, and any physical state, can be described consistently by both equilibrium and imbalance. But how do I feel power, either outside or within myself? Anyone will say as an imbalance, since power tends fatally to be exercised in its totality, although the other way of looking at it, as a permanent dynamic equilibrium and also as a permanent compromise, must be equally true. However, this latter aspect can evade consciousness to very large degrees, since if equilibrium always exists, and the sum of forces is always zero, there is no noticeable change in the balance. Thus, without intending it, fundamental physics seems to lead us to an unsuspected arcanum of power, an arcanum which, curiously enough, would have a blind spot inaccessible to the prevailing physics and mentality.

For as long as there have been stories, we have been told of the struggle between gods and titans, devas and asuras, for the dominion of the cosmos and man; in this respect nothing would be easier today than to establish a narrative that would not be far from reality. However, the truest of these ever-present myths is something that is not captured by time, although it sets in motion the springs of fiction. In the end, neither the one who better deceives nor the one who is not fooled will win, but the one closest to the Principle.

There is virtue in the retroprogressive, a virtue that the fascism of the irreversible makes ever more evident, but also one more elusive virtue that makes the whole wheel of events turn. This virtue does not adhere to anything but allows everything to follow its trace; even power is able to follow it, since it is enough to seek it for it to open a path where it seems to be none.

IRONY AND TRAGEDY IN THE RIEMANN HYPOTHESIS

22 February, 2022

> *Did you know this secret? The worst is that beauty is not only terrible, but also a mystery. God and the Devil fight in it, and their battlefield is the heart of man.*
>
> Fiódor Dostoyevski

Mei Xiaochun published a paper 3 years ago[1] in which he claimed that the Riemann hypothesis does not even make sense because there are already four serious inconsistencies in the 1859 text to begin with. In a later article[2], he uses a standard method to prove that the Riemann zeta function does not have a single non-trivial zero. Zero zeros. Needless to recall that, according to the prevailing mathematical opinion, have been calculated trillions of them.

Mei Xiaochun is not overcomplicating matters. The inconsistencies he talks about are very basic, they infringe even the very Cauchy-Riemann equations that are at the basis of complex analysis. I am not a mathematician and prefer to defer my judgment on the relevance of his arguments, but I think that, at the very least, they deserve an answer; though we are unlikely to find it anywhere. If some mathematician deigned to answer, surely he would say something like that the analytical continuation has principles that the author seems to ignore, but no one ignores that creating new principles out of convenience is the most elegant way of not having them.

Mei is a physicist at the Institute of Innovative Physics in Fuzhou. Looking at a list of his scientific production it is easy to see that he has specialized in the critique of highly speculative aspects of modern science that pass for normal science: general relativity, black holes, the LHC and LIGO projects or even electromagnetism. Regarding the latter, he has a series of papers on retarded electromagnetic interaction that I find particularly interesting, since he links it with temporal irreversibility in certain processes, and we have repeatedly discussed on this page the theory of the retarded potential and also irreversibility in fundamental physics.

Right or wrong, what is clear is that in the West one cannot belong to a university or any official research institution and publish articles like Mei's. It would not even be worth trying, since rejection would be guaranteed in advance. That this is allowed in China only highlights what we already know, and that is that Western science is not assumed there as something of its own, but only as a tool of enormous power that is impossible to ignore.

Otherwise Mei is not trying to "deconstruct the West"; his arguments are perfectly legitimate and right to the point, and what is really revealing is that no Western scientist belonging to the institutions can afford to say these things. Perhaps the last researcher on the payroll who could do so was the Canadian Paul Marmet, and I am not even sure if he was still working at the University of Ottawa when he published *Absurdities in Modern Physics* in 1993.

What can the mathematicians *here* make of Mei's arguments? We will have to ask them. I have been told that the things I have written about the zeta function are "utterly speculative", and it would only be worrying if they were not, when I am not even trying to do mathematics. The fact is that the Riemann hypothesis itself is "utterly speculative" right from the start, and the only worrying thing would be that mathematicians do not realize it.

To begin with, I have never believed that the hypothesis can be solved, which is what apparently motivates most mathematicians; I think that this special function is interesting in itself independently of the hypothesis, even if that naturally includes the zeros on the critical line. At least here we have raised doubts about the correctness of the methods of calculus, doubts that arise even before we enter into complex analysis, with the most elementary differential calculus. We have referred once and again to the work of Miles Mathis[3], who has highlighted the systematic illegal algebraic manipulations to arrive at the known results. We have also talk about the reverse engineering of calculus applied to physics, and what is striking is that in the case of the zeta function the circumstance is the opposite: physics, and in particular many body quantum systems, would reproduce some of the patterns of this function. But, needless to say, all that is no less "utterly speculative". I fully agree with Mathis in his assessment that calculus is just applied mathematics, not pure mathematics; what I don't know is how anyone could claim otherwise. Why else are they still calculating zeros?

The Riemann hypothesis "would just say" that the prime numbers are as orderly as they can be because if they were more orderly they would cease to be prime numbers. That is, it is a more restrictive but still highly uncertain condition on Legendre's and Gauss' hypothesis that later became a theorem thanks precisely to the Riemann zeta function. We put "would just say" in quotation marks since this same thing can be formulated in a thousand and one different ways having as many different resonances in the most diverse fields of mathematics. It is understandable that most mathematicians wish it to be true, since it would provide a criterion of order of immense scope, and the opposite, that the pattern of prime numbers starts to be different at very large numbers, besides being unimaginable, would have devastating effects. Moreover, there is no hope of disproving the hypothesis with the brute force of computation, pathetically impotent in cases like this.

Of course, here we are assuming that these non-trivial zeros of the function exist and that they have a certain and unequivocal relation to the prime numbers and their distribution; but this is exactly what Mei Xiaochun denies. The minds of mathematicians, industrious by nature, will always choose to have something to work with. If an analyst found himself in limbo, heaven or hell, and had to choose between spending eternity idly, or spending it calculating endlessly, he would not think twice.

It does not take much humor to conclude that the proof of the Riemann hypothesis, let alone its implications, is inconceivable, and that its refutation, let alone its implications, is equally inconceivable; and that in the middle of these two inconceivabilities, between zero and one, and zero and infinity, lies the critical line with a real value equal to 1/2.

Well considered, one can see that Mei's just objections do not work to the detriment of Riemann, but unintentionally reveal as never before the genius of the most profound of mathematicians. Like writers and philosophers, great mathematicians hardly do more than exploit and amplify as far as they can the scope of a single idea, by which they are more or less possessed. Everyone does it, but the success of forcing an idea soon finds its limits in reality and in the credit that others are willing to give us.

The 19th century is above all the century of the complex variable and Riemann is its greatest exponent. It is well known that the origin of the complex zeta function is in the theory of functions and not in number theory. Riemann knew that complex analysis greatly simplifies many problems of real analysis, just as he knew that complex numbers provide way more room to move. Riemann sought to better define that still hyperbolically imprecise and unmanageable idea that "prime numbers are as orderly as they can be because if they were any more orderly they would cease to be prime numbers." What he then did was to look for the kind of manipulations that could express that idea in the complex plane starting from Euler's product formula and combining Cauchy's and his own breakthroughs in complex analysis with the indispensable ones of Gauss, Dirichlet and Chebyshev in arithmetic. Riemann was looking for a simplifying criterion and finally came up with one that indeed seemed to be extremely simplifying at the expense of doing four forbidden operations. The criterion was given by his own *sui generis* calculus that was by itself a new modality but which miraculously put the zeros of the function on a critical line. Riemann could not help but be aware of the precariousness of his manipulations and so he dropped his hypothesis as discreetly as possible. But only now does it become clear that the hypothesis was from the beginning the motivation for all his work on the function.

And what could matter four unjustified operations when something so extraordinary emerged? It took time for mathematicians to begin to glimpse the new possibilities opened up by the Riemann hypothesis, but once they got into the game, they found impossible to give up this unexpected and tempting paradise. It is readily admitted that Riemann was not exactly a model of rigor, but no one doubts the boldness and conscientiousness of his approaches. It is impossible that he did not know what he was doing; he simply concluded that it was justified because it gave way to a truth of a higher order. However, the mathematicians who came after him ceased to be aware of this because they already have enough work justifying their additional and increasingly rarefied assumptions.

I have no desire to undermine either Riemann's work or that of all those who have followed in his wake. We can feel sympathy for their ways of proceeding, precisely because we can see them as human weaknesses, and it is foolish to be indignant with our own weakness. I think the Riemann hypothesis is just sublime, and yet I also think Mei is right and the manipulations involved are strictly speaking illegal —even by Riemann's permissive standards. One is tempted to say that the Riemann hypothesis is sublime even if it is absurd, and that even absurdity would prove it right if it could.

Of course, it is not only a matter of human weakness, because if there is any greatness in the life of a scientist, it lies in the fact that his discoveries are not merely due to his intelligence, but to a faith that advances in the dark and to perseverance in the face of adversity. Physics and mathematics have had ample doses of this, although many do not realize it.

I am convinced that what has been said conforms to the reality of the facts because the whole history of calculus gives us all kinds of evidence in the same sense, which is transparent in spite of the triumphalist bias with which its foundations are judged today. Being from its very origins a series of heuristic recipes, by the time Riemann came, analysts had been practicing systematic opportunism for almost two centuries with no other immediate imperative than to extend the domain of the calculus at all costs, repeating at every opportunity that familiar "a rigorous proof would be desirable". In this Riemann was only adding to a practice that by then had become a tradition.

A sagacious mathematician often gets results without knowing how he arrived at them; the Riemann hypothesis would be the most extreme case of this kind of anticipation. The mathematical community has never justified Riemann's procedure, but has accepted it at face value, and accepting it literally, has to justify it later by way of a proof that promises to be infinitely complicated and equally insignificant. However, there are in Riemann's implausible anticipation obscure general reasons and specific reasons that have been erased by that typical forgetfulness that any present needs to attain the illusion of existence.

It is said that the foundation of calculus gave the desired rigor to the great edifice of analysis. But again the foundation was a justification of known results, only with a higher degree of generalization. If previously a right result was arrived at by wrong means, now it was claimed to do so for the right reasons, but how can wrong mathematics work to begin with, and how is it that right mathematics does not change the results of wrong mathematics at all? And of course, the problem with the Riemann hypothesis is that it does not rely on known results, or if it does it is through a connection as elusive as it is problematic.

The very fact that mathematics is expanding in all directions ensures that the gaps everywhere are also increasing. If its knowledge, as in the rest of the sciences, doubles every 15 years, so do its loopholes, doubtful moves and castles in the air. In fifty years they are tenfold, and in 150, thousandfold. More than one hundred and sixty years have passed since the Riemann hypothesis, so that, with a cursory calculation, it can be estimated that many current works on this function could retroactively involve on the order of 6,000 grave violations. This makes serious criticism an almost impossible task, and besides, no one is to blame for this enormous accumulation of knowledge. So what are we talking about when we talk about rigor?

It has always been discussed whether mathematics or the great physical laws are discovered or invented. I have no doubt that they are invented, and the case in point seems to me to be very eloquent; which, rather than detracting from their merit, adds to the work of Newton, Riemann and so many other illustrious scientists. Nothing is discovered without the will to advance in a certain direction. However, the continuators fatally tend to take for granted things that the creators saw with the most extreme and justified caution.

From Galileo's inclined plane to Mendel's peas, the whole history of science is littered with forced arguments in the face of experimental evidence that does not match enough but should. Mathematics is supposed to be the most rigorous of all the sciences, but, once again, and in view of it all, what are we talking about when we talk about rigor? We have had ample occasion to see it everywhere and in this days as never before: rigor is nothing compared to the momentum of research programs and the collective effect they generate. Too many scientists have become specialists in cloaking the wildest ideas in the mask of rigor, opacity and boredom.

*

This may bring to mind that distant book by Sokal and Bricmont[4] on intellectual impostures. No doubt already in the 1990s there was a whole clique of authors and so-called philosophers who were saying ridiculous things, and with some famous names, the derision was even too easy. Otherwise, anyone approves the calls to separate knowledge from political and culture wars, something that in less insane times would not even have been necessary. But in fact Sokal and Bricmont, a mathematician and a theoretical physicist, were also waging their own very timely culture war, diverting attention from the rampant proliferation of inconsistencies in their own disciplines. And this they knew first hand.

Do we want to suggest here that the Riemann hypothesis is an intellectual imposture? The question seems to me of the utmost interest, but not to dabble in the sociology of science and the culture wars, but for the mathematical core of the question. I personally do not believe that the hypothesis nor the current way of calculating the function is an imposture or mere collective illusion, at least for two reasons: because, even if it has arisen from an unsound manipulation, and especially if that were the case, all those zeros still ask for an explanation, rather than a solution; and because its extolled beauty is also problematic in itself.

Much has been said about beauty in mathematics and it is a subject from which nothing can be drawn, not because it is unimportant but because it exceeds the competence of the mathematician. The mathematician cannot say anything meaningful about beauty, he can only, at best, create mathematical beauty, and more for its own sake than thinking of those who can appreciate it. However beauty in mathematics is not just a result, but a generative and productive end. Simplicity is sought, and in the meantime, one tries to simplify continuously, which is one with the process of abstraction, of distillation of heterogeneous materials.

The history of mathematics is fascinating because it reflects things about us that cannot be seen in any other mirror. There is an expressive, physiognomic level to mathematics that can be appreciated without having to know the meticulous intricacies of this science; just as a good connoisseur can judge character, mood, and state of health by countenance without having to know anything about physiology. Mathematics not only has its own physiology but is also unwittingly displaying a historical physiology on its own plane, a logic of internal development of which the "social state" would be, at best, only a snapshot that would leave its only precise figure in chronology. There is an aesthetic plane of the isolated mathematical object, and another aesthetic plane of the relation of all mathematical objects in their interaction at a given moment; and the applications of mathematics to physics are a crucial part of this constellation.

Poincaré judged with good sense that arithmetic is *a priori* and geometry *a posteriori* and therefore a matter of experience; other mathematicians have said that there is no part of this science, however abstract it may be, that cannot come to reflect relevant aspects of Nature. Here we would also like to suggest the opposite: there is no cognition of Nature on our part that cannot migrate and join the compact monolith of arithmetic —provided one knows enough about number theory, of course. This retrograde or posterior path is much less conscious than the descent from arithmetic to geometry in applications through algebra and calculus, and although in theory they should be reciprocal, in practice they are far from being so because of the various contingencies in their development.

Riemann devoted at least as much time to theoretical and experimental physics, especially in the company of Wilhelm Weber, as he devoted to mathematics; Euler and Gauss also devoted themselves to physics to the best of their abilities. Riemann, no one doubts, was a great analyst, a great geometer, and a great number theorist, algebra being his blind spot or at least his most indifferent area. This is very characteristic, and in a way allows us to excuse the liberties he always took in his operations. Being a mind so absorbed in theory, and having a short life and poor health, it is surprising to note the patience with which he attended to the details of repetitive experimental protocols; especially when we note that all his major contributions took place in nine years, between 1851 and 1859, the last of them being, when he was barely 33, the work on the zeta function.

Bernhard Riemann's eight-page paper *On the number of primes less than a given magnitude* was published in November 1859, the same month as *On the origin of species*. That very same year saw the beginning of statistical mechanics with a pioneer work by Maxwell, and the starting point of quantum mechanics with spectroscopy and the definition of the black body by Kirchhoff. Statistical mechanics is the mother of information theory, and, on the other hand, Riemann was also interested in thermodynamics which was in the forge in those same years. I have always felt that the zeta function alone represents a counterpoint, and probably an antithesis, to the triumphant bid at that time to explain everything in Nature with concepts such as chance, entropy or natural selection. Riemann sought in everything the underlying unity, and, with Leibniz's permission, one can venture that never an analyst had such a strongly synthetic orientation. In his writings one can see how his mind oscillated between abstract intuition and abstract imagination, two very different faculties that only occasionally coincide. This highly synthetic intelligence also devoted much time to natural philosophy, the mechanics of audition, the conduction of heat, electromagnetism, light and gravity within a theory of the ether; and although these concerns and works, with an unmistakable hallmark of depth, did not bear the desired fruit, they somehow poured into the only synthesis he was able to achieve —where one could least expect it, in his only work on number theory.

I also believe that in formulating his hypothesis, Riemann was not motivated by the desire to extend the predictive power of calculus but by the desire for simplification or synthesis, and that is why beauty here has a character, never better said, functional, and, at the same time, the least obvious that could ever be imagined. Because, does it really simplify things at all, or are we just looking at another mirror whose relation with numbers never quite closes?

Authors such as those already mentioned seem scandalized by the abusive use of mathematics by some philosophers to construct the most inept metaphors. Here again they are not without reason, and many of the examples they expose to scorn really deserve it; but, again, all this is too patent to devote time to it. However they do not tell us anything interesting about the use and abuse of metaphors in the so-called contemporary hard sciences. It has been said that what would be science without metaphors, and in truth, metaphors are inextirpable, not only in the popularization for laymen, but at the very heart of scientific activity. And it is also eloquent that, in the midst of so much abuse by one or the other, the Riemann hypothesis, so exalted in its beauty, is equally inaccessible to metaphors.

Potential infinity is not a metaphor, but actual infinity is, and yet it is quietly accepted by almost all contemporary mathematicians. In fact, actual infinity is, as Lakoff and Núñez put it, the "Basic Metaphor" of our present mathematics, and it has also something to do with the idea that the Riemann hypothesis must have a solution. Of course, potential infinity is not infinite at all, but only an indefinite finite series. Riemann himself was still thinking in terms of potential infinity, for as we know the actual infinite only began to be contemplated, and not without fierce resistance, after Cantor's work. If actual infinity is so commonly accepted today, it is due to the typical hypertrophy of algebra and its symbolic indifference to the real.

Teratology, the catalog of monstrous and absurd metaphors, begins with the hard sciences themselves, and if in mathematics the aberrant and pathological always remains beyond our sensible recreations, with physics it enters fully into the imaginary. This anomalous use starts historically with the infinitesimal calculus and algebra, which assists it in the application of mathematics to change. It just happens that we have become accustomed to these leaps over reality.

There we have black holes, for example, so well dissected by Mei Xiaochun and many other authors before him; why don't the professionals of skepticism and the artisans of scandal criticize them, when it is so easy to see their absurdity? More than easy, because to pretend that a mathematical singularity can have physical reality is absurd, but to want on top of that that mathematical physics goes beyond a purely mathematical singularity, is epically ridiculous. And yet, we can understand that the fact that every day we have breakfast with the news that two or three new black holes have been discovered has to do not only with the entertainment industry, but it is also the consequence of a logical development of the same theory that comes from far back.

But no matter how much it is the logical consequence of a series of misunderstandings, there is no beauty in a black hole, nor any mystery as to why they exist beyond the mystery of human credulity. Now, once inhibitions against absurdity and ridicule have been overcome, the sense of mystery and depth are guaranteed. Sokal and Bricmont spoke of postmodern intellectuals and fashionable nonsense, but postmodern science, which at the time of their book was in full swing, was already ramping up in the sixties with "the golden age of black holes", and in the particle zoo with entities such as the quarks, whose name alone says it all. Incidentally, the decisive feature of the nuclear force, the relevance of non-central forces in Nature, is totally eclipsed by something as aseptic as the celebrated asymptotic freedom that conceals it.

General relativity was not crazy and inconsistent enough, even with his mother special relativity, and the only remaining theoretical restraint, the veto on black holes, had to be tear down. In a course that does not allow one to go back and seriously reconsider the fundamentals, the only thing left is the flight forward, and giving physical reality to singularities was the ultimate passport to Neverland. This was done by taking advantage of the curvature that Riemann had devised in his daring geometrical generalizations and whose application to physics was in his time a utterly remote possibility. In fact Riemann sketched a theory of matter particles as sinks of ether and considered for them spaces with and without curvature. The cautious and scrupulous Poincaré, half a century later, saw that, if special relativity was assumed, there were two ways left for general relativity: curving the space-time or curving the light. It was much simpler to do the latter, as well as being in accordance with our experience; that this was perfectly feasible is still demonstrated by current theories such as the so-called gauge gravity.

Postmodern science is self-ironic by mere defensive reflex, for it knows very well that it cannot take any idea too seriously. Not black holes, which we take only as an example, but almost everything in contemporary science is destined to a supreme irony, the product of its twists and shortcuts in the double movement of accumulation and rarefaction of knowledge. And that supreme irony will be very different from the small ironies now sprinkled everywhere. The one who has not the slightest hint of irony speaking is Mei Xiaochun, the most unexpected incarnation of frankness.

And something of that supreme irony that is approaching can already be glimpsed in the turns of theory itself in its multiple manifestations. The path of excess also leads to wisdom, if one knows how to get to the very end. Things like the so-called "anthropic principle" in cosmology have long since revealed the degree of disorientation of speculative physics, but here we will only mention the holographic principle because it is closer to our subject, since it arose from the thermodynamics of black holes.

In a black hole, relativistic mechanics, quantum mechanics and thermodynamics —and therefore also information theory- converge. Taking the various theories to the ultimate conceivable limit and trying to see what happens in their contact: this is interesting, even if it is proposed in the demented and pathological environment of an ultrasingularity. The Riemann zeta function is already implicit in Planck's calculations of blackbody radiation, and thus in very basic aspects of quantum mechanics such as vacuum energy levels. On the other hand, this function has been used to regularize the thermodynamics of black holes to obtain finite results, which is only one method among others that does not imply any necessary connection between both.

The holographic principle is already a supreme irony of the theory for several reasons. First, because after having raised the number of dimensions for a theory of gravity to four and for string theories to ten or twenty-six, we end up being told that any physical evolution with all its information is reducible, no even to a volume, but to a surface. This cannot fail to create the greatest perplexity, when it was assumed that processes such as gravity or Huygens' principle of light propagation cannot operate in two dimensions. But that is still a minor irony. The greater one would be that the holographic principle is based on the geometric phase, which is a supplement to quantum mechanics but does not properly belong to it: a loop or curvature added to the unitary, closed evolution of the Hamiltonian. The geometric phase does not belong to the projective Hilbert space, but reflects instead the geometry of the environment which is not included in the definition of a closed system. And this openness of a closed system to its environment is what is now supposed to define the limits of our experience of the world.

In other words, the geometric phase, a mere appendix to a quantum mechanics that claims to be complete but clearly cannot be so, would give us the fine grain of the skin of the universe. The geometric phase, the anomalous shift of the potential, is not part of the scheme of dynamical interactions and is therefore considered entirely secondary. The history of dynamics subordinates the potential to the force. The force is the determinant and the potential, as a mere position, was only understood as a passive auxiliary; but it cannot be that that which has an instantaneous effect, as quantum mechanics demonstrates, is passive with respect to that which needs time to react.

It is as if physics had to some extent inverted the ideas of potency and act; and the truth is that this is far from being the only thing it has inverted. But what is entirely illusory is to pretend that the geometric phase is exclusive to the quantum world, when it is present at all scales and in the best known theories under the misleading name of "retarded potentials" —another of the recurring themes in Mei Xiaochun's papers. Retarded potentials first appeared in physics in 1848 with the electrodynamics of Weber, Riemann's faithful friend and close colleague; and in a different way, also in Riemann's own electrodynamics of 1858 whose publication withdrew himself. If retarded potentials are applied to gravity, black holes become impossible because with increasing velocity the force decreases proportionally.

*

Analyzing the relational mechanics of Weber and some of his more recent continuators such as Nikolay Noskov and André Assis, we come to the conclusion that the so-called "retarded potential" is not such but simply expresses the temporal difference between action and reaction when we leave the implicit principle of global synchronization which dominates all classical mechanics including relativity, and in which the “local causality" of quantum mechanics is embedded. Let us briefly recall, once again, that in a relational mechanics such as those cited, generally compatible with observation, we can replace the first principle of inertia by that of dynamic equilibrium, in the second principle the force ceases to be constant and to depend only on distances, and in the third principle ceases to be simultaneity between action and reaction.

This modification seems at first only a slight deviation, but we soon realize that it implies an idea of time that is very unusual and difficult to conceive. Each thing has its own time, and if there is a "global synchronizer", which as it comes out of Newton and it has always been conceived does not cease to be a mere metaphysical principle, it cannot be in the domain of dynamics and interactions, but in the potential, that which has always seemed passive to us. With respect to something instantaneous, the dynamic interaction can only be a passage from potency to act, but it is the potential that directs the action.

Wolfgang Smith, a mathematician and physicist of Aristotelian orientation, has proposed for quantum mechanics the idea of *vertical causality*. This concept seems to me not only useful but perhaps necessary for our issue, but in a very different sense from the one Smith gives it. Smith adheres to the Copenhagen interpretation, assumes the radical difference between macro and microphysics, and relies on the collapse of the wave function for a deliberately verticalist and transcendent interpretation. By contrast here we do not see that there is any irreducible difference between classical and quantum potentials, and the so-called vertical causality would be a relation of another order between the different proper times that have processes that we normally consider in the same plane, that is, globally synchronized. Such vertical causality could indeed have a relation to the Riemann zeta function.

Needless to say, this "verticality" has nothing to do with the critical line being represented parallel to the ordinate axis, which is a mere convention; rather, it has to do with the non-separability of potentials, the alignment of their respective dynamical equilibria, the connection between different proper time lines and the elementary, irreducible but unspecific relation that arithmetic has with temporality —just as the zeta function has an intrinsic relation with complexity not because it uses complex numbers, but because of its relation to many-body interaction problems.

This idea of vertical causality, which can best be realized along the lines of physical theories that are now forgotten —and indeed it is forgotten that Weber's relational mechanics, which yields non-central forces, made many relativistic innovations unnecessary- could be inspiring to say the least. It should not be understood literally, since the idea that there is a horizontal causality that extends virtually without limits would simply be another illusion. There is neither horizontal nor vertical with respect to a primitive homogeneous undifferentiated medium, which is otherwise the only true infinity; the contrast only arises when we assume that there can be a global synchronization by dynamics itself, which besides being metaphysical involves a contradiction in terms. Vertical causality is an implicate order with its own temporal lines and layers.

Then, we can call this vertical causality a metaphor if we want; but a much more restrained metaphor than the notion of actual infinity, or the metaphysical principle of global synchronization, or the intrinsically contradictory principle of inertia, which asks us to consider a closed system that is not closed. The real difference lies in the fact that we are not accustomed to this new metaphor as we are to the others, and we have yet to delve much deeper into it before we begin to make sense of it.

On the other hand, if the geometric phase and the retarded potential form a feedback loop that we may modulate, as the Weber-Noskov theory would suggest, perhaps there are ways to test not only the holographic principle —which would have to be ubiquitous by definition- but also the relation of these ideas with the Riemann zeta function itself —something that, we should say in advance, has nothing to do with the idea of a "physical proof" of the hypothesis.

Someone might find curious that a physicist as insightful as Michael Berry, the main generalizer of the geometric phase, who has also carried out some of the most influential studies on the relation of the Riemann zeta to physics, has not worked with these or similar ideas. The answer would pass first of all through the fundamental incompatibility of the theories, then through the scarce recognition of relational mechanics and finally through the even scarcer awareness of its temporal implications. Let us recall that according to the informed opinion of Berry and Keating[5], for a "Riemann dynamics" to exist it should have properties such as scale symmetry and a classical counterpart; be chaotic and unstable; be temporally irreversible; be quasi-one-dimensional. Naturally this still leaves all the room in the world for speculation, but it has its value as an estimate of dynamically relevant asymptotic features. Here we have also commented that it should be an irreversible dynamics in a very different sense, but now this would take us too far.

I devote a good deal of space to the relation of the zeta with physics for good reasons: because I think that the hypothesis has no mathematical solution, and because on the other hand I believe that it is a pole of inspiration and attraction for non-standard physics, understood in the most conservative but unexpected sense. Concluding one of his texts, Mei Xiachun says that it should be considered whether or not the study of the distribution of prime numbers should return to the domain of real numbers, although no doubt he is not unaware that the Riemann zeta function was essential for the proof of the prime theorem in 1896 and that the "elementary" proofs, presented much later, proved to be more complex indeed.

But, apart from the fact that Mei does not pretend to settle the matter, perhaps there is no contradiction. The zeta function served to prove the theorem because Riemann already assumed it to begin with when he managed to find out transformations and equivalences of the function. But from there to being a potential game changer in number theory —an enthusiast said that if the hypothesis were solved it would be like shifting from a screwdriver to a bulldozer- there is a gulf. Nothing can justify such expectations. Number theorists may say that their advances have been spectacular, but I think that, rather than being of real use, the only thing this function does is to make other results look in its mirror. These are completely different things. In itself, the zeta function could be effectively as sterile as a mirror, and yet have a singular role as a guideline. The zeta function would not be a shortcut to anything new because it would be the ultimate shortcut itself.

The truth is that, knowing that no major breakthrough has been made in over one hundred and sixty years, expecting the picture to change suddenly and radically thanks to a demonstration seems too much like a fantasy of enlightenment: once on the top of the world, there would be a whole eternity to explore and delve into the infinite variations of all the associated L-functions. But true understanding rather comes when we realize that no enlightenment is possible; this cuts the tree from the root. Since most do not even want to hear about this, one has to devise indirect strategies, in which one imagines that one is getting closer to the truth without harming oneself too much. Of course, analytic number theory cannot give up the complex plane and that is out of question. At its interface with physics, however, there are other very basic issues pending.

Given the nature of the hypothesis, one might superficially think that, even if it were shown to be true, it would not change the landscape of knowledge in number theory at all; and if it were shown to be false, neither would it. And although I believe that neither of these two things will happen, the difference between both possibilities is that more than a thousand theorems, some of them absolutely amazing, have been advanced assuming that the hypothesis is true, but, as far as I know, the idea that the hypothesis is false does not spark ideas of the same scope. That is, the hypothesis that the Riemann hypothesis is false is not quite productive, while the hypothesis of its truth is, and this already tells us something. For Mei Xiaochun the hypothesis is neither true nor false, but plainly meaningless; his judgment stands by the ground of the matter, but something that continually produces theorems of a very different nature does seem to make sense and be a source of meaning.

According to this, seen from the foundation the hypothesis has no sense, the sense emerges beyond the foundation. But this has been happening continuously in the history of science, only with a sense that is very easy to identify, either for the sake of prediction or explanation. Here however meaning might demand another level in description.

An example could be entropy. The interpretation of entropy as disorder, due to Boltzmann's statistical mechanical rationalization, introduces an unnecessary subjective element besides diverting attention from the clear natural tendency to the production of maximum entropy already perceived by Clausius; and yet the view of entropy as disorder still overwhelmingly prevails. If one had to speak of order and disorder, it would be much closer to reality to say, with Rod Swenson, that the world tends to produce order because order produces entropy faster than disorder. The most elementary physical interpretation of the Riemann hypothesis is as a random walk —the discrete counterpart of Brownian motion-, and this allows us to treat both prime numbers and zeros in terms of numerical entropy. But are entropy and the degree of structure of the function opposed here?

The claim that order produces entropy faster seems quite related to the idea that "prime numbers are as orderly as they can be", and yet in both cases we seem to be asking for a better definition of order than either of the ones we have. The backward heat equation has also been linked to the zeros of the Riemann zeta, which is an interesting way to connect it to signal processing and retrodictive determinism, but there still seems to be a problem of definition and description. On the other hand, it is clear that entropy can be defined in other ways without any need to appeal to the Riemann hypothesis, and that is the case for any important concept externally related to its function.

*

One of the tragedies of history is that, no matter how much more perspective we gain, most of the times this broader scope is of no use to change anything. Neither changes the past, nor changes the future towards which we are shot like a cannonball. In real life no one has to convince us that events are irreversible, but in intellectual constructs, such as mathematics and physics, there always seems to be room for making amends in an elegant and bloodless way. Of course, this is only an illusion, since the organic development of the sciences as a whole is completely irreversible. Or is there any question of overturning the foundations of analysis, relativity or quantum mechanics? Not at all, it is only conceivable that they can be transformed forward with always greater complications, never backward. Reflection is permitted on an individual basis, but for the collective development of knowledge, "advancing backwards" would only be to deactivate oneself.

It is only for this reason that things like black holes exist, which express like nothing else the fatal flight forward of theory when looking backwards is impossible and stopping even more so. Truly Faustian is our science, indeed, expanding and striving until the last moment to try to forget that it has long since he sold his soul to the devil.

I write these lines not as a scientist, nor as a historian, nor as a critic, but just as a lover of natural philosophy. It is too easily forgotten that Riemann loved natural philosophy at a time when it was still possible, even though he was mainly responsible for the leap of mathematics to new heights of abstraction in those years. In fact, as we have already seen, 1859 would mark the coup de grace for a natural philosophy already badly wounded since Newton's *Principia;* although maybe it would be unfair to forget Thomson & Tait's 1867 *Treatise on Natural Philosophy*. No doubt Riemann's heart was sharply divided in this respect, but it is easily to understand that the tension between the analytic mind and the search for unity was always the mainspring of his mathematical creativity.

As I do not believe that there is a mathematical solution to the Riemann hypothesis, neither do I believe that there is any kind of "physical solution" to the problem. I believe, without the slightest mathematical foundation, and with the conviction of my own natural philosophy, that the hypothesis is inherently open both above and below, both through infinity and unity; and that it is that very thing that makes it seem impenetrably monolithic. Which need not be a bad thing, for even if we can never encompass it, we can still surround it on all sides as long as we do not pretend to possess it.

It is infinitely easier to make physical sense of the zeta function than to prove the hypothesis; and yet no physical interpretation seems to have even remotely the depth of implications that its simplest mathematical formulation has. It is thus a challenge in itself to find physical systems that can reproduce the behavior of this function or of the associated L-functions, as for example the one proposed by França and Leclerc[6] some years ago with an electrostatic field and its potential; or many other physical analogies that can be modulated, as for example the Brownian motion of electric dipoles in random classical radiation. And it is an even greater challenge to find a physical interpretation that contributes to give a full sense to the hypothesis rather than trivializing it. Riemann and Weber, the close collaborators in theoretical and experimental electrodynamics, would have greatly appreciated these analogies and would have probed them with much more penetration than we have.

It is remarkable the flippancy with which these analogies are disdained with the argument that they are not useful to prove the hypothesis, but everything starts from the fact that those who propose them pretend that they can be used for that purpose. We should approach the subject in a very different manner. The problem is not that they do not serve to prove the hypothesis, which should be taken for granted, the problem is that modern physics is incapable of describing impartially any process, because it has subordinated everything to prediction. Precisely, we speak of natural philosophy when what we seek is neither prediction, as in the physical sciences, nor explanation as in cosmology or the theory of evolution, but a kind of description that is in itself sufficiently complete and eloquent. And from this point of view, the deficit of modern science is truly abysmal.

Incidentally, neither Gauss's electrodynamics, nor Weber's nor Riemann's were "failed theories", as is still commonly read in history books. They are simply approaches that have not found the same development as Maxwell's equations and were prematurely abandoned, but they are not false and rather show another treatment and a different point of view —something that Maxwell, who owed so much to Weber, knew better than anyone else. The same Faust who sold his soul to predictive power has tried to silence his conscience with explanatory sciences such as the theory of evolution or cosmology, but in vain, because only probity in description can find a balance between prediction and explanation.

What we are saying is that in known physics there are already more things than we think, if only we know how to read between the lines. The prediction always leaves us on the surface of things, but the explanation, historically subordinated to the prediction, does nothing more than justify them. We see this with theories such as the big bang or black holes, which try to give us an illusion of depth when in fact they are nothing more than the ultimate extension of the principle of inertia. What happens with the so-called "foundation" of calculus is along the same line of justification.

Riemann had the confidence, repeatedly expressed in his writings, that man has no access to the infinitely large, but that we can know the natural laws thanks to the analysis of the infinitely small. But surely this too is an illusion: the mathematical analysis of physical change cannot have access to the infinitely small either, it should be content with correctly analyzing the physical dimensions of a problem by means of finite intervals. However, it does neither one thing nor the other. Bloch's principle that there is nothing in the infinite that does not exist before in the finite has apparently borne much fruit in complex analysis, but, if we wanted to make sense of it in physics, we would have to travel in time.

To my knowledge, Miles Mathis is the only author who has so far made a relevant critique of the foundations of calculus and its application to physics in more than three centuries; a critique that goes far beyond the well-known objections of finitists and constructivists. The standard calculus conceals from us at least one dimension of the physical geometry of processes, as is proved by the mere fact that the acceleration of a mobile in a straight line is described by a curve. Analysis soon forgot its humble origin in the calculus of curves in order to generalize itself algebraically, and, for the sake of prediction, it definitively broke its commitment to the description of processes. Potential infinity was formalized by the concept of limit, but in reality, if this is not based on a finite interval, it is nothing more than a way of disguising the infinitesimals, and so one can return to them without great difficulty.

Constant differential calculus, having a higher dimension and being more faithful to the physical geometry, makes it possible to deal fruitfully with the question of how to interpret more realistically the use of complex numbers in physics. If this higher dimension does not only concern physical problems, those who are willing to use alternative forms of calculus can check if there is a real zeta function with intermediate values that can be revealing with respect to the Riemann zeta function. Unfortunately for physicists and mathematicians, this calculus is also much more restrictive in its operations, which understandably invites to ignore it. But the big gap between natural concepts and metaphors and those of higher mathematics begins right here. So if anyone wants to close it again, this is the best place to start.

It is not that natural philosophy is outmoded, but that it is older than we think, or, in other words, it is much more ageless than any restricted theory. Despite the incredible expansion of the sciences, we still have a rather provincial idea of what Nature might be. Even such basic concepts as "potency" and "act" hide much more, not only than Aristotle could imagine, but also than today's interpreters of quantum mechanics can imagine. And yet the best way to grasp that unimaginable would be to try to close the huge gap between our experience and our abstraction.

If it is true that there is nothing in mathematics that cannot have its reflection in Nature, we do not fully realize what this means, nor do our theories conduct in harmony with what it implies —for it is intimating to us that Nature and Spirit are only aspects of the same thing. But if we do not wish to get loss in this radical unity which inevitably leaves us speechless, we can always say that we find math in Nature simply because we probe her through it, which is equally true. In any case, there is in Nature a *plenum*, a superabundance not only of complexity, that no theory, barely a spider's web in the bark of this great tree, can ever show us. The Riemann zeta function, for which we have no true physical correspondence, would express something of this *plenum* much better than all our predictive and explanatory frameworks. It would be demanding a different framework by itself.

The permanent expansion of knowledge is not only constantly creating new gaps, but also new rifts that are basically one and the same. Although the Riemann zeta function began as a purely mathematical subject, it is inadvertently but uncontrollably migrating also to physics, chaos and complexity theory, and computation. We could even see it at the crossroads of these four major areas, and yet it remains a purely mathematical question. Why then are mathematicians themselves increasingly appealing to the so-called "physical intuition" or computational arguments to grasp this problem? No doubt because there are no good mathematical ideas, for on the other hand it is more than doubtful that there is today a single physical idea that has not been emptied of any intuition. But surely all these speculations about quantum chaos, Hilbert spaces, operators and so on are just ways of beating about the bush.

Although it is hard to accept, and given that we are talking about calculus rather than prime numbers, the most distinctive mathematical feature of the complex zeta function and its critical line are the infractions committed in its calculation: this would be the zero zone, and perhaps the best way to see through it is to contrast them with the infractions in the calculus of the real zeta function with respect to the more stringent constant differential calculus. This threefold transparency should yield greater depth of vision on the purely mathematical plane, if that is at all possible, and also on the most elementary "physical intuition", and on the relationship between both. There is no substitute for rectitude. Riemann looked for something straighter than the straight line, and almost found it: all the more reason to retrace the shortcuts, of which the history of calculus is the greatest collection. Bringing divide to the heart of the mathematician is the best way to take him also to the heart of the problem, to the origin of its conception; but now he has the great advantage that he can go much further *back* in time than the German mathematician.

It would be good to start here, and then, if one wishes, with a fresher look, to ask oneself in what other calculus could the zeta function make more sense, in what other electromagnetic theory, in what other quantum theory, in what other theory of information, in what other quantum theory of information, in what other quantum thermodynamics, in what other spectral theory, and so on. Today, for example, the question is asked whether it is possible to do quantum mechanics without complex numbers, and some studies answer in the negative[7]; the subject is of great interest but such studies are of very little scope because they are self-limited to the basic quantum formalisms. What is instructive, in order to understand the role of complex numbers in quantum mechanics, is to try to reproduce many of their results with more basic theories that are completely independent of such formalisms, some of them already indicated.

All the areas just enumerated, being entirely focused on prediction, do not even deal with the point that interests us most here. The so-called physical intuition, like any other intuition, is mostly misleading because it is hardly more than the sum of our habits; we would have to regress our intuition to things prior to those habits, and that is why we have considered constant differential calculus, as a case in point of "immediate non-intuitive knowledge", to use the term coined by Jacob Fries.

And if anyone still wonders how an “unsound calculus” like Riemann’s could serve to shed light on other theories than those being contemplated, he only needs to recap history, for all the frameworks that are standard today have likewise been built on an elementally deficient analysis. We know very well how this is possible, what we do not know is how an unsound calculus can reproduce what it does not know.

*

Number theory, with its sovereign uselessness, is an specific continental product; it is only around 1914, when the theoretical corpus has already been comprehensively systematized and even analytic number theory has exemplary manuals such as Landau's, that there begin to be significant contributions from the Anglo-Saxon world. It is from then on that talk about "conquering the Everest of mathematics" begins and the demonstration or refutation of the famous hypothesis takes on the tinge of a sporting competition in select circles. Does the demand for a solution trivialize the problems? Not always, but in this specific and very different case it seems obvious to me that it does: looking for the answer in terms of "yes" or "no" distances us from the question.

Mathematicians themselves tell us that what is missing is an understanding of the relationship between addition and multiplication at the most basic level, and, in good logic, something so basic cannot arise from such derivative, sophisticated and complex ideas as those proposed to obtain a proof. Without an understanding of that so simple and fundamental, seeking a proof *should not even be desirable*.

Riemann's natural philosophy, which was not the German romantic *Naturphilosophie*, may sound too strange for our time; partly because it is unfinished and we only have tentative sketches, and partly because of the resulting conflict of seeking on the one hand "a complete and self-contained mathematical theory", and on the other hand wanting "to go beyond the foundations of astronomy and physics laid down by Galileo and Newton to penetrate to the interior of Nature". This was just a discreet way of saying that he intended to turn the glove of Newton's natural philosophy inside out. Riemann wants to attain the space and the quantitative without taking it for granted, starting from the qualitative, from the inside out. The extent of Herbart's influence on his natural philosophy is disputed, but it is worth remembering that for Herbart reality is absolutely simple and therefore excludes any quantitative concept. Once again we find the German mathematician in the middle of two highly incompatible demands.

The momentous changes in twentieth century physics prevent us from seeing how ambitious this solitary project of a unified conception of Nature was. For it was not only a matter of "unifying" the main forces known at the time, in the style of modern attempts of unification in current theoretical physics, but first of all of finding a previous foundation from below and from within, which was to include all the psychic activity of the soul towards that simplicity of the real which was already completely unattainable for science. Riemann's natural philosophy is an interface between physics and metaphysics: he tried to give back to the great animal of cosmos that soul that mechanism seems to have taken away from it, without renouncing any of the knowledge of physics at that time.

Riemann knew too much to be content with any crude simplification of unity; in fact he knew as much as Helmholtz about natural sciences, and then some, but he was capable of imagining things that no scientist of Helmholtz's style could even dream of. Riemann died at 39 and Newton at 41 had not yet begun to write the *Principia*; but at any rate Riemann's natural philosophy, of a completely different scope, was destined never to be born: not for "lack of information", as today's more superficial reading will judge, but for multiple reasons including internal incompatibilities or the burden of working with the classical formulation of mechanics. So there is nothing to regret here. However, I am convinced that natural philosophy was Riemann's only muse throughout his entire scientific production.

Natural philosophy can provide us with certainties that go beyond the domain of physics or mathematics; it is quite another thing what use they may have for these, which will depend as much on the knowledge of these particular sciences as on the possibilities of application. In any case, this philosophy has a retrograde path that necessarily begins with interpretation, seeks its methods accordingly, and finally arrives at principles. And although it is an reverse path to the usual one in normal science, it is not difficult to see that modern physics itself was established in this reverse order in the leaps that mark its foundation: from Galileo and Kepler to Descartes to Newton.

When physics came to Newton's *Principia*, what took place was a closure through the axioms: the principles as starting points that should not be changed for knowledge to advance and accumulate. But for those for whom the Principle is first and last, and this was certainly the case for Riemann, the principles still have a further evolution and are starting points as well as basic distinctions and foundations of unity. The principles understood as axioms make the fathomless irretrievable: the first mechanical principle of inertia is not definitively closed until the formulation of the third principle of simultaneous action and reaction, which tacitly implies global synchronization. The reformulation of the three principles according to Weber's mechanics, for example, allows us to recover what was buried in the foundation. Inertia is a good example of something fathomless, which for more than a thousand years could not be intuited, and yet is now taken for granted. Transformative basic ideas can only occur at this level. One of these slow sculptures could be reflected in number theory, but our physical intuitions, or lack of them, are preventing its revelation.

My surmise, I have already said, is that Riemann, by putting his previous knowledge at the service of number theory, activated both voluntarily and involuntarily a retrograde method or reentrance of those conceptions into the *a priori* of arithmetic —or at least, pointing in that direction. And the conclusion is inevitable: one can not find in number theory better ideas than those introduced into it.

Today academics employ their best arts in smuggling past or present ignored ideas, but commitment to the dominant trends makes impossible to assimilate them. We know only too well that one cannot ask a discipline to go backwards and reconsider its past; only creating new sciences is viable. We spoke above of a quadruple crossover between mathematics, physics, complexity theory and computational theory. Well, today it is perfectly possible to create a morphology of the most general type in the space not covered by any of these sciences. It is not a product of them; on the contrary, it takes them for granted and tries to distance itself from them as much as possible, with another idea of calculus, causality, generation, equilibrium, and a complexity that does not depend only on the number of elements.

This morphology, which does not yet exist but which may be born at any moment, is precisely the science that shows and describes how and why there are no shortcuts in the forms of Nature; that is, in what sense they are necessary. Calculus cannot know this, not because it is unknowable, but because to date it is a systematic collection of shortcuts. This morphology will largely fulfill Riemann's dream of contemplating Nature both from within and from without, and if it succeeds it will be because it can integrate without difficulty aspects incompatible with current developments —and because it can afford to rectify Riemann's own ideas about calculus, modern physical notions of curvature, and even the very principles of mechanics.

Such a science would not be imperialistic or opportunistic because it neither pretends to envelop the others nor to borrow their expansive impulse, but on the contrary it deals with that which the other sciences have excluded and rejected, seeking to distill its quintessence. We will devote our next work to this morphology, but if we comment on it in this one, it is with regard to the question concerning the expansion of knowledge, one with its loss of value and with the dissipation of its symbolic capital and credit.

It is easy to see that everything that has "added value" cannot depend on the rate of reproduction of knowledge but would have to go rather in the opposite direction. Natural morphology wants nothing to do with such intractable things as the Riemann hypothesis because it has its own object and is based on incomparably simpler notions, but the important thing to remember is that the multiplication of knowledge means nothing if we do not consciously direct it and do not know what we are looking for. Trying to prove something does not at all guarantee our awareness of the question. In science the most difficult thing is not to want to solve the questions that one has not asked oneself.

*

Although Riemann was excessively cryptic in his most famous work, it is known from Siegel's research that he had made an exhaustive previous study with very powerful analytical tools; if he did not write a much larger text it was due to the length requirements of the Berlin Academy. No doubt he would have liked to expand more on the topic, including the technicalities of his calculus, but in the seven more years he lived no one asked him for detailed explanations about them —surely because the few mathematicians who had understood the text were still trying to digest it. What Riemann would certainly not have liked to talk about in public was his natural philosophy, for this was his intimate laboratory, the only place where he was allowed to make all sort of speculations and hypotheses.

In Riemann's grand program, natural philosophy served to connect mathematics and physics with psychophysics, psychology and metaphysics. There is a clear connection between his concept of n-dimensional manifolds and Herbart's qualitative continuum. His idea of constant ether flow in matter obeys other considerations than those of relativity; it arises also in the context of light and electrodynamics, but Riemann does not bother adding physical aspects that only complicate the characterization, such as tension, pressure, heat, color, and so on, because he starts from the idea that there must be a general psychophysical correspondence between quantitative values and qualities. To our age, which has seen the development of so many specialties, this sounds like an undesirable conflation of very different planes, but what Riemann was rather looking for was how the contact between these different planes was possible to begin with. His efforts are best understood along the lines of scientific philosophies such as those of Clifford and Whitehead, with whom he has many similarities; the difference being that Riemann's philosophical ideas had a still unnoticed effect on mathematics.

But perhaps, looking further back, the philosopher with whom Riemann could be most associated is Leibniz himself. One of the many interesting things about Leibniz is that his philosophy is associated with the mechanical time of clocks —with Newtonian-style global synchronization- but in fact his ideas in physics are purely relational, which is something else entirely. Leibnizian pre-established harmony does not point so much to the synchronicity between events or horizontal causality, which would be at most its external limit, but to the vertical causality of their infinite coimplication. The so-called universality of the Riemann zeta function, which would allow it to reproduce an infinite number of times any differentiable curve with any degree of approximation, could remind us the universal concert of monads, sharing this time at least *one* common thread. Some would prefer to speak in this case of "pre-established chaos", but at any rate, order and disorder here tend to dangerously coincide.

Relativity arises to rectify certain asymmetries of electrodynamics; but relational electrodynamics deals with them in a different way. An intuitive mathematician always needs to think with images. Riemann, who undoubtedly also had in mind projective ideas in line with the Riemann sphere[8], on the one hand assumed that there should be an abstract *a priori* symmetry between imponderable or qualitative aspects, and ponderable and controllable quantitative aspects in the relations between matter and space; and on the other hand he must have harbored his own tentative idea, still rather undecided, about the possible symmetries in electrodynamics in both real and complex variables. In any case, in Riemann's electrodynamics masses were mere coefficients for density, and density and frequency are indeed fundamental concepts with multiple applications in arithmetic; this way the transmigration of his natural philosophy to the domain of number theory could have taken place.

It is assumed that the zeros of the Riemann zeta emerge from the difference between a sum and an integral, and on the other hand it is no coincidence that the first ones to deal with Green's theorem, fundamental for electrodynamics and potential theory, were precisely the creators of complex analysis, Cauchy in 1846 and Riemann in 1851; but the German mathematician, to whom we owe precisely the first proof of the theorem, could not miss that this problem may be approached by the discrete method of Riemann sums that he developed in 1853 on the way to the Riemann integral and that are also suitable for counting in number theory. The so called Green functions are really distributions, but the projective analogy that Riemann might have had in mind, considering a transition between point and wave aspects of electrodynamics, would also be related to the Möbius transform, which shows a one-to-one correspondence with the Lorentz transform but has a different temporal connotation in the much earlier theory of the retarded potential.

One could still think of such things as the so-called Wick rotation and its association with an imaginary time derived from the Lorentz group itself - such a rotation is nothing but a mathematical trick, but on the other hand the "real imaginary time" is the global synchronization underlying all present mechanics. In special relativity the extended energy-momentum relation unveils a mass-potential symmetry that in principle could allow sign reversals between matter and antimatter, but both the present annihilation/cancellation theory of matter and our idea of synchronization have a big central hole to be filled. Although Wick rotation as an analytical continuation is just a formal procedure, it also reveals an important connection[9] at the foundations of quantum mechanics, thermal physics or Fourier analysis and stochastic processes such as Brownian motion that admits a reformulation in a more primary context.

In other words, although Riemann had in the foreground the connection between calculus and prime numbers, he also had in mind not only a sort of theory of the ether and the ether of frequencies, but the ultimate limit for his general approach. If it is unquestionable that the motivation of the function comes from the prime counting problem, for which the well-proven method of Eratosthenes was already available, the search for an internal symmetry within the function is due primarily to the inspiration of Riemann's natural philosophy. The adventurous and celebrated hypothesis comes from natural philosophy, and without it, Riemann would never have twisted the arm of calculus to find it. Without Riemann's preconception of a fundamental symmetry, mathematicians today would frequent the complex zeta function without even suspecting the existence of a critical line.

*

But may be this is not just archeology of knowledge. As we have already said on other occasions, there are at least three absolutely fundamental equilibria in physics and in electrodynamics in particular: the zero-sum dynamical equilibrium, the ergoentropic equilibrium between minimum energy variation and maximum entropy, and the equilibrium of densities with a unit product, which are hardly contemplated in current frameworks and when this is done they are not properly related. Of course, that Riemann conceived a symmetry in the zeta function does not at all mean that it should find a translation in physical reality, but if that is really the case, as so many studies now suppose, and we try to guess what that unimaginable symmetry responds to, it would be desirable to study it in the light of these equilibria and their relation to vertical causality.

Henrik Stenlund has recently noticed[10] that the ratio between a zero of the Riemann zeta function and its conjugate in the negative half of the plane, besides a certain phase has a unity absolute value, instead of being singular as one would expect. This, which is a fact for all zeros on the critical line, fails immediately outside of it, though the deviation increases gradually as one moves away from the line. His basic idea is to substitute the zeta function in the functional equations by more elementary functions, but it is also the sort of thing that Riemann himself might have taken into account in looking for a critical line in the function, and furthermore it makes one think that along with the analytic aspect of the function there exists an absolutely simple and unitary aspect; and the most basic link between the analytic realm and the irreducibly simple is the constant differential calculus based on the unit interval.

If the Riemann hypothesis makes sense at all, the most essential point about the zeros on the critical line of the function is still unknown; but the most essential, be it the most ambiguous or not, is always that which most directly links the divisible with the indivisible. There should be something much more basic than anything so far considered, but it cannot be recognized due to the order of conception created by the very history of the development of calculus as applied mathematics. Riemann's *sui generis* calculus would be a cut through various strata that have constituted themselves seeking also the shortest path. The best way to go deeper into pure mathematics should be to delve into applied mathematics, for ultimately mathematics is always about Nature, and this was still the most profound conviction of Euler, Gauss or Riemann. But it is easy to see, or at least to feel, that if calculus has been modeled by the shortest paths, *a posteriori* we always lack the background to know for what is a simplification. It is only *natural* that deep down we do not believe that mathematical physics deals with Nature, since this has been trimmed everywhere.

The surmise that there is a vertical causality that we cannot now conceive of is also based on the certainty that both calculus and physics, contrary to the claims, are determined from the top down and from the global to the local. As we have already seen on other occasions, this has been the case since Newton, both in the actual proceeding and in the inverted interpretation at play with the reverse engineering that makes prediction possible. From then on, the inversion of the relation between the global and the local advances by leaps and bounds: in analytic mechanics, in the same number theory, in complex analysis, in the passage from the macroscopic to the microscopic in statistical mechanics and quantum mechanics, in the holographic principle, and so on. This is the general procedure, but since a foundation is given, the claim is that all this is built bottom up.

We see for example that fractional calculus and fractional dynamics show long-range spatial and temporal correlations and non-local dependence on history; however classical calculus, in which this does not happen, is only a particular case. We spoke earlier of Brownian motion in dipoles, and, although the fractional dynamics of this type of motion is generally local at the macroscopic level, it would be of interest to trace the correspondence with the microscopic level when using non-relativistic mechanics entirely based on retarded potentials for transitions between low and high velocity motions. This would not only be different from the study done by Mussardo and LeClair[11] on Brownian motion and the Riemann zeta but would have a distinct motivation, namely, to try to give content to the idea of vertical causality and see if it makes sense at all. The underlying idea is that the global synchronization that has dominated mechanics since Newton prevents us from seeing another type of homogeneity, much more general in Nature, which affects our whole conception of space, time and causality, matter, form and motion.

Martin Gutzwiller noticed[12] long ago the correspondence between the zeta function and the delay of an electron with the resulting phase shift in scattering problems, the zeros corresponding to the values of the momentum where the time delay changes abruptly. The smooth quantum chaos would have here a classical counterpart, but to go from locally smooth processes to globally chaotic ones begs not only for a more appropriate mathematics, as he points out, but also for a different mechanics and interpretation. However one can have all three without innovations, filling in the gaps in the line that goes from Weber to Noskov.

The phase of the Riemann zeta function along the zeros of the critical line has been previously associated with the geometric phase of a quantum potential and with the inverted harmonic oscillator[1]3, in which the frequency of the oscillator is analytically continued to imaginary values; the inverted oscillator allows an almost canonical passage from linear systems to nonlinear and chaotic systems with the possibility of feedback, both at the classical and quantum level. The inverted harmonic oscillator can be hidden in the most unsuspected places in Nature: in the same vertical stance of the human being, which is nothing less than an inverted pendulum that needs continuous adjustments in order not to fall. Yes, quiet standing implies a balancing random walk without translational motion. The balancing feedback of the inverted oscillator can also be associated with torque and torsion and generalized to an ergoentropic reformulation of mechanics[14] and vortices in all kinds of natural environments.

Leibniz believed in infinitesimals but conceived physics in a relational, top-down manner; Newton used fluxions in a sense closer to the modern concept of limit but his physics, also constructed top-down, claims the opposite through a foundation not only axiomatic but also metaphysical; Riemann was closer to Newton in matters of calculus but closer to Leibniz in his ideas about physics; Mathis, who does not even find a use for complex analysis, has finally shown as no one else to what extent calculus inverted the physical geometry of problems, but still believes that physics can be built from local causality.

The contours of this constellation suggest that historically we still live in the first phase of calculus or analysis, that which is dominated by the movement of resolution from top to bottom; and that this phase will only finish when it is reflexively understood that it has always been a descending movement since the beginning. That is to say, knowing that it is not so much an ascending construction as a descent will mark the end of the very descent of calculus with respect to the natural world. And after this, there would still be two stages to cover, one of ascent after the unavoidable inflection, and one of final descent.

The second phase of analysis, after the indispensable awareness of its true situation, will be dominated by the ascent to the global by new methods of abstraction. This can hardly be possible without assuming Mathis' critique, a critique less motivated for the sake of controversy, as many will be tempted to think, than by an undoubtedly positive argument that is also in harmony with the demands of a sound natural philosophy. In parallel there is another motivation for a mathematical abstraction of a new kind, and as we have already suggested, it is to be found in the possibility of a general morphology that starts from very simple notions of physical geometry. The conjunction of both aspects, critical and generative, will make it possible to create, among many other things, new kinds of dimensional analysis and measure and dimension theories.

The third phase of analysis may not even deserve that name anymore, and is simply an entirely different dimension and conception of applied mathematics. This second descent, starting from a new position within the always obscure notion of the global, may perhaps show us if a "Riemann dynamics" makes sense in the physical world, but in any case this would be only a particular case within the much more general subject of downward causality and vertical causality in Nature.

Thus, all known analysis and its indefinite continuation in the same direction does not even reach the end of the first phase of applied mathematics. If such phases are to be conceived as a historical development, then it is impossible to know what periods of time separate them, whether lapses like that between Newton and us, or much longer ones like that from Archimedes to Leibniz; but they can also be seen in a perfectly timeless way, as movements of the spirit that are often already present even within the proof of a simple theorem.

The demon of the endless calculus and the *daimon* of finite analysis could not be more different, but by insisting that they are one and the same, the spirit of circumspection is confused and the evolution of applied mathematics, no matter what its growth, is hopelessly misdirected. *Peras* and *Apeiron* have very little to do with the modern notions of limit and infinitesimal, for if the latter was an idealization since the beginning, the concept of limit was from the start pure rationalization. It is extremely dangerous to conflate both because it leads one to think that there is no more room for anything else, and yet in the midst of their busy commerce there persists an axis inaccessible to arbitrary manipulations.

It is to be hoped that this adamant axis of constant differential calculus will also permit a much more natural reentrance from the finite to the homogeneous which is the background of all comprehension. The future belongs to simplicity, and it is even more so because of the nature of the present complexity. André Weil was asked if he believed that there was any path left to explore in the work of the great creators of 19th century number theory, and his answer was a categorical no. Weil's judgment is undoubtedly competent but typical of a specialist; number theory is less and less isolated by the day, which has both good and bad aspects. At any rate, the development of number theory is a peculiar case within modern mathematics, and the order of conception that dominates the whole science is, as in physics, that of calculus or analysis. And in both analysis and physics we know positively that some valid options have been ignored that even today may give rise to a completely different perspective. There is no history of science less interesting than the one that assumes that the possibilities of the past are exhausted.

The infinity of present mathematics is above all the infinity of complexity; that of future mathematics will be an infinity of simplicity, and even then it will be in a certain sense only an intermediate stage. Detecting the homogeneous outside is a way for the heterogeneous to contrast with the homogeneous inside, which is also undetectable but is the support of discrimination. To level both aspects is tantamount to crossing a threshold that in the present circumstances is not even considered. Today there are two things that systematically block the possibility of syntheses of a higher order: the first is the elementary deficiency of analysis, of a foundational and obstructive character; the second is the absence of an abstractive synthetic imagination for the morphological aspects intrinsic to Nature that are already present at the level of phenomena. But today we already have anticipations enough to see what can respond to these gaps, anticipations that must contrast sharply with the excess of superfluous complexity of analysis and with the excess of simplification that inevitably accompanies it and that erases the natural context from which it emerges.

Needless to say, this outlook has nothing to do with those of the mathematicians about the future of their discipline: its interest lies precisely in the fact that there is nothing inevitable about it, in that it is a latent possibility independent of any chronology. In the end it matters very little whether in two or three generations computers can prove theorems better than humans, what matters is what man can understand and even more what he wants to understand. The future of mathematics is a reflection of the future of the spirit among us, and the spirit is not just the intellect but the union of intellect and will.

At the bottom, math is the only true science; and yet it cannot define its relation to reality, which remains entirely open. It cannot even find a single criterion on how to conduct itself in its own internal affairs, those wich seem to be beyond any possible application. But in order to take possession of herself, and not be at the mercy of external interests, as she is now to an extreme degree in the midst of her fictitious freedom, she should be able to reconsider at all times both the criterion of application and the degree of receptivity she shows to Nature; for the two things always go hand in hand.

In the strange case of the zeta function, long before considering whether the hypothesis can be proved, without doubt it would be desirable to understand the role played by complex numbers in physics and Nature; the relation this may have to time; the meaning of the zeros of the function; and finally, how we should interpret vertical causality, which, it is worth remembering, is but the antithesis of the tacit principle of global synchronization that dominates all physics. These four questions are closely related, and of course a large number of mathematicians will see them as purely philosophical; however we have the advantage that at least three of them can also be posed in much simpler and general cases than the one addressed by Riemann. If these three were understood, it is likely that the fourth would be understood too, provided that the zeros have some meaning. We are just talking about understanding, not about proving anything.

It is constantly said that the Riemann hypothesis is strictly a problem of pure mathematics, but if we agree that calculus is applied mathematics, it is undoubtedly a bicephalous problem that necessarily concerns both spheres. To conceive it only as pure mathematics already miss the mark. This reveals the misplaced Platonism that still underlies current mathematics, even when it is assured that it is just a human construction; and the same can be said of current physics, so often pure mathephysics. What is not understood is the supreme importance of the criterion of application, which decides the relationship between the time and the timeless aspects of mathematics. This means infinitely more than any attempt at demonstration that does not even succeed in posing the question.

Without duly rethinking their own foundations, the sciences are serfs of the accumulated debt that they are unable to cancel and that drives the enormous inertia of their career. Some will not understand that a science like mathematics, which should need little more than a pencil and paper, may have debts issues, but she has them like any other, as of course we are referring to internal debts, to all that she owes to herself. Of course, one can always ignore that issues to keep in the same direction.

We talked about the future of simplicity, but in order to properly appreciate the simple, we must first educate our taste —we must develop an acquired taste that we do not have right now. Certainly, the taste for complexity we have developed enough, and although one may come to think of it as a spoiled taste, it is also a necessary condition for the formation of a much more conscious one that presupposes it, but which we would like to leave behind as much as possible. The Riemann zeta function could be one of those intermediaries between extreme complexity and simplicity that could help in the reeducation of taste, at least among the most recalcitrant complexity addicts, but for that one should first reconsider the fundamentals; and even then one cannot rely too much on such an issue. For one can get simplicity through simplification and through simplicity itself.

That the Riemann hypothesis fully affects the criterion of application of calculus has nothing to do with hypothetical practical repercussions such as those sometimes considered; for example, the fantastic associations that have been made with respect to its impact on cryptography, when no one has specified how it might lead to faster methods of factorization. And yet fantasies of this kind, together with the widespread surveillance on the web and the kind of careers that many researchers lead, not to mention the very fantasy of an imminent demonstration, only tend to increase the opacity, secrecy and mystifications in this minoritarian but nonetheless collective quest.

Consciously and unconsciously, as in everything else, the ceremony of confusion is promoted. The hypothesis would conceal something absolutely basic, but its understanding could only be reached through the most incredibly complicated detours. And it is added for the avoidance of doubt: any naive attempt of elementary proof is doomed in advance. But in all these statements it is already assumed that the only desirable thing is that otherwise almost unattainable demonstration —the rest being only philosophy or semantics. A narrative is constructed to channel the "young talents" so that the unwary and the deluded will look in the same direction without daring to consider any other. And in this way, no matter how much we are told about mystery and grandeur, everything tends to the lowest possible level and nobody questions the criterion of application of mathematics, which is what matters above all else.

Being well assured that there is no proof around the corner and no risk of cryptographic meltdowns should help more free collaboration among mathematicians. And by openly rejecting the bait that a proof is desirable without first understanding the much more basic things that would need to be understood, the quality of that collaboration will increase beyond quantification.

What certainty is that which natural philosophy has always conveyed? That the Principle is not only the starting point, but also the aim of all our inquiries. Not much, but sufficient for those who know how to adhere to it. With modern calculus begins an immense work of reverse engineering of Nature that nevertheless is irremissibly moving us away from her. And in the midst of this dangerous journey, emerges the Riemann zeta function, of which we still do not know whether it represents an even greater diversion or a portentous maneuver of rapprochement with yet highly inadequate methods. There should be simple ways to answer this question, ways that do not require a thousand years of preparation.

Against a reverse engineering that separates us more and more from Nature, we have ventured the existence of a retroprogressive method, that is, of a progress towards simplicity, a progress in the direction of a Principle that has little to do with any mythical origin situated in the past. I would like to think that Riemann's hypothesis points in the same direction, but I do not yet have the slightest indication that would allow me to affirm it. In any case, here too the retrograde method can be applied from ends to means to principles with an incomparably lower time cost and a much greater impact, from the moment we begin to ask our own questions.

My appeal to Nature is anything but rhetorical because she is still the great absent in this equation. It is already significant that physics and mathematics are only in their comfort zone precisely when they can evade the question of their relationship with Nature, and if we want to know why this is so, we only need to interrogate the entire history of calculus in terms of excess and defect, and guess of what it has had too much and what it has lacked; and as a counterpart, find out what this has meant in terms of abstraction and mathematical imagination over more than three centuries.

*

The "Riemann moment" of 1859 stand alone in the history of science, completely apart from all others. Not even its clear precedents in Euler and Gauss shed light on its true nature. Numbers have generation but no causality. Physics is supposed to have causality but this becomes irrelevant at the level of the so-called fundamental laws. The classical methods of calculus, which analytic theory applies to numbers, further obscure this question of causality, besides the relations within the number line. How is it that prime numbers, being so important for arithmetic, have practically no relevance in Nature? This is a question that is never asked and that we should ask ourselves. Here we are referring, of course, to the natural world of directly observable phenomena; as, moreover, this point is in sharp contrast with the multiple echoes of the zeta function in physical systems by well known and unknown reasons.

Mathematicians try to reformulate the Riemann hypothesis in multiple equivalent statements; and on the other hand a great number of physical phenomena can be associated with the zeros of a function. What is not done is to try to reformulate the principles of mechanics in equivalent forms to see what new kind of interpretation it allows of functions of complex variable; for, to begin with, some of these reformulations can dispense with the use of complex numbers in the description of both classical and quantum systems.

We should place more value on the attempts to proof the hypothesis by the most elementary means, for only in this way can we attempt to answer the question just asked. Some of these approaches bring us closer to the essential points than impressive displays of mathematical weaponry. We will cite only three of them, associated with natural numbers instead of complex ones, the Möbius and Mertens functions, random walks and Denjoy's probabilistic interpretation: that of K. Eswaran[15], that of Spencer Brown[16] and that of Henk Diepenmaat[17]. Spencer Brown reformulated the hypothesis a la Legendre in the cumulative terms of the Möbius function relative to its previous values. In this way the function would have a self-regulating negative feedback balance and would not be random at all. Diepenmaat's approach has also common ground with Eswaran's, while including a recursive component like Brown's. Diepenmaat offers an original view of prime numbers as a driving pattern in social development in what he calls "recursive perspectivism" although probably the major difference between society and Nature is that we perceive the former from within while we tend to believe without the slightest justification that the latter is only out there.

Needless to say, the mathematical community disdains these approaches. Rather than attempts of proof, they may be seen as attempts of assimilation, but they are valuable nonetheless. It has rightly been said that if Kepler had had more complete data on the trajectories of the planets he would never have advanced his laws of elliptic planetary motion. Today the specialists on the subject know too much on the one hand, while on the other hand there are too many things they prefer to ignore. In the long run we are likely to learn more from attempts of assimilation, from the desire to incorporate the problem and make it our own, than from any kind of proof whatsoever. Other researchers[18] have also noted the scale invariance in the distribution of primes and its relation to power laws; the presence of the latter being as frequent in social phenomena as in natural ones.

Ultimately, any effort to demonstrate the hypothesis based on complex analysis is an attempt of assimilation because its principles, like those of mechanics itself, are based on convention. Which of course is not to say that they are merely arbitrary. What Riemann was doing in his original paper was already an assimilation, although no one doubts that complex analysis offers another perspective for innumerable problems.

However, the prime counting problem is just the handle to introduce a much broader subject —if there was a way to justify it. As for the prime numbers themselves, already countable without error margin in the classical manner, the German mathematician could expect a fair behavior, just as in the probabilities of tossing a coin infinite times; but this kind of ideal balance, the discrete expression of the random walk, would only be the most basic form of equilibrium on a scale extending to the other numerical fields and which gives the function its immense richness of implications.

In any case, if this function may be considered the expression of a natural philosophy, such a philosophy could only rest on the idea of equilibrium, an idea *truncated* right from the start in the conception and foundations of modern physics. If dynamic equilibrium underlies the idea of inertia, its unfolding as action-reaction is also above the plane of global synchronization with the fiction of the instant and the simultaneity that is also at the heart of classical calculus, and the retarded potential would be an index of that unknown scale yet to be interpreted and assimilated. But on the other hand, the existence of natural finite intervals at the core of calculus suggests that there is a theory of proportion hidden in analysis, that this should be connected with the theory of equilibrium in a homogeneous relational mechanics, and that both aspects would be intimately linked with the "deep structure" of the function.

If the hypothesis is true, the most direct implication is that the prime numbers have no deep structure yet to be known; why then should it be disturbing that different physical systems tend to reproduce the patterns of the function?

Epilogue (June 2022)

For those who take for granted the usual way of calculating the function, that is to say, almost everyone, there is probably no simpler explanation of the critical line and its zeros than the one offered by Bertrand Wong in this brief survey[19] instantly understandable by any layman with some interest. Evidently his argument in terms of optimal equilibrium cannot prove anything, but at least it shows clearly that the idea of proportionality still remains essential even in classical analysis. From this purely mathematical point of view, there is not the slightest mystery in the hypothesis, nor is there room for mystification; any proof is supposed to aspire to this, although it can hardly be achieved with demonstrations of two hundred pages. The mystery, if any, would be in the physical resonances of the function. If the simplest is also the broadest and most universal, then the physical parallels of the function should also be understandable in terms of equilibrium and proportionality, and the problem here would be in recognizing how our physics and calculus have truncated these concepts.

Scot C Nelson discovered in late 2001[20] that the logarithmic spirals of plant growth —sunflowers, daisies, pine cones, etc. — "serve as a simple and naturally efficient prime number sieve". Like everything related to the continuous proportion and its associated number series, this has received barely any attention and seems relegated in advance to the ever-growing section of anecdotal coincidences. And yet this is the first basic connection that has been found between prime numbers and these ubiquitous patterns of phyllotaxis, which should have told us something. In light of Nelson's finding, there appears to be a "central symmetry of prime numbers within three-dimensional objects", and vegetal growth would have natural prime number generating algorithm in its becoming. The same passage from the number line to the unfolding of these patterns on surfaces and in three dimensions should be a thread for the geometric intuition of the fundamental theme of arithmetic. It is not the same to try to link arithmetic with modern abstract "geometry" than to link it with a natural geometry. A mechanical analogy comes to mind: the parts repel each other like magnetic dipoles with a minimization of the energy between them, and as the plant grows the time delay between the formation of new primordia is reduced. One can think of it in ergontropic and information entropy terms as well.

Needless to say, when linking primes with complex variable Riemann always had harmonic analysis in mind. Since the zeta establishes such a deep connection between discrete and continuous aspects, this should not only be contemplated as an industrious abstract elaboration but also in its more basic association with waves. In fact an Australian team recently showed[21] that the simple superposition of optical waves can generate the sequence of prime numbers with no need of dynamical chaos or ad hoc factorization constraints, "encoding them holographically" into the propagating wave-field. What is really surprising is that such simple things have not been noticed before; findings like this show there is still room for elementary discoveries in the area. And after all, the idea that there is a destructive interference or cancellation of the primes that tendentially affects their distribution is inevitable and absolutely natural. Besides, it brings back the suspicion of a certain projective analogy and a certain "projective arithmetic" in a different sense from those considered today under such terms, as well as of a space-time geometrodynamics which need not be four-dimensional, and which would restore de Broglie's notion of phase harmony, with its duality between the energy-momentum and the periodicity of space-time, to the perspective of the retarded potential prior to the relativistic description. Which would allow to connect the so-called "external symmetries" of classical physics and the "internal symmetries" of particles in a much more natural way than by the more popular methods of algebraic deformation and deformation quantization. If de Broglie particles are elementary clocks reciprocating mass and time, the crucial question about the time parameter of the hypothetic Riemann dynamics would find its meaning in a relational mechanics that can be expressed with and without complex numbers. Today it is admitted that the constants emerging in detailed calculations in physics are almost the same as those arising in number theory, but to see the other side of global synchronization one should first go out of its domains. Both the primes and this special function could finally break into our conception of Nature if we give them the proper place in our representation.

References

Bernhard Riemann, *Riemanniana selecta*, Consejo Superior de Investigaciones Científicas, 2000

Mei Xiaochun, *The Inconsistency Problem of the Riemann Zeta Function Equation*, 2019

Mei Xiaochun, *A Standard Method to Prove That the Riemann Zeta Function Equation Has No Non-Trivial Zeros,* 2020

Miles Mathis, *A redefinition of the derivative —Why the calculus works, and why it doesn't,* 2003

J. Neuberger, C. Feiler, H. Maier, W. Sleich, *Newton flow of the Riemann zeta function: Separatrices control the appearance of zeros,* 2014

Notes

(Links available in: Miguel Iradier, *Irony and tragedy in the Riemann hypothesis,* www.hurqualya.net)

1. https://www.researchgate.net/publication/334668577_The_Inconsistency_of_Riemann_Zeta_Function_Equation
2. https://www.researchgate.net/publication/339529915_A_Standard_Method_to_Prove_That_the_Riemann_Zeta_Function_Equation_Has_No_Non-Trivial_Zeros
3. http://milesmathis.com/are.html
4. https://monoskop.org/images/5/53/Sokal_Alan_Bricmont_Jean_Fashionable_Nonsense.pdf
5. https://empslocal.ex.ac.uk/people/staff/mrwatkin/zeta/berry-keating1.pdf

6. https://arxiv.org/pdf/1407.4358.pdf

7.https://www.researchgate.net/publication/357075014_Quantum_theory_based_on_real_numbers_can_be_experimentally_falsified

8. https://www.researchgate.net/journal/New-Journal-of-Physics-1367-2630/publication/266858082_Newton_flow_of_the_Riemann_zeta_function_Separatrices_control_the_appearance_of_zeros/links/6229d1243c53d31ba4b5f726/Newton-flow-of-the-Riemann-zeta-function-Separatrices-control-the-appearance-of-zeros.pdf

9. https://arxiv.org/ftp/arxiv/papers/1804/1804.05204.pdf

10. https://www.researchgate.net/profile/Henrik-Stenlund/publication/325567892_On_Studying_the_Phase_Behavior_of_the_Riemann_Zeta_Function_Along_the_Critical_Line/links/5b1632a7a6fdcc31bbf53cdb/On-Studying-the-Phase-Behavior-of-the-Riemann-Zeta-Function-Along-the-Critical-Line.pdf?origin=publication_detail

11. https://arxiv.org/abs/2101.10336

12. https://pdfs.semanticscholar.org/b81b/8dc760ad55188667aa06365354487-ca5972a.pdf

13. https://arxiv.org/pdf/chao-dyn/9406006.pdf

14. https://www.researchgate.net/profile/Mario-Pinheiro/publication/318710701_A_reformulation_of_mechanics_and_electrodynamics/links/59790a60aca27203ecc632f8/A-reformulation-of-mechanics-and-electrodynamics.pdf?origin=publication_detail

15. https://www.researchgate.net/profile/K-Eswaran

16. https://arxiv.org/pdf/2010.13781v3.pdf

17. https://henkdiepenmaat.nl/wp-content/uploads/2018/07/A-dicey-proof-of-the-RIemann-hypothesis-def.pdf

18. https://mdpi-res.com/d_attachment/computation/computation-03-00528/article_deploy/computation-03-00528-v2.pdf

19. https://www.researchgate.net/profile/Bertrand-Wong-2/publication/333583029_A_SURVEY_OF_THE_RIEMANN_ZETA_FUNCTION_WITH_ITS_APPLICATIONS/links/5cf5f7304585153c3db1a098/A-SURVEY-OF-THE-RIEMANN-ZETA-FUNCTION-WITH-ITS-APPLICATIONS.pdf?origin=publication_detail

20. https://www.freedomsphoenix.com/Uploads/129/Media/A_Fibonacci_Phyllotaxis_Prime_Number_Sieve.pdf

21. https://arxiv.org/pdf/1812.04203.pdf

MORPHOLOGY AND INDIVIDUATION

30 March, 2022

If mathematics, physics, complexity theories and computer science continued to expand and cross-fertilize at the same rate as today for a thousand years, they still would not find the key to morphology; not even in two thousand years either. Perhaps this may give some idea of its value, although we all know that nothing can be found without actively searching for it. That is the point: the mentioned sciences already have their own momentum and inertia that nothing can change, only a creation of a new science from scratch could overcome without obstruction the deficiencies of its predecessors.

Introduction

The morphology of organisms has its origin in Goethe's 1790 study on the metamorphosis of plants, in which he describes in detail the continuity between different forms that we have called homology since Owen. The term "morphology" was coined by Goethe himself, but the notion of homology, sporadically noticed, would go back at least as far as Aristotle. The concept of homology was later subsumed within the theory of evolution, but as Ronald Brady points out, in it it becomes a static similarity of explanatory order, not a dynamic transformation in time self-contained in its description. Morphology is not comparative anatomy.

The situation has not changed to this day. Basically, we have "hard" predictive sciences, such as physics, and explanatory sciences such as cosmology or the theory of evolution. And mediating between them there are other disciplines based on probability, such as thermodynamics or statistical mechanics, used more than anything else as mathematical tools that in no way serve to describe phenomena on the same level they appear. That is to say, between the predictive sciences and the explanatory sciences there is only a connection at an abstract level, but not at the proper descriptive level, which is what we need to make the world intelligible, to give it more and more accessible layers of intelligibility.

Goethe, who was at the antipodes of mathematical talent, did not fail to express some hope that someday his ideas would find their Lagrange; that is, that mathematics would finally be able to stick to the phenomena themselves, besides assisting in the foundation of physical hypotheses and theories motivated by prediction as it has always done. But there has been plenty of great mathematicians since then, and yet the distance between the phenomenal world and the queen of sciences has not ceased to grow. And the alibi has been precisely physics, with which it is assumed that the real world is sufficiently accounted for.

In biology there have been some notable exceptions to this general mismatch, such as the work of D'Arcy Thompson, Waddington or René Thom on the morphogenesis of organisms, who prolonged a certain Aristotelian line of thought using increasingly modern mathematical tools, but their efforts have had no continuity and today are seen as little more than occasional deviations from the mainstream of this science. Even before them, in 1892, Otto Snell had formulated the classical allometric equation, but allometry, which studies the relationships between size, shape, anatomy and physiology, although neatly developed, is still just another analytical tool within the much broader trend of applying statistics and data analysis to biology. Current ontogeny and developmental biology connect directly to the basic questions of morphogenesis but are abducted by the causal and explanatory models of embryology, molecular biology or cell differentiation, without even considering the possibility of more general principles that might mediate between the organic and the physical.

And, of course, there is today a thriving area known as mathematical morphology, but since it is basically concerned with digital image processing, not only is it not a general study of shapes, but it is not even a branch of mathematics, being just another application of computation. Waiting for who knows what that might give it its long-awaited birth certificate, morphology as a science on its own remains in the limbo of its pure possibility. More than two and a half thousand years after the birth of geometry, morphology remains in no man's land.

What does morphology need to be born? Obviously, first of all, the desire for it to exist. Then, to have a clear awareness of the kind of gap that exists among the present sciences and that none of them cares to fill, for this gap is bound to be its natural place. Beyond this, there is no need to speculate too much because we already have an accomplished anticipation of what morphology can be in the exploratory work of Peter Alexander Venis.

Vortex morphology and the Venis transformation sequence

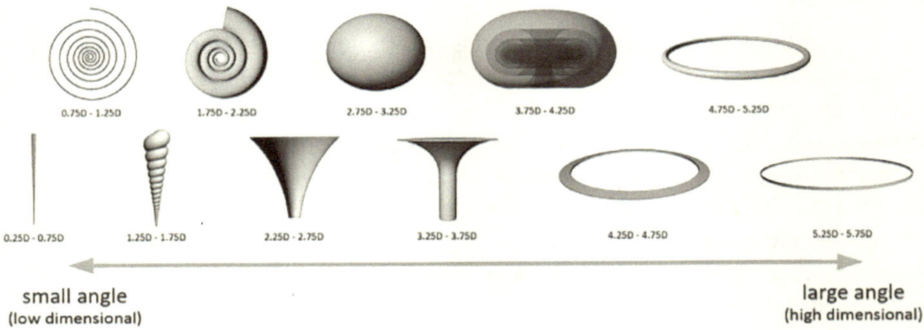

Image by Peter Alexander Venis

The simplest change of shape or transformation that we can observe in a continuous medium is a wave. What we see above is a series of sections in fractional dimensions of the evolution of a wave-vortex according to Venis' reconstruction[1]. Venis describes this evolution in six spatial dimensions, but it should be kept in mind that he does not speak simply of whole dimensions such as length, width or depth but of the continuum of all the "fractional" dimensions, in fact dimensions with any real number. It should also be noted that within this sequence the term "vortex" includes both rotational and non-rotational flow movements.

We cannot even intuit well the evolution of a spherical wave in 3 dimensions, so why try it in 6? Because there is something in six dimensions, or at least in a certain way of understanding them, that is better intuited than in three. But for a physicist or a mathematician the most unusual thing about the new meaning of the word "dimension" introduced here is not the possibility of fractional orders, something that is already often used today, but the fact that they are also linked to material density: dimension 0 and dimension 6 of the sequence would correspond to maximum and minimum extremes of density as well as to changes of size or scale.

Even physicists continue to separate space and matter in spite of the sophisticated field theories they now hold. For example, general relativity assumes, in the terms of differential geometry, that "space tells matter how to move, and matter tells space how to curve"; and this theory is after all very similar to fluid dynamics, with an extra dimension for space-time. It is an attempt to get out of a dualism that has nothing to do with Aristotelian hylemorphism, but with the inheritance of the atomistic vision, which was also Newton's, of corpuscles of matter moving in the void. Quantum mechanics, too, which has so little to do with ancient atomism, has adopted this view in spite of itself and in spite of admitting that the particle is meaningless without its field. Modern physicalism tries to get rid of the dualism of Plato and Descartes by importing it into its conception of physical reality; which is even more paradoxical if one considers that both philosophers, like Aristotle, were monists with respect to physical reality and explicitly denied the separation between space and matter.

In the unfolding in six dimensions of the Venis sequence, the three dimensions that we consider our ordinary space would be only the midpoint of equilibrium between the extremes of the sequence, which in terms of vorticity, would correspond to the most formless point, that in which any polarity or axis of rotation is neutralized. From this morphological perspective, our three familiar dimensions are not just one possible dimension between the point and infinity, but the center between the extremes of our perception. It is very probable that the number of dimensions of the "One Field" is infinite, but the fact that its projection seems to be reduced to 6 would be due to our own perception, which, as we know, always involves an intellectual component in its organization.

One can look for many concomitant reasons for this hexadimensional arrangement, but what is really important is the relocation of spatial experience with respect to oneself. In our geometrical conception of space, which is something so different from our experience, one represents oneself as a point within coordinates ruling for an infinity of different and interchangeable points. Even if one stands at the origin of those coordinates, it is only as a pure abstraction, not as a center in the middle of pairs of opposites, between full and empty, contraction and expansion, dense and subtle, stress and strain, birth and death; and it is these attributes what colors all the time our experience of reality.

We will never appreciate sufficiently the scope of this reinterpretation of space, which once again places man at the center in the midst of conditions by virtue of his own nature and of the larger Nature that envelops him. After all, the famous Copernican turn into which we are still thrown off is just an intelectual outburst that cannot last long, while the alignment of forms with perception and of this one with the formless that gives it its being is inscribed within a timeless return.

It is thus understandable why Goethe did not find "his Lagrange" before, and why he will never find him if we insist on seeing things from the angle of established specialties. Venis' is the inference of a naturalist, not that of a physicist nor that of a mathematician in possession of a great arsenal of analytical methods; and certainly Lagrange himself, the father of analytical mechanics who plugged the great holes in Newton's celestial mechanics, contributed like few others to the divorce between the mathematical and the natural. Goethe made his remote appeal to mathematics within his discourse on color, but all his observations on Nature are morphology as they consciously deal with the boundaries of phenomena.

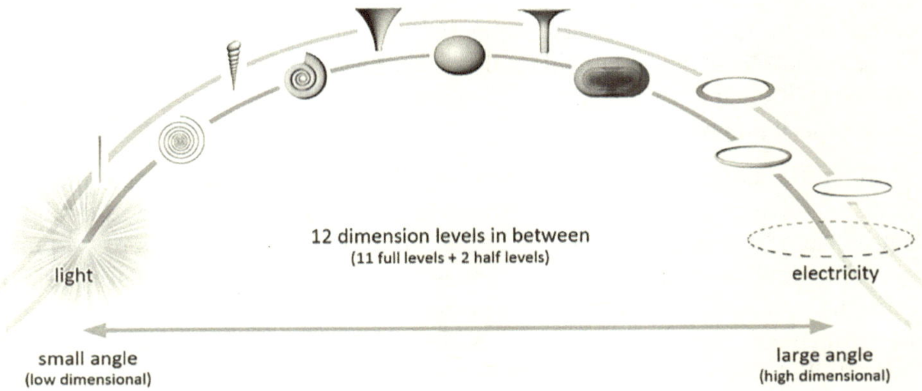

Image by Peter Alexander Venis

Although Venis does not seek a demonstration, his sequence of transformations is more eloquent than a theorem. It is enough to look at it carefully for a while for its evolution to become evident. It is a general key to morphology, regardless of the physical interpretation we want to give it. Venis does not rely on formalisms at all, but he has detected for the first time something that has an irresistible attraction for mathematicians and physicists: a new type of dynamic symmetry. However, this symmetry does not fit in any obvious way into the dynamical variables that physics use to handle.

For Venis the appearance of a vortex on the physical plane is a phenomenon of projection of a wave from a single field where dimensions exist as a compact whole without parts: the "One Field» as he puts it is just another way of speaking of the primitive homogeneous medium as a reference for dynamic equilibrium. It is clear that a completely homogeneous medium can be characterized neither as full nor as empty, and it is equally valid to say that it has an infinite number of dimensions as to say that it has none.

Here, however, the homogeneous and undifferentiated is as much in the background that surrounds the contour of the forms as in their central point of equilibrium. For us, three-dimensional space is synonymous with depth, but here it seems that everything moves on the surface, and it is precisely in three dimensions that the amorphous reaches its maximum. «A vortex is the only part of a wave within the One Field that intersects our physical world»: strange definition for anyone who ignores that here projection is everything. But what is projecting it?

We said that Analysis became too abstract and that biological ontogenesis has never been able to go beyond the particular. Well, the morphological sequence of Venis is at exquisite equidistance between the most basic mathematical physics and biology, between the concreteness of the latter and the abstraction of the former; and this is already something, when one admits the striking inadequacy of mathematics for living processes. Venis' wave-vortex is form and process indissolubly linked, but its natural place is in the projective space free of measurement, which could only descend to the quantitative through affine geometry and metric geometries. The forms we observe in Nature, from mushrooms and spirals to rings, bulbs, horns, chameleons, clouds, galaxies and whirlpools, typically involve a definite portion of their evolution, but we will never find the complete matrix exemplified in a phenomenon before our eyes.

Imagen de Peter Alexander Venis

One has the impression of contemplating an interdimensional ouroboros, a hysteron proteron, a womb that turns itself inside out as it forms, a still distilling itself, that which gives its form to the matrix of all forms. It is true that there is something inevitably subjective in any appreciation of forms, but this matrix seems to be expressing an irreducible reciprocity between the subjective and the objective.

Venis calls his explorations an “Infinity Theory», not so much for the pretension of «having a theory» as to apologize for not finding more justification for it. But the truth is that if we stick to his procedure there is no more need of justification beyond its correspondence with phenomena, for as it emerges it is undoubtedly a phenomenological morphology inseparable from our perception of appearances. The whole subject relies entirely on the primary plane of the aesthetic, and this cannot be seen as a limitation but as proof that we are dealing with something broader than the always limited domain of measurement.

«The human being himself, to the extent he makes sound use of his senses, is the most exact physical apparatus that can exist»; this statement by Goethe can only be understood as another way of saying that the most important thing for the human being is already summed up in an unsurpassable way in what he can perceive by himself. The flood of ultra-sensory data that all kinds of devices provide us with today cannot change this, but they do help us to forget it. When the German poet speaks of the experiment as a mediator between the object and the subject, he is not so much looking for a bridge between the two as for the needle of the balance that helps us to remember their unity. There is in the fidelity of description a Middle Way along which science has barely walked.

In our writings we have spoken of a certain retroprogressive method of return to the Principle and of an infinity as simplicity that would be at the antipode of the present infinity as complexity. The construction of any number of dimensions with orthogonal axes and points seems infinitely simple, but the determination of their possible contents and evolutions becomes infinitely complex. On the contrary, the variety of forms of Nature seems arbitrary to infinity, but starting from its global conception, a simplicity irreducible to the quantitative can be attained.

If one attends to Goethe's experimental method, two stages can be observed: the analytical one that goes from the unlimited complexity of phenomena to simpler ones that in turn seek to converge in their principle, and the synthetic one that proceeds in reverse order seeking the relationship between the principle and the complex phenomena. Venis also proceeds in a similar fashion; but since Leibniz and Newton the reverse engineering of calculus, instead of determining the geometry from the physical considerations, in order to derive from them the differential equation, what does is to first establish the differential equation in order to seek therein the physical answers. These two procedures need not even be equivalent, which authorizes us to think that there must be a physical geometry very different from the one considered today by Analysis, both in the interpretation and in the relation to the Principle; and this is what really matters to us.

Morfhology y causality

Physicists and mathematicians would never have detected the symmetry of the Venis sequence since they move in more defined fields and with other historical lines of motivation, but as soon as they become aware of it they will not cease to make questions trying to circumscribe it. It is something desirable, and on the other hand to be feared insofar as there is the the risk of throwing the baby out with the bathwater. In any case, Venis' great achievement is that since the appearance of his sequence, morphology can now pose questions to physics and mathematics standing on its own ground. Venis himself tells us that before starting his journey of discoveries he was a game programmer working with graphics and algorithms, and no doubt that has left an imprint on his style of exposition, which is as clear and logical as something that has infinity in its background can be. And the truth is that programmers determined to follow this Ariadne's thread of visual logic could often be more helpful in developing and completing this theory than those who want to subdue it with the usual arsenal of quantitative analysis methods.

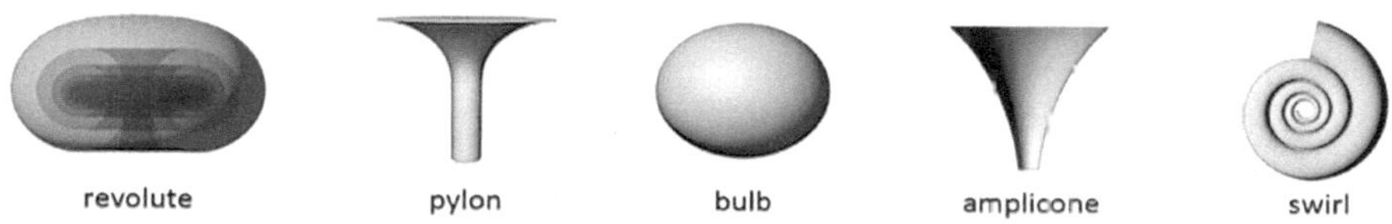

Image by Peter Alexander Venis

In the transformation sequence there are 5 basic types or modes of vortices mirrored, but the interplay of their variants, combinations and metamorphoses has a richness that aspires to be up to the challenge of the inexhaustibility of phenomena. In any case this is not a mere classificatory typology like those so abundant in the life sciences, but it is imbued with a logic that connects the modes causally, and this causality is in a certain sense stronger and more univocal than that found in the fundamental laws of physics, not to mention complex systems in general, as here causality coincides with the direction of the flow.

The notion of physical causality has long been in crisis, and it can even be said that this crisis is one with the relentless expansion of physics since Newton's time. As is well known, the fundamental physical laws, insofar as they depend on variational principles such as the Lagrangian or the Hamiltonian, admit an infinite number of causes to explain them; which is the main reason why physicists have ceased to concern themselves with them at all. Already in the 21st century it was claimed that with the date deluge "correlation supersedes causation"[2]. This is plainly false and does not even need a refutation, since the more the mass of data increases the more necessary are interpretations and models; but this does not matter because it is the very inflation of causal models that causes the devaluation of any idea of causation. If physics, from gravity to electromagnetism to entropy to quantum mechanics to cosmology, was the first to dispel faith in causes, lately it has been the excess of information and complexity.

In physics we use to associate the word «cause» with the well-defined notion of force, but the truth is that in relativity force ceases to be primary and in quantum mechanics as well although for different reasons. And in spite of this, the order of conception of physics still prevails and we cannot help thinking in terms of forces, and specifically of controllable forces, since merely measurable forces are not within the competence of this science. Physics continues to interrogate Nature with and through controllable forces, which thus, if only because of our unavoidable anthropocentric procedure, get in our mind the status of causes.

However, in the new morphology emerging with Venis, other meanings of causality and generation appear. On the one hand there is the apparently univocal causality of flow, but on the other hand there is the generation of the wave in the One Field and its projection as a phenomenon in the form of a vortex; being the appearance a dimensional flowing intersection of the wave. One of the most basic distinctions Venis finds is between central wave/vortex and peripheral wave/vortex; central waves are transverse, like those of electromagnetism, and peripheral waves are longitudinal like sound waves. This distinction is very important from the descriptive point of view although we will see that it still hides much more than it suggests.

But before contemplating the possible connections of this general morphology with physics or mathematics, a difficult and slippery subject that presupposes an adequate knowledge of the tenets of the former and of the descriptive shortcomings of the latter, we may ask how and to what extent the transformations in the Venis sequence are representable —directly or indirectly representable. Representation and description are, for what matters here, practically equivalent; but rather than taking it to the always pertinent levels of philosophical discussion, we prefer to keep these notions in their original context: computer graphic design and our limits in the representation and perception of different dimensions.

Venis, who is well aware that his approach is just the beginning of a long journey, attempts to arrange the basic motions in a series of charts involving the number of dimensions, the apparent shape of the vortex motion, and the motion that might correspond to the wave in the One Field prior to the projection. For motion the most basic distinction would be that of translational and rotational movements; and it is already remarkable, to begin with, that physics has treated such motions almost always separately and has avoided a concrete view of their connection. Let us not forget, for example, that, at the statistical level, the virial theorem eliminates the spin or rotation[3] in its balance of kinetic and potential energy, so that the derivation of a supposed moment of inertia is just one more of the many accounting tricks of physics.

What would we see in a graphical animation of the transformation sequence in a virtual reality application? It would obviously depend on the program and the concepts we introduce, since Venis does not intend to settle the matter and is more concerned with appearance than with its possible recreation; while, on the other hand, the requirements of physics at the descriptive level have always been secondary. Classical mechanics, for example, routinely ignores Heinrich Hertz's distinction between a material particle, which is an irreducible physical point in space at a given moment, and a material point as it appears to our senses and which as a volume can contain any number of material particles —a planet, a star or a galaxy. Nicolae Mazilu reminds us that this purely classical distinction also partially coincides with the wave-particle duality of quantum mechanics.

Reducing a body or a volume to a point, either in quantum mechanics or in classical mechanics, is a way of omitting three dimensions, not in the ideal empty space but in the ordinary space-matter continuum. If in quantum mechanics the description of extended particles is not possible, it is, first of all, because special relativity does not admit them, in spite of being a macroscopic theory. Maxwell's equations are only useful for portions of the field with extension; special relativity is only valid for point events, and in the field equations of general relativity point particles are again meaningless. However, in a Weber-type relational mechanics, which can also be extended to fields by integrating over the volume, one can work with both point and extended particles, which allows it to cross the gulf between the micro and macroscopic without the need of curvatures or pretending to convert time into another dimension, among many other contortions.

The electromagnetic transverse waves of Maxwell or Hertz do not indicate that the electric and magnetic components are perpendicular in the literal geometrical sense but in the sense of a statistical average between space and matter, and this is why such waves have resisted all attempts of sound geometrical representation. However, it is enough to understand that electricity implies charged matter separated by space, and magnetism space separated by matter, for the irreducibly statistical character of the geometrical attributes of the wave to become self-evident.

Weber's relational mechanics, an inexcusable reference for the latter Maxwell's electrodynamics, raises since 1848 the problem of the retarded potential and its justification of the conservation of energy. Following Weber's logic, Nikolay Noskov[4] postulated a longitudinal vibration internal to bodies in motion to cover the difference between the kinetic and potential energy of the retarded potential. Since Weber's mechanics can contain terms proportional to the square of velocity, one cannot distinguish always between these energies and the internal energy of the body —and one would have to dwell on the link between this indistinction and the quantum indeterminacy relations. This internal wave coincides with de Broglie's famous «matter waves»; the Schrödinger wave equation itself is a mixture of different equations describing waves in a medium and waves inside the moving body, but as we know this ceases to make sense in the relativistic Dirac equation.

However much the differential calculus has accustomed us to believe it, a point has no physical reality, but is a purely mathematical concept, or as Leibniz put it, a simple modality. Going from a point to a sphere with volume adds three more dimensions to the physical characterization of a particle, just as allowing freedom of rotation of an oriented point with inertia we have six physical dimensions instead of three. The rotation of a point in all three dimensions allows us to generate vortices and volumes along its translational motion. If we wanted to visualize the internal wave of matter in an extended particle, it would not be something as simplistic as a corpuscle traversed by a longitudinal vibration, but a configurational process in six dimensions that would include the spin in its wave-particle field.

Returning to the graphic representation of the transformation sequence, if I surmise here that, starting from a poin vortex, there must be a continuous movement of folding and unfolding, of mutual immersion and eversion between the three dimensions of translation and the three of rotation, it is not for any sound evidence of it or because Venis puts it in those terms, but simply because I suppose that an ideal primordial wave that emerges from the undifferentiated whole to return to it must exhaustively cover all the combinations, all the possibilities of differentiation. But of course the multiple intersections of a single wave allow many other variants. At any rate, the simplest way to describe that combination of translation and rotation, immersion and eversion is with the concept of torsion. A morphology of vortices is by necessity a morphology of torsion.

We propose the possibility of virtual representation of the sequence not as a mere computer application or an interactive exploration but trying to find its own place within our imagination. The starting point is the idea that each organism recreates the whole as much as it can; this double movement outside in and inside out is reminiscent of the afferent and deferent processes of the nervous system, and therefore they should be deeply related with the specialization of the cerebral hemispheres. If, in addition, the two aspects of torsion can be conveniently connected with our own interacting cycles of action and perception, our ability to conceive morphogenesis and morphodynamics would be greatly enhanced.

Venis himself assumes that the wave-vortex must have a close relationship with the Hopf fibration, a well known mathematical object relevant for the two-dimensional harmonic oscillator, fluid dynamics, magnetohydrodynamics, quantum entanglement, the Dirac monopole, the geometric phase of a potential and many other aspects of physics. Its analysis requires complex variable, but before approaching the connection with physical processes, it would be necessary a careful study of the mathematical correspondences starting with the calculus itself. It goes without saying that here we must content ourselves with just some remarks.

For example, passing from projective or synthetic geometry to a synthetic differential geometry that includes motions, the descriptive as well as analytical deficit of ordinary calculus cannot be ignored. Without good analysis there is no good synthesis, and standard analysis does not even correctly analyze the physical geometry of problems, as evidenced by the fact that a mobile accelerating in a straight line is represented by a curve. Miles Mathis has shown conclusively[5] that classical differential calculus, no matter whether it is based on infinitesimals or limits, always has at least one dimension less than the physical problem, and this cannot be overlooked when we seek a reasonably complete representation. Calculus had a humble origin in the study of curves but with the algebraization of analysis the primary relationship between calculus and geometry was buried forever. Mathis proposes a constant differential calculus based on a unit interval to put an end once and for all to instantaneous velocities and fictitious points. Otherwise, a synthetic differential geometry starting from points may be discretionary but in any case unspecific.

Can a virtual graphical representation help us to visualize higher dimensions or complex hypersurfaces? This question has a catch and the catch is the essential point. Since any number of dimensions is «just» representation in motion, but on the other hand what we want is to rearrange that representation and bring it closer to our consciousness. On the one hand, it would be perfectly illusory to think that morphology can illuminate that secret chamber of Nature of which Kant spoke and which would always elude our senses, since nothing that is based on concepts and representations can ever do so. But the search for an accurate description brings us closer to the balance between explanation and prediction, arithmetic and geometry, participation and distance, analysis and imagination; since, moreover, imagination already springs directly from that same chamber which is also in each one of us. The balance zone between these two extremes is often crossed by a great deal of data traffic, but it is also potentially a zone of silence.

As for hypersurfaces in the complex plane, although this one does not need any justification in mathematics, in physics it should, and in a morphodynamics of a descriptive nature they are completely necessary; and it is obvious that the most direct generator of complex numbers and non-commutative properties is the rotation of vectors. This would bring us back to the dynamics of the oriented point in 6 dimensions. The use of complex numbers in physics is nothing but a way to have more degrees of freedom and extra dimensions without the need to justify them. What we don't use to ask is wether we are able to conceive balance in terms of dimensions, and yet this is at the heart of the transformation sequence.

If we think of morphology in terms of projection and intersection, two opposing but almost equivalent ideas emerge: the absence of causality, and vertical causality. The term «vertical causality» has been proposed by Wolfgang Smith in his perennialist, neo-Aristotelian interpretation of quantum mechanics. Although it may surprise many, reading quantum perplexities in an Aristotelian key is nothing new, and already Heisenberg wrote that the quantum domain, «between being and non-being», corresponds quite well with what has been understood since Aristotle under the concept of potentiality.

*

Smith simplify things and affirms that the world that physics and quantum mechanics in particular are concerned with is not the world of bodies, but that of matter under the sign of quantity —physics is concerned with the measurable, which would be a completely different ontological domain than the one we perceive with our senses. Returning to the traditional body-soul-spirit triad, Smith concludes that the corporeal is limited in time and space, the animic only by time, and the spirit by neither of them; the three would be like the boundary of a sphere, its interior and its center.

Smith's argument may be acceptable for quantum mechanics, since no physicist claims it is on the same plane of reality than the world of our perceptions, but for classical mechanics and cosmology does not hold water. One cannot simply appeal to «substantial forms» while ignoring the formative capacity of forces, motions and potentials. Few physicists doubt the real character of hydrodynamics or astronomical processes, even if their causal status becomes more problematic as we move away further in space and time.

In any case standard quantum mechanics, that of the Copenhagen interpretation and point particles without extension, is clearly incompatible with all our notions of macroscopic reality, and therein lies the whole force of Smith's argument, which does not need to create alternative realistic or causal descriptions, such as that of de Broglie-Bohm. He does not need it because there is simply no such thing as a «quantum reality,» but only a potentiality that we actualize from our macroscopic corporeal world composed of matter and form. After all, and with a very elementary argument, Smith would like to bring us back to realism.

But there is another totally different way of contemplating vertical causality in Nature, which does not involve separating matter and form beforehand, as Smith does, in order to reunite them later through a mere concept. In addition, physicists have never said that hylemorphism is wrong, but rather that it is of no use for modern physics. Here we have repeatedly observed that, not only quantum potentials, but physical potentials in general, being independent of any velocity of transmission, can only be pure act, while on the contrary the interactions with which dynamics is concerned, being limited by velocity, always require time. Physics, understood as dynamics, subordinates potential to force, but it cannot be that what is instantaneous must be subordinated to what needs time to react.

It is legitimate to consider the existence of a vertical causality over and above the efficient or horizontal causality of dynamic interactions, since potentials are not reducible to forces and involve an implicate order of a different kind. Bohm's famous implicate order is the resurrection of Leibniz's pre-established harmony in the wake of de Broglie; the great difference is that the theory of the retarded potential allows us to see that this also holds at the macroscopic scale. But, since a potential cannot be retarded because it has no transmission velocity, what is not simultaneous is the action and reaction between forces, blowing up the global synchronizer on which all modern physics is tacitly built. The implications of this are hard to conceive.

We have already seen in other places that Weber's relational mechanics allows a reformulation of the three principles of classical mechanics that do not pass through the law of inertia, nor through the constancy of the central forces, nor through the simultaneity of action and reaction or global synchronization. In this sense, relational mechanics, which has its precedent in the ideas of Leibniz, would be simply an acausal phenomenalism. Thus it is not necessary to understand vertical causality literally since the idea that there is a horizontal causality that extends virtually without limits would be simply another illusion. There is neither horizontal nor vertical with respect to a homogeneous and undifferentiated primitive medium, the only true infinite; the contrast only arises when we assume that there can be a global synchronization in dynamics, which besides being metaphysical involves a contradiction in terms. Vertical causality is an implicate order with its own temporal lines and layers.

The idea that the phase shift of the "retarded potential" cannot be such and actually implies a proper time would be indirectly supported by the cosmological implications of Venis' own morphology. Venis speaks of time branches within a hierarchy of scales in which only larger scale vortices can affect the time branch. Even anomalies such as the deceleration of deep space probes would be due to a transition of scales. For Venis the reality of these temporal branches could only be verified by traveling great distances, leaving the sphere of influence of a star or a galaxy; but following the thread of relational mechanics it can be interpreted that each phase shift of the potential entails a certain deviation. The difference between the phase shift of the «retarded potential» and that of the so-called geometric phase is that the former is already included in our equations, while the latter is not and has to be considered as an additional curvature.

But, of course, the geometric phase is not unique to quantum mechanics, as physicists themselves often claim, and exists at all scales, and can be clearly verified in animal locomotion or in simple experiments with vorticity on the surface of water. Interference affects the flow and the simplest way to conceive the geometric phase is as an interference phenomenon. But, being a universal phenomenon, depending on the case and description we can interpret the geometric phase in countless ways: as a parallel transport, as a self-induction, as a curvature or flow of the symplectic form, as a conical intersection between potential energy surfaces, as a transition between dimensions, as a transition of scales, as a torsion or a change in density, as a phase transition, as a degeneracy point, as a retarded potential, as the difference in the Lagrangian, as a resonance, as a holographic interference, as a loop, as a principle of enslavement, as a hole or singularity of the topology of motion, as a conversion between spin and orbital angular momentum, as a proper time or timeline, as a memory, as an interface, or even in other ways that need not be exclusive.

Only the degree of fidelity in the description will allow us to decide which of these interpretations best fit the case; but several of them fit perfectly in the logic of Venis' morphology. It may well be said that, if variational principles allow for an infinite number of causes, the geometric phase gives us infinite ways of seeing the absence of causality. And this absence is the only light we can shed on the so-called vertical causality; but in any case it is something that is open to both mathematical description and experimental verification.

In the mechanics of contact and controllable forces we associate force with deformation; but in the so-called «fundamental forces», as in gravity, there is no deformation when motion occurs, as in free fall, but only in the potential situation, as in solid bodies on the surface of the globe. If we attend to the form, it is as if there were a third state of rest; but mechanics based on inertia can draw no lesson from this striking circumstance. In relational mechanics even the shapes of ellipses, whether in planetary or atomic orbits, depend crucially on the so-called retarded potential. Potentials have a definite effect on shape; we can imagine them as the landscapes and valleys through which matter travel.

Continuing with the idea of projection, we have seen before the supreme irony that the holographic principle, arising from the most extreme conditions that theoretical physics can conceive, tells us, after having raised the number of dimensions for a theory of gravity to four and for string theories to ten or even twenty-six, that ultimately any physical evolution with all its information is reducible, no longer to a volume, but to a surface; when it was assumed that processes such as gravity or Huygens' principle of light propagation cannot operate in two dimensions. However, the greatest irony is that the holographic principle is based on the geometric phase, a supplement to quantum mechanics that does not even properly belong to it: a loop or curvature added to the unitary, closed evolution of the Hamiltonian. The geometric phase does not belong to Hilbert's projective space, but reflects the *PHYSICAL GEOMETRY OF THE ENVIRONMENT* that is not included in the definition of a closed system. And this openness of a closed system to its environment is what is now supposed to define the limits of our experience of the world.

If the geometric phase is equivalent to a torsion, it is an entirely trivial truth that the torsion of a surface is sufficient to describe any conceivable form. Since the holographic principle is universal, to speak of the universality of the morphology of torsion would be equally justified; the question is how many degrees of simplification or synthesis this general principle admits, which ultimately only tells us that we know nothing except with light and in light. There has been a great deal of speculation as to how could be possible to check the holographic principle experimentally, but if it really is universal and we have it everywhere, all that is needed is to pull the slipknot that should exist between the «retarded potential» and the geometric phase; for what this lag tells us is that there is a feedback loop that depends on the environment. In fact, if the retarded potential is applied to gravity, the black holes that brought the holographic principle to the foreground become impossible to begin with because as velocity increases the force decreases proportionally.

*

It is almost an assumption of Venis' theory that all waves would have their origin in a single process, changing only the circumstances of projection or intersection. Here we share this view, the truth of which would have wider consequences than any treatise can summarize. Even today there is a general consensus among physicists that the longitudinal waves of sound and the transverse waves of electromagnetism and light are completely different processes with no possible connection. However, in October 2021 Shubo Wang's[6] team in Hong Kong demonstrated the existence of transverse sound waves created with a metamaterial that couples the spin of the wave and its momentum.

This discovery probably marks the beginning of the end of the preconceived ideas we still have about waves, and soon will find its due counterpart with the verification of longitudinal electromagnetic waves. In fact, a few months earlier a St. Petersburg team[7] had arranged another metamaterial to easily produce longitudinal electromagnetic waves. Longitudinal electromagnetic waves are not entirely unknown, as was the case with transverse acoustic waves, but their theoretical status is still debated. It should always have been obvious their link to Maxwell's equations since these are only a special case of Euler's fluid equations as well as Weber's —even if Weber never predicted electromagnetic waves of any kind. There are of course Noskov's longitudinal waves associated with the retarded potential, which the Russian author always linked directly to sound and the wave mechanics of gases and fluids; Shubo's team considers the scattering field of the waves as a spherical projection of the incident field in which it induces non-commutative rotations and «leads to geometric phases that account for the spin to orbital angular momentum conversion".

We have mentioned these recent discoveries because they open a new way to contemplate the unity of Nature, a way that does not go through the «unification» of the known fundamental forces but through the exploration and reconstruction of the unity of form when trying to represent the physical geometry of processes. Of course, this physical geometry can and must include statistical aspects. This search of unity has nothing to do with the umbrella project of unification of the fundamental forces, for to approach the characterization of this primordial wave is to approach an acausal perspective and the vertical causality implied by the projection. And since acoustic holography offers a complete analogy with optical holography, and the geometric phase allows controlled wavefront shaping, it should be much easier to study the critical connection between "normal" retarded potentials and "anomalous" geometric phases, that in addition provides us with the key link between wave mechanics, dynamics and morphology.

From the relational point of view there are no causes in the mechanical sense but relations, and the co-implication of these relations has a depth still entirely to be explored. A new synthetic perspective opens up, but the physical geometry of which we speak is obviously more than kinematics; this would be the major limitation of a relational mechanics such as that of André Assis, who on the other hand raises an issue of great depth by replacing the principle of inertia by the zero-sum dynamic equilibrium of forces.

With an approach related to the Duhem's attempt to join analytical mechanics and thermodynamics, although based on an argument suggested by Landau & Lifshitz, Mario Pinheiro[8] has proposed an ergontropic reformulation of mechanics, defined by a balance between minimum energy variation and maximum entropy production. Since these are the two most fundamental tendencies we can observe in Nature, it is just reasonable to treat them together integrating consistently rotational vortical motion and translational motion.

Entropy, conceived by Clausius as a tendency to a maximum, already entailed a spontaneous finality; what physicists have never answered is what is the reason behind the blatantly teleological component of action principles governing physics. With a formulation of equilibrium between both, this second part seems more understandable. Boltzmann created a formidable confusion by equating entropy with the subjective concept of disorder, but, as R. Swenson already said more than thirty years ago, «the world is in the order production business, including the production of living beings and their capacity for perception and action, because order produces entropy faster than disorder.»

If entropy is not a by-product of mechanics, but is within the same equilibrium conditions of the latter, the meaning of physical processes changes completely. An internal equilibrium formulation such as Pinheiro's assumes that there is always free thermodynamic energy in the environment, an «external» disequilibrium; by the same token, mechanics itself would emerge from an open and irreversible background. In the ergontropic framework, systems are subject to external forces but react to external constraints with vortical rotational motion and dissipation, with a direct conversion of angular motion into linear motion via topological torsion. This is therefore of interest both at causal and morphological level.

Pinheiro revisit Newton's rotating bucket experiment and gives us a rendition that is neither the absolute space interpretation of the author of the Principia nor the relational one along the lines of Leibniz or Mach, who attributes the formation of the vortex to the frame of distant stars. Since we believe that the principle of inertia is entirely dispensable, we have always leaned toward the relational interpretation; however Pinheiro's interpretation adds a necessary element, since in a case like this we cannot simply dispense with local causality. *«What does matter is the transport of angular momentum (imposing a balance between the centrifugal force, pushing the element of fluid to outside) counterbalanced by the fluid pressure."*

The transfer of matter to regions of higher pressure is already an indication of the second law of thermodynamics, and the fact that it takes time for water to acquire its concavity speaks eloquently of the presence of friction. But, on the other hand, the vorticity within the water cube also reminds the hypothetical case of an extended particle, with intrinsic rotation while being traversed by a wave according with the "retarded potential". What we now call local causality can never be understood more geometrico, but, like the same electromagnetic waves and so many other physical processes, somewhere halfway between geometry and statistics; local causality would be the non-deterministic reaction to global causality, which is another name for vertical causality or unity. Even today, an experiment as simple as that of Newton's cube allows us to make a different reading of the connection between central forces and non-central or peripheral potentials.

We have frequently spoken of three fundamental equilibria in physics and in electrodynamics in particular: the zero-sum dynamical equilibrium, the ergontropic equilibrium between minimum energy variation and maximum entropy, and the equilibrium of densities with a unit product. Such equilibria are rarely contemplated in current frameworks and when one of them is eventually considered it is not properly related to the other two. These three such basic equilibria should be aligned with the morphodynamic balance in the midst of the Venis sequence, which in a sense is reminiscent of the so-called «potential flow» of fluid dynamics. Naturally, this central equilibrium has ramifications in a wide variety of different circumstances.

We said that Venis mentions as a possible mathematical approach to his morphology the Hopf fibration with use of complex variable. In a morphology of flow and torsion that keeps the emphasis on description the relation between the use of complex numbers and rotation, of these two elements with time and the retarded potential, and of all of them with timeless projection and vertical causality should be investigated. Of course, any idea of causality is part of a complexion of concepts, but once one begins to delve into morphology as knowledge of figure and configuration, this will affect our ideas about space, time and causality; matter, form and motion.

Individuation and Singularity

In Venis' hexadimensional wave-vortex the most concrete and the most abstract merges. Concrete, because it is possible to follow the evolution of its form in a way that is both intuitive and logical, and because it is based entirely on the simplest notion of flow. Abstract, one should perhaps say in quotation marks, not because it would involve a hypersurface in six non-integer dimensions, nor because there always seems to be something in its movement that eludes us, but because as a whole it does not correspond to any static «thing» or object, but only to processes. Indeed, nothing could better illustrate something usually so unrepresentable as the philosophy of process and becoming, from Heraclitus to Whitehead or Simondon.

It is surely no coincidence that it was Whitehead the first to formulate a point-free geometry, which was later followed, for different motivations, by point-free topologies. Moving away from set theory, the mathematical theory of categories finally fulfilled Aristotle's program of making the categories of concepts explicit. Synthetic differential geometry was developed, and, in general, the modern mathematical concept of topos coincides surprisingly well with that defined by the Greek philosopher: *the boundary of the enveloping body according to where it touches what it envelops*. René Thom also moved closer and closer to Aristotelian concepts in his «sketch of Semiophysics»; these are just examples of how the attempt to think forms and the organic, our most immediate tangible reality, has generated a movement of mathematical abstraction at the most fundamental level —of new concepts that nevertheless connect in depth with the first thinker of the organism. The now ubiquitous language of homology and cohomology has great affinity with morphology as it is natural to treat a vortex as a topological defect, but the extreme algebraic abstraction of the specialty has succeeded in separating it from its intrinsic connection with directly observable formative processes.

Our notion of projection ideally starts from a point, but we have already seen that the idea of point in differential calculus is dissociated from physical geometry and that such dissociation has a clear solution. Synthetic differential geometry emerged as an attempt by Lawvere to give an axiomatic basis to continuum mechanics but it leaves intact the dimensional analysis of classical calculus and cannot access the physical geometry that morphology demands.

The notion of the point particle, which as we have already seen is prescribed by the constraints of special relativity, is one example among many of physical pseudoindividuation of an entity. The Copenhagen interpretation of quantum mechanics believes in the identity of individual particles, but the very characterization of its aspects inevitably passes through statistical criteria. Thus we see that an irreducible mathematical concept, such as the point, has a deficient physical and differential application that is also allied with the congenital nominalism of all modern science. From a configurational point of view, focused on the physical geometry, there should be no contradiction between geometric and statistical characterization, as rather both imply each other.

To put it more directly, for us to be an individual is to be singular, but one is not irreducibly singular without the contribution of the idea of point, which implies a very long and abstract evolution of concepts. The same is true in our cosmology, with the idea of initial and final singularities, not to mention black holes. These singularities only seem possible to us not only because of our notions of point in space and time but also due to taking the principle of inertia and central forces to their ultimate consequences, and we have already seen that in a relational mechanics there are no such evolutions. Physicists who speculate about black holes speak of irretrievable information, but what is not realized is that the same principle of inertia makes irretrievable aspects of physical geometry that contain information of a morphological order.

The last great theorist of individuation was probably Gilbert Simondon. Simondon made an exhaustive critique of the Aristotelian principle of individuation and in general terms his contribution is of great value, although one may sometimes doubt if the Aristotle he criticizes is really the Greek philosopher or rather his scholastic derivations. In any case we will not discuss this since what matters to us now is the alternative he offers to this principle with his idea of the process of individuation and the ultimate limit of its contour in this our age of extreme nominalism.

Simondon, not by chance either, is one of the thinkers who most clearly aspires to situate himself in a middle ground between physics and biology; at least that is where he wants to situate physical individuation, since he clearly sees psychic and social individuation as processes with great similarities but of a different order. For the Frenchman, the individual cannot be thought of without equally contemplating the pre-individual and the transindividual. The individual would emerge as a change of phase in the potential; this formula is a way of passing from physical causality to the cause of individuation, that is, to its very process. Simondon does not directly relate this gap to the geometric phase shift, not least because in 1958 even physicists did not use the term. Some have questioned whether his application of physical concepts is legitimate, but what is certain is that the philosopher proves to be more aware of the importance of this phase shift, not only than the physicists of his time, but also than those of present times.

It is necessary to insist, given the deep-rootedness of our preconceived ideas: that the geometric phase closes the contour of any form is, in principle, as trivial a fact as saying that every form must be circumscribed by its environment. But this would only be trivial if the physics with which the system has been defined had not been constructed from the start obviating the influences of the background environment. That is why phenomena such as the so-called Aharonov-Bohm effect, which is by no means exclusive to quantum mechanics, created such perplexity among physicists. In a relational mechanics like that articulated by Assis reformulating the three principles, neither the forces are constant because they always depend on the environment, synchronization depends on the phase of the potential, and the principle of inertia is replaced by dynamic equilibrium. The principle of inertia is contradictory in itself, since it asks us to consider «a closed system that is not closed». And yet this contradiction is at the heart of the dynamism of modern physics, of its «creativity» for better or for worse. But within the inertial context of modern physics, in which also quantum mechanics remains embedded, it is impossible to appreciate the temporal differentiation of each entity. The principle of inertia demands that everything is out there: it is the basis of the construction of the Law as pure exteriority. But in itself there is nothing necessary about it, it only transcribes the phenomena to the most external interpretation possible. According to this principle, nothing can have a proper time, and therefore its form must be at any rate accidental.

If this is well understood, it is much easier to understand Simondon's various approaches to the concept of individuation, both physically and morphologically. For example: «The individual is the partial and relative resolution that manifests itself in a system containing potentials and involving certain incompatibility in relation to itself, an incompatibility composed of forces in tension as much as by the impossibility of an interaction between extreme terms of the dimensions». Or: «Individuality is a resolution of an initial incompatibility rich in potentials». Or: «one could say that the only principle by which one can be guided is that of the conservation of being through becoming». Or his vision of the individual as communication between different orders of magnitude, macro and microscopic; all these verbal formulas are perfectly in tune with the descriptions of the transformation process in Venis.

Venis, always cautious in his judgments, is the first to provide indications that the transformation sequence may not be complete. He also rightly points out that if the vortical morphology does not translate easily to the domain of organic matter, it is due to the presence of several simultaneous equilibrium states. The coexistence of these equilibria and their compromise could be, perhaps, the most important key to seemingly irreducible biological complexity —and also to a general theory of development and aging.

Today biomathematics is developing exponentially but, like so many areas, not because of theoretical advances but because of the brute force of computation and massive data mining. Massive biological data analysis can in no way be confused with theoretical biology, for the latter must aspire to understand the problem already posed by Aristotle, namely, how life is possible as the existence of autonomous agents. The theory of evolution does not deal with this, but rather takes life for granted. Neither does molecular evolution, which revolves around self-replication of DNA or RNA: for not only short peptides, but even vortices of inorganic matter can replicate[9] under certain conditions of stratified flow. These vortices are so robust that they can survive indefinitely even under subsequent turbulent conditions.

Even helical amino acids have a degree of vorticity, can be subjected to torsion and exhibit defined geometric phases: that is, they have a critical sensitivity to their environment which is the only thing that can explain that the same enzymes can create different proteins depending on the environment. But we have already seen that Pinheiro's ergontropic vortex or systems with retarded potential have a capacity for self-adjustment or feedback that is not contemplated in either classical or quantum mechanics. This is prior and much more basic than self-replication. The idea that matter is blind is untenable: matter sees through form, so to speak. And in a sense, it sees much more than we do.

Matter itself would be only a configuration. If today it is a cliché to contrast physicalist reductionism, which conceives observable properties as resulting from the interaction of particles of matter, with emergentism, which, descending in a direct line from Aristotle, affirms the appearance of irreducible configurations and that the whole is always more than the sum of its parts, it is clear that morphology will always be on the latter side. And if we distinguish between a «weak emergence», in which the higher level cannot modify the lower one, and a «strong emergence», in which descending causality is possible, it is not hard to support this second thesis from the point of view of a vortex morphology.

The so-called downward causality, still in the midst of controversy, does not necessarily have to be synonymous with vertical causality, although the two can easily overlap. Perhaps the main difference is that vertical causality is at the moment a rather negative and too open-ended term, whereas when we speak of downward causality we want to pay more attention to the details and positive aspects of mediation between different physical strata. If what we understand by vertical causality is rather the implicate order and the intrinsic synchronicity of the potential at all scales, *one and the other can coincide without the slightest difficulty since in any case the downward causation also takes place through the mediation of the environment*, which it is not to say that there is only one means of causation, but rather the opposite. As we said, if the variational principles of dynamics admit an infinite number of causes, the geometric phases admit infinite ways of interpreting the absence of causation, that is, of interaction; everything depends on whether we attend to the controllable force or to the potential. That quantum mechanics is not complete is conclusively demonstrated by the geometric phase; what we do not know is how far the loop of its feedback can be extended.

Complexity researchers such as Stuart Kauffman or Danko Nikolić stress that downward causation cannot be consistently defended without a deep respect for the details and circumstances of the «system» or organism in question. Recourse to downward causation might allow substantial, if not decisive, advances in specific areas and problems, whether in chemistry, biology, psychology, neuroscience or sociology, and indeed the concept was first proposed by Donald T. Campbell, a social scientist, in the context of hierarchically organized biological systems —although already Kant postulated that organisms are cause and effect of themselves. But there are also well-known and robust physical processes, such as Bénard convection or the self-replicating vortices mentioned above, that would exemplify it in the most obvious way. That a vortex can survive turbulence is no less worth of respect, a respect that also demands attention to the details of how this is possible.

*

In dealing with the principle of individuation, Leibniz makes another great concession to nominalism and settles the problem by affirming that in reality only individuals exist. Long before postulating the existence of monads or simple substances, Leibniz also contemplates the physical and mathematical aspects related to the surfaces of individuation of bodies in the context of a fluid plenum to reach the conclusion that forms lack real entity and always have an imaginary component inextricably connected with our perception. The position underlying Venis's morphology essentially coincides with this, even though he tries to avoid metaphysical statements. However, Leibniz's position does not deny the objective component either; as we have seen, in relational mechanics there is always feedback with the environment, so that one can rightfully interpret action-reaction cycles as action-perception cycles. Each physical potential is not merely a position, but an authentic perspective: the circumstance of the environment from the agent's perspective.

The individual implies the pre-individual in the same way that a form presupposes the environment or the phase shift a previous potential. This surplus on course in the individual seeks to transcend itself both on the horizontal, or transindividual, plane and on the vertical plane, which has been so often called transcendent; and yet both would be projections on the self, the homogeneous and undifferentiated medium. Aristotle appealed to an entelechy to explain the autonomy of the organism; Leibniz and Goethe stick to the same idea. Simondon prefers to shift the question to the process itself, but in Venis's morphology the very symmetry of flow suggests a reflexive principle or self even in the midst of the formless. This would be the most immanent way of grasping something that is both wholly immediate and entirely metaphysical, or, if one prefers, transcendental.

At the heart of the individual, as at the heart of consciousness, is the simple and undivided, rather than the singular; any singularity or form can only exist in contrast with this background without qualities. If in a homogeneous medium with a unit density we imagine the appearance of two separate volumes with a reciprocal change of density, this does not seem possible without the emergence of a torsion connecting them. The fact that in this background or homogeneous medium it is impossible to distinguish the full from the empty, nor consciousness from matter, makes it easy to speak of a «neutral monism», but of course this is only a philosophical position, which, like any other, becomes complicated in the way of a more precise definition. For that it is better to think that emptiness is form and form is emptiness, and stick to it.

Every form is mind communicating with a much more minute and vast one, but from the point of view imposed by inertia, it is not possible to «recover the information». All mind is spirit because spirit is intellect and will, but for intelligence it is not possible to recover the will that has been invested in it.

A vortex is a self-contained process, which seems to have its own entity and mass while remaining connected to the whole —for it cannot exist without permanent contact with the background from which it emerges. It opens and stretches, dilates and contracts always with its own rhythm, much like the heart, which in reality is a spiral muscular band acting like a spring regulating the flow of blood also in helical trajectories. But all animal and plant organs can be considered as different types of frozen vortices in different states of evolution and equilibrium.

Undoubtedly, the transformation sequence can be seen as an archetype of the individuation process and its stages, an ephemeral configuration within the homogeneous environment with which it never loses contact. However, the correspondence between the evolution of the sequence and the process of growth, maturity and aging remains to be interpreted. It becomes necessary to distinguish the abstract from the concrete in the sequence of the vortex before reuniting it, and the same should be done for the development of the organism in terms of flow and obstruction, in order to be able to appreciate their common ground and differences.

This may seem an extremely complex problem and yet it contains a very simple principle. Modern medicine, which has accumulated such mountains of information on diseases, has not even bothered to find a functional definition of health, such as Ehret's efficiency principle, which says that vitality is equal to power minus obstruction (V = P — O); power P being synonymous with pressure and energy, and obstruction O with tension and matter, having at a constitutive biomechanical level, the boundary of deformation between both. Since these variables fall squarely within the most elementary logic of flow, they can also be associated with the transformation sequence without much difficulty.

There is a fundamental principle of development, including economic and social development, which, for some good reason, is permanently ignored: the principle of increasing restrictions for complex systems, which can also be related with the efficiency principle and the transformation sequence. And other basic aspects such as the principle of minimum action understood as movement along paths with least curvature or restriction, the principle of maximum action in size or scale, the principle of maximum entropy production, or the energy density flow per mass, which as Eric Chaisson points out is a much simpler and more reliable indicator of the degree of complexity than entropy production. Although these are very different aspects, all of them and some others converge in *the evolution of an individual entity as a flow process.*

Surely the most essential aspect of this process is something very simple and perfectly intelligible to anyone. Surely it is also a qualitative, inner law that can be understood without the need for measurements, physical quantities and mathematical apparatus. If we mention such late and elaborate physical concepts, it is because, at best, they also can benefit from a convergence in a much simpler complexion. The internal law of individuation is reflexively simple but allows us to connect very complex aspects of organisms as well as abstract mathematical structures; that is why the Venis sequence suggests a true *symplectic morphology*, taking this last word not in the mathematical sense Weyl gave it, but attending to the complexion of the causal in the phenomenon of form. It was Goethe himself who said that causality had not only length but also width; now we give it height, and it would be this unfolding of the implications of the causal dimensions that makes it possible to simplify the complex.

But what is an «internal law»? To date, still under the influence of the law of inertia, all physical laws are considered external. However inertia is not a law, but a mere principle that can be replaced by others, such as the principle of dynamic equilibrium: this demands a zero sum between all the forces acting at any point in any state of motion. It is not the same to see any process, even organic aging, as being governed by inertia, as it is to see it like an equilibrium between factors.

External law is and always will be odious to the living. This is the critical aspect of religion that is still perpetuated in its entirety in science and in the social. But the «inner law» of the living of which we speak is so inner that it is not even possible to internalize it, as has always been done with religions, laws or instruction. In fact, it may well be said that it is not even internal, but that it is intimate, which is another way of saying that in it the interior and the exterior cease to be different. It is this law alone that interests us, because it is the only one that connects us directly with the formless of the Principle. The external law binds us and appears to us as such only to the extent that we do not know our own law, that which the Sanskrit language called *SWADHARMA*.

A Middle Way

Venis' morphology, a sort of nondual hylemorphism, sees the flow in terms of yin and yang, but the dialectic of pairs of opposites exists in any culture and is only one method of approximation among others; Goethe himself speaks of systole and diastole, syncrisis and diacrisis. What is really important is to seek a balance between explanation and prediction by means of adequate description. We insist that the great deficit of science is still at the level of description of phenomena, as our mathematical models have accustomed us to replace phenomenal infinity by a mathematical infinity with a very poor correspondence.

The dialectical method proceeds by assimilation and in this sense easily allows us to bring a problem to the level of experience; it also has the advantage of allowing us to notice early on the fundamental ambiguity of any question posed between extremes. There is no important question that cannot be simplified to the extreme while maintaining a patent nucleus of significance that usually coincides with its fundamental ambiguity; an ambiguity that nevertheless soon escapes the threshold of attention of consciousness.

Morphology lies in a no-man's land between mathematics, physics and natural philosophy, and thus also in their junction, something from which different knowledges can benefit. Richard McKeon[9] defined with great clarity the four basic methods of philosophy —dialectical, problematic, logistic and operational- and noted that there is no question arising in any one of these positions that cannot be translated into the terms of the other three. So far philosophy has scarcely been able to draw lessons from this internal architecture, but for a general overview mathematics may find it more useful than the conventional division of its domains into arithmetic, geometry, algebra, and analysis.

There is certainly nothing imaginable that mathematics cannot translate into its language of equivalences, equalities and identities; the problem is rather the opposite, that mathematics has left far behind the world we are able to perceive and represent. If it would only devote a small part of its forces to faithfully account for appearances, science would be a very different issue. Even postulates that seem as unscientific as that of Goethe's polarity of colors, with its sound perceptual basis nevertheless, would have found their corresponding «analytical continuation»; for color is between light and darkness as light and darkness are between space and matter. Man can only perceive his own world, but in his world everything else is already implicated; any attempt to transcend his own limits by ignoring them or trying to overthrow them only leads to the loss of meaning and the dissolution of his own sphere. And so, almost everything in current science is actively and passively a force of dissolution.

If mathematics, physics, complexity theories and computer science continued to expand and cross-fertilize at the same rate as today for a thousand years, they still would not find the key to morphology; not even in two thousand years either. Perhaps that may give some idea of its value, although we all know that nothing can be found without actively searching for it. That is the point: the mentioned sciences already have their own momentum and inertia that nothing can change, only a creation of a new science from scratch could overcome without obstruction the deficiencies of its predecessors.

Something that has such strategic potential will not go unnoticed for long by the powers seeking to capture the symbolic initiative in any field. And since today it is impossible to have the slightest illusion about science and what it serves, it is obligatory to keep in mind the worst and the best that can happen to morphology in this last journey, since morphology, as a theory of individuation, is really the most finished product of nominalism as well as its turning point.

Spengler, the first proponent of a morphology of cultures, noticed a century ago that we are basically indifferent to the idea of causes and predicted the advent of a general «physiognomic science» which, beyond the millennium, would finally envelop any particular knowledge, transcending the petty ideas of causality; but his words were soon forgotten and no one knew what to do with such a prophecy. And yet, as we have already seen, the gradual erosion of the idea of cause in the sciences themselves has continued its unstoppable process, a process that began to roll down as early as the theory of universal gravitation and its subordination of description to prediction.

But Spengler was pointing to something both anterior and posterior to the idea of causality. The reciprocal arrangement of space, time and causality is undoubtedly a late achievement, but the sense of inner direction and irreversibility has always existed and will always exist. Spengler opposes History to Natural Science and blood to calculus, but in the morphology of flow and torsion causality in the most elementary sense and internal direction coincide by necessity.

Spengler's physiognomic science, that last Faustian unfolding now freed from the idea of causes, would be a comparative historical morphology in which physics plays no part. Perhaps what Turchin baptized with the name «Cliodynamics» has something to do with this; although chronologies, statistics, sociometry, ecology, evolutionary biology, etc., would constitute, at best, only a preliminary phase in trying to access the interior of the historical heartbeat, that which the German philosopher called «the soul of cultures». But, given that these transdisciplinary attempts always have a precarious basis — like any other theory of complexity- if they also aspire to predict the future, they can hardly be situated within a process, since prediction requires stepping out of the current and abstracting from the given context.

Although Spengler opposed physiognomics to the natural sciences, the rise of correlation and its paracausal character can easily be extrapolated to what is now happening with computing, data mining, expert systems, artificial intelligence, surveillance or personalized marketing and medicine. What machine learning networks do today, with individual databases and their exhaustive and systematic survey, from medical data and images to fingerprint or facial recognition —often with the assistance of mathematical morphology- would fall squarely into the preliminary phase that we could call «analytical physiognomics,» «statistical physiognomics,» or «soulless physiognomics.» But it is clear that, in this relentless process, no soul will be born by spontaneous generation of statistics.

Thus, although no one speaks of «physiognomics» today, it is evident that within the great traffic of data there is an overwhelming tendency towards algorithmic and biometric control that wants to descend from the instances of power to the innermost recesses of the individual. The same computation feeds back and reshapes itself in trying to reproduce the various semantic fields of the individual, which, understood in Simondon's way, can go far beyond human biological individuals. Naturally, if all this tendency could already be foreseen by Spengler himself, it must be part of the *FATUM*, and therefore has nothing to do with the symplectic morphology we are dealing with. This one is not a convergence, but an aside from all that came before, and even a conscious distillation of what has been rejected by the other sciences.

There is also something in China that gravitates intensely towards both morphology and physiognomics, with a still very wide margin of indecision. Chinese culture has a marked interest in the process of individuation that largely follows from its extreme nominalism and from the fact that, geometry and morphology being largely antithetical, the historical lack of development of geometry in China will tend to be compensated spontaneously by a great development of morphology. On the other hand, it is just one of those mirrorific counterpoints that the origin of the binary code can be traced back to Leibniz's logistic interpretation of the Chinese hexagrams of the *Book of Changes*.

We have said recently that the limits of a science such as mathematics are given by its criterion of application and its degree of receptivity to phenomena; but our idea of causality also depends on both. The basic criterion of application is given by our idea of calculus or analysis; we have also seen some of the reasons why it is conceivable that we have not even reached the end of the first stage of analysis and general application of mathematics, which would be the realization that differential calculus descends from the global to the local without any real foundation. If we succeed in getting down to the bottom, the end of this phase would be followed by another one of ascent in abstraction from the physical geometry of phenomena, followed by a final one of descent according to its knowledge of vertical and downward causality.

If these phases are intended to be conceived as a historical development, then it is impossible to know what spans of time separate them, whether such as that between Newton and us, that from Archimedes to Leibniz, or even way longer periods; but they can also be viewed in a perfectly timeless way, as movements of the spirit that are often potentially present even within the proof of a simple theorem.

By way of contrast, we can compare these 3 phases with the three levels of civilization according to the Kardashev scale. This is a scale of development based on the amount of energy a civilization can extract from the environment: a type I civilization would use all the energy available on a planet —something we would still be far from-, a type II civilization would use all the energy radiated by its own star, and a type III civilization would use all the energy available in its own galaxy. It is a shame that the original scale does not include the energetic harnessing of black holes, which is where this kind of logic invariably ends up, although I understand that later cosmologists have corrected this shortcoming. And as nothing is too much, others have added a type IV and a type V, which would control universes and "collections of universes».

In any case it is clear that a scale based simply on extraction of resources is the ultimate scale of the poor devils. It is unlikely that if we had a deep knowledge of Nature and of ourselves we would think of interstellar or intergalactic travel, and in fact the latter can be taken as good evidence to the contrary. Higher knowledge, including technical knowledge, learns to dispense with detours and intermediate steps —with mediations – and even mathematics attests to this, although on the other hand mathematics such as the present ones are incapable of containing their own expansion in all directions.

It is doubtful that the archons of each solar system are willing to see how the interplanetary space junkyard grows, but neither will they need to take violent measures: the increasing restrictions of scale of a civilization as an individual phenomenon are extremely effective to contain it by themselves, and these restrictions reach their limits of contradiction and incompatibility on the intellectual plane earlier than on the material plane. Already now it can be seen that the internal logic of these limits imposes an alarming inefficiency in the rational use of our limited intellectual resources.

Other lovers of speculation have proposed different metrics for such scales. Based, for example, on information, although it is doubtful that this is a less barbaric category than energy. Or Barrow's inverse metric, based on the mastery of decreasing, microscopic scales rather than increasing or macroscopic scales. All this is supposed to be of interest in order to ponder our own anthropological case, but I don't think that is the point: all it reveals is the limitations of the current scientific knowledge. Because not even our idea of limit as a basis for analysis is sound to begin with.

A symplectic morphology like that of Venis, even if it is in its earlier stages, gives us at least other different ideas about the relationship between macro, micro and mesocyclic scales: ideas that remain faithfully linked to the phenomena we perceive. The very idea that going beyond phenomena, with the mediation of machines and high mathematical technology, leads us to transcend our limitations is the basic fantasy that most distances us from knowing our place in the Cosmos. Venis, for example, does not believe that gravity dominates the universe on a large scale, but considers it a mere local phenomenon of what he calls orbs or spherical regions within a flow regime, and even within them its action diminishes perceptibly; the motion of stars in galaxies escapes the domain of those regions and is due to a vortical motion like that of the hurricane although on a larger scale. But the best example of transmutation of meaning without changing anything of what we know is the interpretation of the cosmological redshift: within Venis' dimensional logic this cannot imply the expansion of the universe, as our comical egocentrism surmises, but simply the contraction of our local group.

Naturally, Venis is not in a position to quantitatively justify his judgments, but he is much more likely to be right than wrong. The whole of Western science is built on an inordinate and unwarranted extension of predictions that in reality have been a reverse-engineered copy of known results that no one knows how to justify. This generalized reverse engineering without guarantee or justification, and leaving in the shadows most of what that does not fit with it, is what creates the persuasive illusion of «the unreasonable effectiveness of mathematics». And that absolutist extension without guarantees, that speculative inflation to infinity is what creates the price, not the value, of its symbolic capital.

We see for example how today the geometric phase is routinely used in control theory or in attempts of quantum computation when it places before our noses the umbilical cord that connects an individual being with its environment, the very thing that denies that we can consider a system as inert and closed. Between the living and the dead, we choose the dead without hesitation. We want to control Nature but not to know how it regulates itself, and if we exclude this, we cannot recognize our own law either.

We have already seen before that calculus has oscillated between the idealization of infinitesimals and the rationalization of limit theory; the unwary might think that by reaching both extremes nothing could be missing, but the most important thing that remains in the middle still is being ignored. We have also spoken of the rights of the description of phenomena to find the balance between prediction and explanation, which is but another angle of the same question. Without the proper assessment of phenomena, we will never find our own vertical. Finally, others, like Kant himself, have also believed that there must be an internal law as surely as there is an external law, but the truth is that there is neither, and that both thinker and thought are only the activity of thinking, the Logos we have always been looking for. Thus there are some good reasons to think that there is a Middle Way in the sciences and scientific knowledge, and that we need not to look very far, but only cease to adhere to a willfully unbalanced conception.

As we said, a vortex is both a very concrete and abstract process, in a different sense to the one we usually give to these words. And this has many consequences if studied in the proper perspective. The Indian samkya, for example, calls *vrittis*, vortices, to any mental modification, that is, to the ephemeral par excellence, thoughts. It may seem a figurative way of speaking, but if we start from the idea of a still medium without modifications, the comparison is inevitable and implies much more than what we associate with such vague terms as «mental waves» or «brain waves». The point is not that an interference of waves can be seen as a vortex, but that the degree of abstraction in the way a vortex is considered can echo the degrees of abstraction of thought itself. Moreover, mental representations require discrete states for there to be cognition: to have clear and distinct cognition the first thing is not to have it, and this also has a correspondence in terms of vortices and even their possible self-replication.

We have already seen that in the third quarter of the last century there was a one more revolution in the fundamental degrees of mathematical abstraction with category theory, algebraic topology and other developments; latter there were even pioneers, such as Robert Rosen, who tried to understand the autonomy of the organism by applying these new mathematical categories. It is true, as Kauffman has said, that the world is not a theorem and that the mathematics of set theory falls short to describe life, but it is also true that there is life in mathematics beyond set theory. However, here again a middle ground between the concrete and the abstract has been lacking so that neither goes too far and admit the rectification of an agile, "gentle empiricism», so different from the positivism on the industrial scale of refutation and specially confirmation of hypotheses.

Mathematics is the science of pure forms, in the purest intellectual sense; morphology can only be a science of the forms of pure phenomena. This defines the extreme affinity and contrast between both ways of knowledge. But the most universal of phenomena is change, and the branch of mathematics that studies change is calculus, its main application, through which mathematics ceases to be pure form at all. Yet it is clear that the calculus only deals with change in a very definite sense that eludes descriptive requirements. «What moves does not change, and what changes does not move»: taken at its symbolic extremes, and with due license, one could understand by «motion» translation, and by «change» rotation. Since physics has been modeled by an idea of differential calculus based on a baseless difference, we should see what happens when we use calculus as a function of non-difference, as constant differential calculus does, and apply it to the symplectic morphology of torsion.

We have alluded to two extreme forms of nominalism: a Western nominalism that is intellectually overdetermined, and an Eastern nominalism that is underdetermined but more receptive to phenomena. The former greatly overestimates the value of theories by trying to exploit them to their maximum conceivable extent, while for the latter everything that is theoretical pretension already seems to be an excess of systematization. But the truth is that morphology opens a free path to the knowledge of forms, ideas or essences, inasmuch we abandon the Platonic pretension of a separate, immaterial and immutable reality. Given the one-sidedness of its analysis, physics has not been able to get rid of its own misplaced Platonism either, just substituting metaphysics for mathephysics. There is another analysis and another object of analysis that can change our idea of zero, infinity and unity; of number, measure and equilibrium.

Seeking reintegration into unity has nothing to do with the elucidation of intellectual aspects unless we start from an intellectual overdetermination; in the same way that simplicity would not be interesting if we did not start from a circumstance marked by complexity and differentiation. Knowledge in the first person by itself is not the same as that same knowledge examined from the third person, but no analysis of the latter can recover the immediacy of the former. However, we have seen that the constant differential calculus is in close correspondence with the execution of immediate actions whose procedure cannot even be explained by those who perform them, as for example catching a flyball in the outfielder problem. In this respect we have spoken of the possibility of a fourth-person knowledge or an immediate non-intuitive knowledge which is at the basis of all our assumptions, the aim is to connect third-order knowledge with this other knowledge in the most direct way possible, avoiding with the greatest care distractions and detours. The only certainty that natural philosophy has is that the Principle is not only the starting point, but also the end of all our inquiries.

There are two ways open to the scientist today: to continue on the descending path as a microserf under the protection of the great technocratic powers, or to take the ascending path towards new forms of intelligible simplicity. The first way, in which young dwarfs carry old giants on their shoulders struggling to advance one centimeter towards nowhere, though socially rewarded has increasingly insignificant yields; the second does not promise technical applications and when it does, one must know how to elude them, but it allows a free advance towards what is most significant for man, provided one knows how to stick to the most basic and does not expect a particular return. No matter how many advantages it grants me, knowledge as mastery over external things always distances me from the mastery of myself.

The social individual is a type of socially induced contraction and reaction; the biological individual has a more extensive physical background but is also a reaction to the environment. The individual personality is a mixture of biological and social individuation but is still reactive. The person perceives the individual and his personality, but they do not perceive the person, just as I perceive my body and my thoughts but neither my body nor my thoughts perceive me. The center of the person is not a singular consciousness, but an impersonal simple consciousness that is nevertheless aware of the singular in the person and the individual; just as the formless envelops the form and at the same time is hidden in the balance of its manifestation process. From the degree of alignment between our perception of phenomena, the individual, the person and the impersonal consciousness depend the degrees of abstraction and concreteness that our consciousness of being may attain.

The Middle Way in science is like the axial mountain of ancient cosmographies, which despite being at the center of the world also surrounds it and defines its contours. The way of extremes in science as we know it today seems on the contrary to be all-encompassing and yet blind to the essential. Science can be a mountain of opacity or a mountain of transparency depending on the point of view we adopt, but for most the historical perspective we have reached is of no use. Those seeking the mountain at the center have two ways of approaching it, either by asking for its location, or by wondering how it became possible that they cannot see it.

References

Peter Alexander Venis, *Infinity Theory*

Ronald H. Brady, *Form and cause in Goethe's Morphology* (1987)

Miles Mathis, *A redefinition of the derivative —Why the calculus works, and why it doesn't* (2003)

O. M. Dix and R. J. Zieve, Vortex simulations on a 3-sphere (2019)

Nikolay Noskov, *The phenomenon of retarded potentials*

Mario J. Pinheiro, *A reformulation of mechanics and electrodynamics* (2017)

K. T. Assis, *Relational Mechanics and Implementation of Mach's Principle with Weber's Gravitational Force* (2014)

Gilbert Simondon, *La individuación a la luz de las nociones de forma y de información* (1958)

Philip S. Marcus, Suyang Pei, Chung-Hsiang Jiang, and Pedram Hassanzadeh, *Self-Replicating Three-Dimensional Vortices in Neutrally-Stable Stratified Rotating Shear Flows* (2013)

Shubo Wang et al., *Spin-orbit interactions of transverse sound* (2021)

Notes

(*www.hurqualya.net)*

1. http://infinity-theory.com/en/homepage

2. https://www.wired.com/2008/06/pb-theory/

3. http://milesmathis.com/virial.html

4. http://milesmathis.com/are.html

5. https://www.nature.com/articles/s41467-021-26375-9

6. https://arxiv.org/pdf/2103.10205.pdf

7. https://www.researchgate.net/profile/Mario-Pinheiro/publication/318710701_A_reformulation_of_mechanics_and_electrodynamics/links/59790a60aca27203ecc632f8/A-reformulation-of-mechanics-and-electrodynamics.pdf?origin=publication_detail

8. https://arxiv.org/pdf/1303.4361.pdf

9. https://richardmckeon.org/content/e-Publications/e-OnPhilosophy/McK-PhilosophicSemantics&Inquiry.pdf

NAME, CONSCIOUSNESS AND NUMBER

31 May, 2022

A well-known mathematician said once that it would take a million years to understand prime numbers well, if ever. One should have asked him what he aspired to understand about prime numbers that could take that long. As far as I know, no mathematician has even asked the question why the prime numbers, being at the core of arithmetic, do not seem to have any importance in Nature —when yet the Riemann zeta function, so intimately linked to their structure, seems to be reflected in many different kinds of physical systems.

Learning to count does not require counting to a million. Prime numbers only appear in retrospect as a reflection on the order involved in their succession. This is in itself eloquent, if we agree with Poincaré that, unlike geometry or logic, arithmetic follows from a synthetic a priori principle: the recursive definition of the sum or product is irreducible to the logical definition. And what this means, if one thinks well, is that the retroprogressive method has in mathematics as much course as its forward advance by formal generation.

Poincaré comes to say that one demonstrates by logic whereas by intuition one invents; although for demonstration, as opposed to mere verification, which is purely analytic, intuition would also be necessary. I quite agree with his position; but there is still potentially another double retrogressive movement that points beyond logic and intuition. We have called it fourth-person knowledge and also, following the terminology of Jakob Fries, immediate non-intuitive knowledge. Yes, intuition invents because it is synthetic and creative; but all this is under the sign of the impulse of formal generation that is connatural to mathematics, and here we would like to point to something beyond it. For it has long been the case, according with the general tone, that mathematical invention is too speculative, and formal logic too exploitative.

In his reflections on mathematical discovery and invention, Poincaré speaks of a conscious ego that does the arduous work of preparation and of an unconscious or subliminal ego that would make possible combinations on the basis of the previous work. We can verify this not only in mathematical work, but in all kinds of problems with which we sufficiently occupy our mind. But these two egos only indicate the activity of the outermost and intermediate layer of our consciousness; the most basic level does not even deal with forms. This pre-formal level is responsible for our sense of identity and identification, it is just the Self without any other quality than its capacity for identification.

This coincides with the three *avasthas* of nondual Vedanta: the waking state, the dreaming state and deep sleep state, which are respectively concerned with the external world and the body, the internal world and the mind, and that intimate which is neither internal nor external and which we usually call

«I» or “self”, although this very I tends to identify itself with the body and the external experience of our conscious life. Anyway this ternary scheme still remains useful to overcome the dualistic oppositions such as objective/subjective, body/mind, conscious/unconscious, and in any case provides a route to delve into our present theme.

This «third I» at the bottom of consciousness is indeed the most primary, and if we don't acknowledge it as such is because this same I without qualities is the very principle of identification that commonly adheres to qualities. Nevertheless this self is beyond the domain of form, including mathematical form, and this is the main reason why it has not attracted more attention from this science or any other, such as, for example, psychology. If for Poincaré it was already a very delicate matter to discern the part of the subliminal ego in knowledge, and his ideas, such as the central place of convention in physics and mathematics, have never been popular, something more subtle is still needed to detect the role of this naked self in knowledge beyond the objective and subjective.

*

Elsewhere[1] we discussed the basic concordance of these three moments of the self with the three categories of Peirce's ternary logic, as well as with the three principles of classical mechanics, and we also advanced, without much justification, that there is a certain correspondence between fourth-person knowledge and the so-called «fourth principle of mechanics,» which comes to play with a reformulation of the three classical principles in terms of a relational mechanics without inertia valid for open systems with a feedback loop including the geometric phase. Of course, Poincaré already showed that the principles of mechanics are a matter of convention —which, of course, does not mean that they are simply arbitrary; though the same could be said of other branches of applied mathematics, not only geometry, but the very same calculus.

If the conventional character of mechanics has hardly been admitted, much less has been done with the principles of analysis, despite the fact that various forms of non-standard analysis are well known and used. The discovery of non-Euclidean geometries is considered a critical moment in the history of mathematics, yet non-standard forms of analysis have had nowhere near the same impact. And yet, if instead of contemplating only the new forms of infinitesimal calculus, which are still new rationalizations of the original idealization, one were to take into account an elementary redefinition such as the constant differential calculus based on the unit interval discovered by Miles Mathis[2], this would affect much more closely the immediate object of mathematical thought which is the natural number. For, as we will point out later, and although Mathis himself does not consider it, his redefinition potentially implies the rebirth of the theory of proportion at the heart of analysis.

Let us return to the role of the «third self», which is actually the first one, in knowledge in general and in mathematical knowledge in particular. Mathis rightly speaks of how the problem that his calculus tries to cover has not only defied any solution, but "has defied detection". And this alone suffices to give us a correct idea of the role of the most basic self, even if Mathis here does not concern himself in the least with questions of discovery. Poincaré's subliminal ego performs a work of selection. The Frenchman invokes the random collisions of the kinetic theory of gases, but, as he admits himself, this is just a crude simile; it would be more appropriate to speak of the overlapping or superposition of the innumerable possible options. The fundamental I, the same one that operates identifications, is also the only one capable of inhibiting itself from realizing them, and can thus detect both new questions and new truths, which are completely different things. And also, since this I is simple in the sense that it would be absurd to attribute a composition to it, it is responsible for detecting the most common, basic and recurrent facts: the simplest facts, which exist in innumerable ways but which are also the most difficult to recognize since they are veiled by the multiple structures to which we adhere and with which we identify ourselves.

If the primary I is the source of identification, it is also the source of the objectification of thought, and by the same token, it alone makes possible an objectivity without objectification that permits the detection of new questions within the most familiar. Of course Mathis does not reach this point by equanimity alone, but by sheer rectitude in reasoning and by the prior conviction that there must be a much simpler way to justify calculus; but all this facilitates that equanimity which is the condition of recognition "by lifting a single curtain". Classical calculus is full of recipes and artifices, but no artifice can ever replace rectitude. As for simplicity, which in itself is a sign of the degree of truth and its quality, the challenge is there for anyone who wants to accept it: to show a way for calculus simpler than his.

Of course, if the mathematical community ignores such an amendment in full, it is not because it lacks relevance but just the opposite; at this point, it cannot afford a revision of the fundamentals at such a basic level, which should affect all the main branches of this science. But this tells us enough about the economy of knowledge. Today's mathematicians use to work at the tactical level, and even when they propose bold connections in their most ambitious programs, for example of geometry with number theory, their evolutions do not go beyond the operational level. The real strategic level is forbidden to them because, as it is easy to understand, going back to the fundamentals one is very careful not to touch the foundations.

The constant differential calculus also has a surprising correspondence in its procedure with the way we act when we try to catch an object describing a parabolic trajectory. Although someone running after a fly ball cannot explain how he catches it, it has been shown[3] more than convincingly that he simply moves in such a way as to maintain a constant visual relationship, naturally without making any complicated time estimates of acceleration or anything of the sort. Incidentally this expose the computational paradigm, many preconceived ideas about cognition and other associated issues. However, here too, Artificial Intelligence experts have made heroic efforts to ignore and minimize these simple facts.

If the purely formal procedure, of simple numerical analysis, of constant differential calculus corresponds to immediate actions involving clearly informal knowledge, we are reconnecting third order knowledge with first-person knowledge, and the mere fact that this is possible justifies the claim of the existence of fourth-person knowledge and thus of non-intuitive immediate knowledge.

*

Pythagoras, Plato or Euclid still seriously considered the contemplative value of number and proportion. However, contemplation has never disappeared, and it is still easily perceived in the fact that even today mathematician can isolate themselves from the world without a sense of sacrifice and with the best fruits for their activity. Regardless of the era, the mathematician is a spontaneous recluse and a natural ascetic.

Poincaré distinguished three purposes of mathematical work: mathematics by and for itself, mathematics that studies Nature, and mathematics as a philosophy that tries to delve into basic concepts such as number, space, time or logic. He also rightly insisted that mathematics should not be navel-gazing since much of its development is due to what its application raises, and anyone can see to what extent physics, hand in hand with analysis, has favored its growth. But in this classification of goals it can be seen that the primary plane of mathematics is an aesthetic plane —just like the logical category of firstness in Peirce, only the other way around: not with a minimum, but a maximum intellectual component. This is where the autarchy of mathematics, this self-sufficiency that invites contemplation, comes from. Mathematics applied to Nature would be secondary in Peirce's logical sense, and the philosophy of mathematics tertiary in that same sense. Also for Peirce, however far he was from intuitionism, mathematics had an irreducibly aesthetic component in this relatively immediate sense.

In this very fact that mathematics, an eminently intellectual third order activity, always returns to the first plane as its own, lies not only the autarchy of mathematics but also its indefinite capacity for synthesis. Yet mathematics has expanded too much for anyone, least of all its practitioners, to be seen as a self-contained whole. It is this uncontainable expansion that prevents us, above all, from contemplating mathematics in the same spirit as Plato. Now, our intention here is not to recover the Platonic spirit for this activity, since in fact it has already preserved it more than one would think, and not only in mathematics but also in physics. Rather, we would like the Platonism in which we still move to close the full circle of its transformations and misunderstandings in order to open a new cycle after more than two thousand five hundred years.

The unstoppable expansion, both cumulative and accelerated, of mathematics begins with modern calculus. As we use to remind, if knowledge doubled regularly every fifteen years, today the body of mathematics would be four million times larger than in Newton's time. There are two thousand years from Euclid to Newton, but although minds were anything but idle during all that long period, one cannot speak of a remotely similar growth. The great difference lies in the emergence of calculus.

With calculus there is a radical inversion of priorities with respect to the external world that at the same time disrupts the primary plane of mathematics. In this first attempt to reverse engineer the change in Nature from known results, instead of determining the geometry from the physical considerations, in order to deduce from them the differential equation, the differential equation is first established and then the physical answers are sought in it. This, which seems the definitive overcoming of Greek mathematics, is in reality only the misplaced maintenance of some of its illusions. In fact Mathis rightly attributes the basic misunderstanding of calculus no longer to Newton or Leibniz but to Archimedes himself.

And this way we have passed from Platonic metaphysics to present day *mathephysics* with its mirages and rationalizations. Analysis, which soon forgets the humble origins of calculus in the study of curves, replaces physical geometry by another parallel reality. And finally, as a consequence of the attempt to found analysis by arithmetic, the very idea of number dissolves and is replaced by set theory, a new branch of logic. The new mathematical continuum, irreducible to the perceptual continuum that had motivated it, also puts an end to the primacy of number.

Modern mathematics has swallowed too much to be able to digest it properly; not only in the total mass of accumulated knowledge, but in each of its parts, which contains too many operations that are far from being obvious when they are not openly illegal, as Mathis shows even with the most elementary functions of calculus. And just as knowledge in every field multiplies, manipulations of this kind increase in the same proportion. How is it possible that in such conditions, so far removed from evidence, the present-day mathematician is still capable of losing the sense of time for long hours while absorbed in a problem? Poincaré gives us again the answer when he observes that it is not only the order that matters, but the unexpected order —the renewal of identity through different things.

*

The excess of expansion and the fading of mathematical evidence is one with the primacy given to prediction over description; an imbalance that has been compensated by a justification or foundation that has never gone beyond rationalization. If calculus had been adjusted to the physical geometry of the problems, our idea of Nature today would be way too different, who knows to what extent. Mathis applies his finite method to the most direct solution of a great number of crucial problems of physics, but this is only one side of the question, and in fact the most difficult thing in science is not to want to solve problems that one has not asked in the first place.

In its relation to Nature, the extremes that define the range of action of mathematics are its criterion of application to external problems, which affects in particular the calculus, and its degree of receptivity to natural phenomena. The quality of the connection between both depends on the fidelity of description, which in modern physics is entirely subordinate to prediction. The characterization of phenomena in physics is always much more tentative and approximate than in mathematics, more qualitative as it depends on the domain of measurement —if it is true, as Hegel wanted, that measurement is the synthesis of quality and quantity. But at the same time the intimacy of the physicist with the methods of calculus has created a pre-selection of the phenomena most susceptible of *controllable* measurement.

Peter Venis's projective morphology[4], which is a phenomenology of Nature, gives us a modern contrast of what a truly qualitative approach to the domain of Form can be without the quantitative pre-selection from which physics starts. Unlike Mathis, Venis does not try to solve famous problems, he discovers new questions in need of a more rigorous formulation. But in what sense rigorous? In the same sense that we have been applying calculus to physical problems while ignoring their physical geometry? Let us hope that Mathis' warning will do some good. Mathis' critique of calculus goes far beyond the well-known objections of finitists and constructivists, and this is why it is ignored. Mathis and Venis perfectly illustrate the two extremes of the criterion of application and responsiveness to phenomena, and the quality of the connection established between both will also define the quality of their development.

As we have seen before, the physical meaning of dimensions in Venis' morphology has yet to be clarified. But Mathis makes it clear that analysis has never been able to correctly analyze even the number of physical dimensions of the most basic problem, as shown by the mere fact of representing a straight-line acceleration by a curve. The question is not only the representation since in any case the analysis of change still depends on the given point and instant. According to Venis, however, all geometric manifestations, whether points, lines or planes, have reality only as projections and sections of a homogeneous field or medium that can have zero or an infinite number of dimensions.

Venis conjectures that the study of the wave-vortex will require the use of complex numbers and techniques such as those derived from the Hopf fibration. Mathis on the other hand sees no need for wave mechanics in quantum mechanics, nor for the use of complex numbers in physics, not to mention that he also shows that even the application of the most elementary trigonometric functions is not understood. The perspective of the two authors is diametrically opposed and yet should also be complementary. Venis' morphology puts forward another idea of the continuum that reconnects with the perceptual continuum, but this perceptual continuum need not coincide with the treatment of continuous media in physics, nor can it refer to the classical use of calculus in them.

Mathematics is the science of pure forms, in the purest intellectual sense; morphology can only be a science of the forms of pure phenomena. This defines the extreme affinity and contrast between both ways of knowledge. But the most universal phenomena is change, and the branch of mathematics that studies change is calculus, its main application, through which mathematics ceases to be pure form at all. Yet it is clear that the calculus only deals with change in a very definite sense that eludes descriptive requirements. «What moves does not change, and what changes does not move»: taken at its symbolic extremes, and with due license, one could understand by «motion» translation, and by «change» rotation. Since physics has been modeled by an idea of differential calculus based on a baseless difference, we should see what happens when we use calculus as a function of non-difference, as constant differential calculus does, and apply it to the symplectic morphology of torsion.

If we go back to the role of the three egos in mathematics, the conscious ego is on the level of form, the subliminal ego on a pre-formal level, the primary self on the formless level. In India the manifestation of any impermanent entity is seen in terms of *nama-rupa*, the concurrence of name and form. In the West we do not usually contemplate this pair, which has no dualistic connotations, but we do speak of mind and matter, which for us does; a more akin translation would be that of essence and accident. Mathematical form is detached from matter, but the form contemplated by Venis' morphology is one with it. This view of morphology cannot be dualistic because form depends on a continuum of non-integer dimensions and these dimensions depend on perception. However this does not exclude an objective description because each physical potential is not merely a position, but a perspective: the circumstance of the environment from the perspective of an agent.

Put another way, number as a mathematical form is detached from the Word, from language, but the most qualitative aspect of forms, which in physics has always been considered contingent, is a direct expression of the language of change, that same change that has always escaped like water from the web of artifices woven by the calculus. The question is how it could not escape. This can only be explained by the continuous change of dimensions and by the shift of the potential. For the former there is already the fractional calculus, widely used but lacking credible physical interpretation; for the latter we have in physics the geometric phase and the theory of the retarded potential, aspects undoubtedly linked by their dependence on the environment but which physics has not yet been able to connect.

Morphology opens an indirect way to the knowledge of forms, ideas or essences, precisely insofar as we abandon the Platonic pretension of a separate, immaterial and immutable reality. In this wave-vortex morphology, the substantive is the wave emerging from the homogeneous medium, while the adjective is the cut or dimensional section that produces the appearance of the vortex. That which gives name can only be that which identifies, and only that which is formless and lacks identity can identify itself with something. Physics has been unable to grasp the process of individuation of an entity, but that is precisely the question of morphology; answering that question, it overcomes the nominalist reduction of everything possible to the individual and undertakes the way of return.

*

Economy of thought is essential in any science and even more so for mathematics, but the degree of economy of thought depends crucially on the right choice of terms, and precision presupposes prior indeterminacy or ambiguity. Accurate naming allows one to assume things that would otherwise be much less tractable, and Poincaré sentenced that mathematics was the art of giving the same name to different things. The importance of this maxim should not be underestimated, even though the relationship of mathematics with names and with the art of naming is anything but fortunate: one only needs to realize the absurd naming of such important and honed concepts as the «real numbers», «imaginary numbers", «complex numbers», or "topology» —which actually deals with that which is independent of position— to be convinced of this. One more symptom is the unstoppable proliferation of proper names that strip the new concepts of their very conceptual contribution.

The economy of terms, the search for synthesis through the name, remains an almost entirely untrodden path for mathematics. Analysis and synthesis are supposed to be reciprocal, but in practice analysis has overwhelmingly predominated in modern mathematics. This, however, holds a sort of promise, since this overwhelming predominance has nothing to do with the quality of analysis, on the contrary, if we duly appreciate the fact that analysis does not know how to analyze even the dimensions of the simplest changes. Since without good analysis there is no good synthesis, we should be concerned with the quality of the steps, instead of the quantity as a technified mathematics, blinded by the achievement of results, has done.

Today emphasis is put on the unity of mathematics above the multiplicity of its manifestations, applications and branches. In this sense, we would then speak of arithmetic and geometry, of algebra or analysis more as a tribute to the history of the development of the subjects, the contingency of their appearance and the present state of knowledge, than out of deep necessity. But on the other hand, mathematics without its form is pure indeterminacy and potentiality. Mathematics in act is pure form, but without form there is no objectivity. Science must be objective if it wants to have a social function and not fall into mysticism, but the subject cannot stop reading continuously between lines and planes, displacing the value of signs.

To follow any mathematical argument one needs to know the underlying structures, which partly depend on the contingencies of historical development and partly are subject to progressive rearrangements. Mathematics is thus eminently historical not less than eminently timeless, but the relation between the synchronic and the diachronic within is subject to constant redefinition. In Philosophic Semantics and Philosophic Inquiry[5], Richard McKeon characterized with great clarity the four basic methods of philosophy —dialectical, problematic, logistic, and operational— and suggested that there is no question raised in any one of these positions that cannot be translated into the terms of the other three. So far philosophy has scarcely been able to draw lessons from this internal architecture but from a material rather than a formal perspective mathematics may find it more useful than the conventional division of its domains into arithmetic, geometry, algebra, and analysis.

McKeon's modes of inquiry —assimilation, discrimination, construction, and resolution— are themselves a recapitulation in a historical key of Aristotle's famous four questions at the beginning of the second book of the *Posterior Analytics*: whether it is, what it is, what sort it is, and why it is —or questions of experience, of existence, of that which is, and of being. As it is well known, these four questions became the four *constitutiones* of Roman rhetoric, thus entering into legal and political philosophy; however, this foundational work of the Aristotelian canon is the first explicit attempt to formalize the method of scientific knowledge at a time when it was not formalized at all, and no doubt the geometrical analysis of contemporaries such as Eudoxus is at the basis of his reasoning. Aristotle does not do geometry but he expounds what geometricians do.

The four cardinal questions of the *Posterior Analytics* —to which is added the knowledge of what the name means- do not seem to have an internal symmetry, but the modes that McKeon derives from them can be conceived themselves as the four angles of the activity of knowing: the knowledge, the knower, the known, and the knowable. The modes of inquiry thus "serve to unravel the tangled history of the methods of induction and deduction, analysis and synthesis, discovery and proof which first emerged from distinct modes and were variously merged with each other and inverted.»

The four questions and their associated modes are on the intermediate, pre-formal plane of semantics and heuristics, between the formal and the formless, between logic and metaphysics, between what is known by definition and what the name hides. But if Aristotle, reasoning within an unformalized science in which almost everything quantitative was still expressed verbally, tries to descend towards the undeveloped formal plane, today we can use the intermediate plane to ascend the scale of generality of the mathematical concept, from the unnameable in actual mathematics to the unknowable in the name.

Everything in our modern age was born out of a pronounced imbalance, lives out of imbalance, and will end with imbalance. Much has been said about the separation between «the two cultures» of the sciences and the humanities, but little notice has been taken of the fact, and McKeon barely does it, that the modern sciences have given almost all their preference to the logistic and operational methods, vehemently rejecting the dialectical method exemplified by Plato and the problematic method inaugurated by Aristotle, the latter two have been relegated to the verbal realm of politics and the humanities. Even the life sciences, from biology to medicine, have been reduced in their theory to the combination of the first two, leaving the other two just for «the discussion of their social and human aspects" —for what is now understood as «rhetoric».

Accordingly, it could be thought that the first two methods are more suitable for quantitative analysis, which is so predominant today, while the second two would be better adapted to the subtleties and nuances of the qualitative. However, mathematics is a complete whole that includes the quantitative and the qualitative. The question is similar and is linked to the purely conventional distinction between «pure mathematics» and «applied mathematics»; mathematics always tends to go full circle between theory and application, their conflation being only a matter of time.

Poincaré said that there are not solved and unsolved problems, but more or less solved problems, and history proves him right since the very idea of what is an acceptable proof or solution changes insensibly with the times, just as other key notions of mathematics change. On the other hand, any important mathematical concept points in multiple directions at the same time, and the four questions and the four modes would be a synthesis of the possible fronts from the point of view of the contents.

It has been said that every great culture has had its own concept of number, but ultimately every mathematician and every person has his own concept, which results from his perception of the whole of mathematics and its unity, or rather his lack of it. This perception, and the will to mobilize it in one direction, is what determines the spirit of a research program —if we understand by spirit the unity of intellect and will. The four questions and the four modes can be used to choose a formal course of investigation, or else to find a more neutral and equanimous perspective, an «objectivity without the will to objectify». The first is a descending path directed towards terms and solutions, the second is an ascending path that tries to distill the ideas enclosed in the conventions liberating the will locked in the contents.

The four modes admit as many variants in their principles, methods and interpretations, as well as selections, so it is not difficult to elaborate a notation of their 4x4x4 combinations, which in any case should be seen as ideography or mnemonics rather than as a formalism. Applying these schematism to modern mathematics may seem an eccentricity at this point, but, if so, it would rather be due to the fact that it is this mathematics that has had an eccentric and unbalanced development. Justified by specialization, there is an interest, essentially unintentional but no less jealous of its forms, in keeping the methods of the sciences and the humanities isolated in such a way that they can never communicate or translate meaningfully.

The still recent mathematical theory of categories, with its successive elaborations and extensions, has become a natural language of mathematics at both the most basic and the highest levels of abstraction, and yet it is still the fulfillment of Aristotle's old project of making explicit the content of the categories of concepts. In fact his book on the categories or predicates is much more metaphysical and remote from mathematics than the *Posterior Analytics*, but the idea that persists from this foundational text basically boils down to the trite scheme facts→ induction → principles → deduction. McKeon himself applied the modes to the historical approaches in modern physics but avoided to deal with mathematics, and yet there would be nothing «out of place" in it but for the places where modern developments have led. Apart from the fact that theoretical and applied mathematics always go full circle, we argue here, moreover, that it is only by delving into its application to the physical world that mathematics yields its ultimate philosophical value, including in it that of the philosophy of mathematics. And it is precisely the failure to take this seriously enough that most limits the present mathematical theory of categories.

The mathematical theory of categories, according to its initiators a metamorphosis of Klein's Erlangen program, is a deepening of the common basis of logic and geometry as spaces of transformations. Of all the advocates of this program, it has surely been Lawvere who has endeavored most to bring the viewpoints of mathematics and philosophy closer together, having tried to model even dialectical logic and the identity of opposites in calculus and physics. Lawvere has always tried to connect categories and physics, especially continuum physics, but if we duly ponder the «selectively unbalanced» development of modern sciences, physics included, it is easy to see why these attempts have not gone far despite the permanent growth in the applications of the language of categories.

*

To what extent do the four questions and the four modes of thought define an axis or common ground for the selected topics, beyond the nature of the topic itself taken in its first generality? This would have to be investigated with a sufficiently large series of examples to come close to an answer. The questions have no apparent symmetry, but the modes are not entirely exclusive and admit a common ground. It would be desirable for mathematics to explore this in a heuristic fashion, not so differently from what it has done in its theory of categories, as this last one had a pronounced heuristic profile right from the start. If the mathematical theory of categories acquired relevance in its day because many of the concepts that mathematicians handle were far from being explicit, the semantic and research schemes are particularly relevant because since the scientific revolution two basic modes of investigation have taken precedence at the expense of the other two, and it would be necessary to bring to light what has been left in the shadows in order to have a broader and more neutral perspective.

Any subject or problem should be able to admit four cardinal formulations, but the most important thing is what their superposition may illuminate. Today, for example, the analysis of dualities is not just a method but almost a general principle of research in physics and mathematics that has allowed great advances in territories that would otherwise have remained obscure. The contrast between the four modes may bear some superficial resemblance to the analysis of dualities, morphisms, homologies, maps, correspondences and transformations in general but leads to a different domain of possibilities.

There are not a few who, like Dingler, consider Poincaré's conventionalism «naive»; one readily concedes that principles such as those of mechanics are not directly dictated by experience even though they are drawn from it, but it is not admitted that there can be any other possible result. We have however good reasons to think that it is the latter assumption that is naive: we have repeatedly seen a relational alternative to the three principles of mechanics that for the most part allows the same predictions as Newton's principles, even if it must inevitably yield discrepancies as well, which are otherwise not difficult to justify. The overlay of modes of investigation should show both the discrepancies and the common ground; but in contrasting them we are also contrasting developments that tend to exclude each other in time.

Despite its understandable limitations, Aristotle's method in the *Posterior Analytics* is essentially retrogressive, since it tends to go from conclusions to premises and principles; and yet it remains analytic as its title promises. The main interest of explicitly contrasting modes of investigation is heuristic and intuitive, but as we note, this can be understood in a double sense and not only in the currently predominant sense of «solving problems». In fact also here it is possible to go back from our intuitions, always conditioned by habits and acquired knowledge, to a «non-intuitive» base that is momentarily excluded but which is actually broader.

The pertinence of this synoptic approach can only be verified through the study of concrete cases, cases in which the material implications of a problem leave ample room for ambiguity. Almost all applied mathematics falls into this category, but also concepts and problems of pure mathematics. Just as in the sciences, beginning with physics, the constructive and the operational approaches have taken hold, so in mathematics the presence of analysis and algebra has grown at the expense of arithmetic and geometry which were at their origin; one need only see that in such vast areas as current algebraic geometry or non-commutative geometry any figurative notion of geometry was lost long ago.

Both well-defined solved theoretical problems and unsolved ones can benefit from this unfolding and overlapping of the four modes of investigation. Although contrasting the modes requires a conscientious work of preparation and composition, what emerges again can be seen as a distilled form of paralogy if not analogy. Viewed from the outside, Aristotle's four cardinal questions are a way of systematically present the results of scientific investigation; but the constructive interference of the four modes of thought contains something that is too internal even to the very activity of investigation and tends to be reduced or rejected. This «something» is something cognitively relevant and points beyond systematic organization: it points to the inarticulate arcana at the origin of language and name.

At the other extreme, this breakdown can be applied to problems which resist univocal definition and where application and theory, the quantitative and the qualitative, are totally mixed and muddled: a glaring example of this is money and the monetary system which regulates it. No field can be less pure than this, and yet the directed technological evolution, the politics behind the technology, and the juncture of global geopolitical confrontation place this instrument of domination at the most acute of crossroads. However, the digitalization of money, which is only one more episode in the digitalization of consciousness, albeit a key episode, also runs, as a presumed technological solution, within a publicized but secretive meroscopic or logistical-operational axis that blocks the holoscopic, dialectical-problematic axis. Naturally, those who wish to increase the control of the ruled have to argue that up to now a fair monetary system has not been possible due to technical difficulties which are now definitely overcome, while trying to obviate the long history of collusion between money and debt which is still reinforced by the new tools. However, the simultaneous contrast of these four approaches should reveal something quite different from a mere historical analysis.

Technological solutions to the money problem always tend to optimize control of the population by a tiny minority. But much of what is conceived as the best alternative solution is also a hostage to technological solutionism in one way or another. Surely the best options are those that depend less on the degree of technological development, since those that defend the logic of the irreversible only want to consolidate a capture operation. We believe that the contrast of the four modes suggests at all times a certain timeless axis, a possibility and a way of evading the mechanisms of coercion based on the supposed irreversibility of progress. Needless to say, forcing an irreversible framework only bring us closer to the catastrophe.

According to the myth, human language has undergone successive stages of degradation, and propaganda is the best proof that language can not sink any lower. In addition, the quantitative and statistical has always been one of the most basic tools of such endeavor. Economic theory and its analysis is the supreme exponent of this radical perversion of means by ends using quantity as an alibi, of an ideological elaboration deliberately constructed to divert and conceal, but it would be very naive to think that the rest of the sciences that rely on quantity have not exploited its enormous potential for duplicity; when economics began to make intensive use of mathematics physics had already been doing so for more than two centuries.

*

Logic and geometry can always be reconciled, but physics always takes us beyond extension and motion, and it is this intensive aspect that gives it all its interest. The intensive is no less subject to quantity than the extensive, but its nature is not reducible and for us it is like an adjunct to our representation. Yet even natural language seeks to reflect in its words such intensive qualities as color, density, or tension. There is something profoundly physical in the name, and it is precisely that which escapes the domain of form and extension.

In science and in physics in particular, besides deduction and induction we have abduction or hypothesis, but this is only a veiled form of analogy to which an attempt is made to give a mathematical expression. Analogy plays on the other hand a genuinely functional role in modern physics since the Lagrangian and Hamiltonian variational principles, which are at the basis of the fundamental laws, are nothing but exact analogies.

The so-called fundamental laws and the hypotheses that have given rise to them ignore almost everything about the simple fundamentals of mathematical proportion. Their elegant-looking equations not only ignore the principle of homogeneity of physical proportions, but over time they pile up more and more heterogeneous quantities, as many knots to be untied but that are never untied, remaining at the basis of the inexplicable numbers and constants that must be entered by hand for calculations to work. An analogy without proportion is just a black box.

Mathis' constant differential calculus, being the natural form of the method of finite differences, reintroduces the possibility of a theory of proportion into the analysis of change. The theory of proportion was basic to Greek mathematics, and without it number and quantity are disconnected from the Cosmos and the idea of the good. On the other hand, it has been reproached to Eudoxus that his theory of proportion, which passed to Euclid and Archimedes, only succeeded in separating geometry from arithmetic, blocking the development of algebra and calculus. But today we can interpret all this in a new light.

When Aristotle says that a problem of geometry must seek its principles in geometry and not in arithmetic, on the one hand he is only vindicating his own philosophical mode which demands that each problem have its specific reflective principle, unlike, for example, the global principles of Plato's dialectical system, or the simple or actional principles of the other two prototypical modes; but on the other hand he is also echoing the dominant idea of the time, endorsed by Eudoxus, that the two are categorically different sciences.

Today there is an emphasis on the unity of mathematics but only on a large scale, since at the most basic level analysis has become completely dissociated, much more so than in Greek times, from geometry. If anyone ever succeed in bringing them together again, it is Mathis, and yet he does not consider the implications of his work for completely recasting the theory of proportion in Analysis —when in fact he has found so many elementary transparencies and correspondences for the «mysterious numbers» of physics. Certainly no one is obliged to take these explanations for granted, but it must be recognized that they are much simpler than most speculations, and that means something.

Mathis also takes pleasure in showing multiple examples of how physics is incapable of identifying the causes of even the simplest questions, things like why a sail allows one to sail, and it would not take long to see that his basic criticisms fully affect not just one, but all the four cardinal questions Aristotle poses within the context of geometry and astronomy. Since, unlike the Greeks, here we are not only convinced of the unity of mathematics, but believe that the degree of unity of its concept directs reasoning at higher levels of elaboration, it would be valuable, rather than embarking on large programs connecting vast areas, to attempt to connect the four major branches of mathematics and the four modes at the most basic level, where decisive divergences and complementarities already arise.

What can be seen in the current "grand unified programs" of mathematics, such as Langlands', is a radical separation between extremely elementary questions and extremely sophisticated theories; and such a lack of middle ground, so necessary for science, forces one to doubt not only the theoretical subtleties but also the choice of basic elements. If applied mathematics exhibits holes such as those revealed by Mathis, one cannot expect theoretical speculations of the nth degree to have more consistency, but rather the opposite. One must search for the minimal matrices of meaning that are representative of a totality attending the foundation, not ignoring it.

*

The Aristotle of the *Posterior Analytics* was already doing metascience and metamathematics, as the modern category theorists would do again twenty-three centuries later. And in both cases we have a struggle between the search for the organic and the attempt to organize vast theoretical panoramas. For also the mathematical theory of categories emphasized from the beginning the idea of natural transformations; but within the context of the great mathematical industry, it is difficult to see what can be «natural» and what «machinery». All present-day "grand unified theories,» whether in physics or mathematics, have a clearly conservative function, exhibit common ills, and are expressly designed not to rethink the fundamentals except where it is harmless: they are the escape route for the manifold contradictions of modern research. Mathematicians have been enormously perspicacious in exploring the possibilities of their science, but only in a definite direction —the one that preserves the achievements and limitations of a model that, although it seems to transform continuously, has not changed in essence since Newton.

So the «natural transformations» of mathematics have at least as much of history as of nature, a history that is not only made by its «impressive successes" but must necessarily have a counterpart. We have seen one way to unravel, at least partially, these compromises. Another way is to apply from the outset the natural transformations of concepts to a framework that deals as directly as possible with the very forms of phenomena, such as Venis' projective morphology —a morphology of the vortex that has been formulated in terms of pairs of opposites. This morphology is still awaiting proper formalization without the historical connotations that analytic and algebraic tools have acquired, but on the other hand it has a far superior analogical potential, a potential that could still be much better realized if calculus and differential geometry can naturally incorporate the theory of proportions. Vortices, as topological defects, are a natural object for the modern language of homology and cohomology, but in projective morphology they acquire an intuitive value they never had before.

Venis' morphology naturally incorporates the idea of morphodynamic equilibrium, and a proper idea of equilibrium is necessary to connect and give physical content to the theory of proportions. Of course, equilibrium by itself is such a general concept that it cannot but be present in a thousand ways in physics and mathematics; but again we meet with truncated modalities of equilibrium from the very beginning —from the same three principles of Newton's mechanics onward. Equilibrium is as principal in open systems as are conservation laws in closed ones; but, contrary to what is supposed, we have reasons to think that the latter emerge from the former. So we have insisted on the importance of replacing the intrinsically contradictory principle of inertia by the principle of dynamic equilibrium, with the consequent modification of the other two; just as we have emphasized the importance of ergontropic equilibrium, and the equilibrium of densities with a unit product. The connection of these four equilibria is already a whole research program, and quite different from the present ones; the golden scale of Zeus has nothing to weigh in closed systems.

These other forms of equilibrium play a fundamental role in the morphological process of individuation, and thus it is possible to connect equilibrium, proportion, form, proper name and transformation. This would in turn create a new constellation for the relationship between quantity and quality, language, logic and number. The untruncated study of balance in Nature and in human activity, including individuation processes, points to an understanding of purpose and time that transcends teleology and dysteleology, our prejudices about the perfection and imperfection of individual phenomena of any order, be it a wave, a sun or a civilization.

Whereas the great "unified programs" point toward an indefinite unity, here we consider that what directs intuition is always behind judgement, in the minimal matrices capable of condensing a balanced totality. The balance of the four questions and the four modes are some of these matrices, but even the three principles of mechanics are the extreme contraction of the three basic modes of the principles as starting points, basic distinctions and foundations of unity, which determine common limits of global closure.

Mathematics self-exam is revealed in a particular way in its diagrammatic dimension, something that Peirce rightly emphasized. As a visual icon, the diagram reverts reasonings of higher level towards the most primary dimension of thought, and this retrograde movement is essential not only for analysis but also for new syntheses; Peirce wanted to use it above all for logical analysis and had to assist with resignation to the most involutive effects of this process. The last decades of the twentieth century have seen a resurrection of diagrams with the study of Riemann surfaces and their topological and group correspondences, but this too must have a diminishing and limited return, because, as Alfonso De Miguel Bueno claims, the first thing that should be translated to the basic plane of the diagram are the most elementary aspects of numbers, groups and models of physical systems.

We start from the obvious but little recognized fact that current mathematics is full of leaps into the void, sleights of hands and the most varied subterfuges. In these conditions of rarefied abstraction, diagrams must be, to use Peirce's terminology, degenerate icons. For diagrammatic reasoning to reach an optimum of iconicity —of relation between the sign and the object— it should try to meet those conditions that are now so much despised: descriptions with homogeneous quantities, correct physical dimensions, analysis with finite quantities, and so on. When diagrams are sufficiently immediate, as are some of Bueno's, there is no need to attend such demands.

In any case, since the icon is the most primary thing in mathematics, and is opposed to the arbitrariness of both the tertiary symbol and the secondary index, the search for an optimum of iconicity supposes a maximum of suchness, of participation in knowledge, and an approach to that which we call fourth-person knowledge and non-intuitive immediate knowledge, with the possibility of connecting elaborated knowledge and immediate participation. The intelligibility of the real depends on continuity. The homogeneous medium cannot be the ultimate interpreter insofar as it rests in itself; but to the extent that demands the maximum degree of continuity between its most distant moments of intelligibility such as Peirce's three categories.

It should be emphasized that diagrams with high iconic value are much more than heuristic aids and even reverse the sense of the current research. If it is true that each culture has treasured its own intimate sense of number, and only in the final phase of civilization does an interest in the «mathematical results» of other developments arise, the conversion of its great motifs into icons of high purity is equivalent to the moment when the seeds of a tree come back into contact with the primordial soil. For example, number theory has been shaped by the contributions of three very different cultures, represented by the names of Euclid, Diophantus and Gauss or Riemann; what would it mean to find critical diagrams within these motifs, as Bueno seeks to do? It would mean going beyond the search for results, beyond the will oriented in a certain direction; it would allow the sense of the language of Nature and the meanings of our languages to coincide in ways hitherto forbidden by the very direction of development. In a retroprogressive sense, diagramatology can be both science and technology, although opposed to the present trend towards complexification.

Surely the only open problem in mathematics that seems to have by itself a potential to change our basic idea of numbers is the Riemann hypothesis. And this for several more or less obvious reasons: because it is the most basic relationship between addition and multiplication; because it directly affects the integers, the relationship between natural and prime, rational and real numbers, and by extension the other number fields; because it has multiple physical and mathematical connections; because it invites us to interpret the relationship between real and complex numbers in the physical plane; because it establishes the most irreducible link between chaos and order, chance and necessity, simplicity and complexity; and because it is the only possible unifying motif, together with its large family of associated functions, for a theory, that of prime numbers, which would otherwise be the least unitary area of mathematics. The only kind of unity that it is reasonable to expect from the prime numbers are asymptotic laws, otherwise of the most diverse orders, for which there can be no more general common motif than the Riemann zeta function.

Faced with a question like this one, some strive to achieve a proof and others seek a better understanding and intuition of the problem. Arithmetic and geometry continue to be basically dissociated, even though the efforts to connect them are stressed today. Thus, for example, there have been attempts to build bridges between the theory of the zeta function and the most sophisticated algebraic geometry, which has so little geometry in it after all. Since it is a commonplace that «all geometry is projective geometry», that the latter is simpler than affine or Euclidean geometry, that it shows a persistent tendency to reappear in unconnected fields, and that it provides an undoubtedly intuitive basis of the widest range for asymptotic motifs, one may wonder why this function has not been successfully linked to the immense possibility of the projective in general.

The way to establish the link would have to come from physics, and electromagnetic theory in particular. It is well known that multiple correlates of the Riemann zeta and the zeros of its critical line have been found in electrodynamics, from the potential of electrostatic fields to far radiation patterns; as it is also known that electrodynamics has resisted every attempt of geometrization and that there is an irreducible statistical component in it. Delphenich observes[6] however that the metric theory of electromagnetism may well conceal a more general geometrical question of complex projective geometry commensurate with the transition from point mechanics to wave mechanics, and his arguments are worth pondering. On the other hand, it is assumed that the zeros of the Riemann zeta emerge from the difference between a sum and an integral. The Lorentz transformation in electrodynamics shows a one-to-one correspondence with the projective Möbius transformation, but the same can be said of Weber's electrodynamics with retarded potential, which in addition gives us a different interpretation of time.

In a word, if there are electrodynamic correlates of the zeta function, there should also be non-trivial correlates associated with the projective geometry of waves, and this, which in principle has nothing to do with any kind of proof, does have the potential to alter our qualitative conception of numbers and their use in the physical world. Obviating harmonic analysis, this longitudinal cut would make it possible to connect the function with other questions more or less unrelated: the fundamental irreversibility of radiation, equilibrium in open systems, Weber electrodynamics and its relational mechanics, the relation between retarded potentials and the geometric phase, the shaping of a wave front, the projective morphology of the Venis wave-vortex and its transition between dimensions, the theory of proportion in differential projective geometry, and so on.

For identical reasons to the holographic principle, the torsion morphology can trivially produce any surface or pseudo-surface, while according to

Voronin's conditional universality theorem, the Riemann zeta function can reproduce any analytic curve with any degree of approximation an infinite number of times. The zeta function is already an extreme numerical synthesis that is looking for another synthesis in its extreme counterpart. And there would be a «projective arithmetic» quite different from those that have been developed since the time of von Staudt, with its theory of proportions and correspondences, its algebra and its own "science of the balance".

Riemann's express motivation was prime numbers counting but the origin of his intricate elaboration is in the theory of functions; as for the prime numbers themselves, which can already be counted in the classical manner without much difficulty, the German mathematician could only expect them to behave fairly, just as in the probabilities of flipping a coin an infinite number of times. But the great mass of implications of the function cannot come merely from this ideal balance, but from its fusion with complex analysis, and the converse question is whether by recomposing the idea of the function and the differential one can draw lessons for the very idea of number. Projective geometry is purely qualitative, and evidently, the interest of finding a projective, as well as a physical, analogy for the function, consists in distilling the qualitative aspects of a purely analytic numerical expression. The concept of function should naturally correspond to that of process, and yet, due to the rupture of analysis with description, there is a huge divide between them. If truly the Riemann hypothesis can tell us something qualitatively new about number, this would be the best way to investigate it.

The simplest interpretation of the hypothesis as infinite balance must maintain an essential continuity with the countless less obvious, deeper interpretations; but since according to the present standards it does not serve to prove anything, it cannot be taken into account. Here, on the contrary, we believe that the proof is secondary to our capacity of understanding and assimilation, and this increases by maintaining the same principle at different levels and observing the transformations of its application.

We have seen that it is difficult to justify[7] the critical line and the calculation of the zeros from the very criteria of the Cauchy-Riemann equations, however much the mathematical community has considered them unproblematic; we have also conjectured that the same identification of a critical line and its zeros could arise from a projective analogy in the context of the densities of an electromagnetic theory with retarded potential such as Riemann's. And it is inevitable to think that if mathematicians have accepted Riemann's calculations it was not so much because of their analytical rigor as because without them there would be no axial motif in number theory.

But even if these connections on a large scale were achieved, and took place at a much more intuitive level than in the present macro-programs, their qualitative impact on such things as our idea of number would be limited to what we are able to assimilate in the minimal matrices of meaning and in the maximum contraction of principles —they depend on the degree of harmony and solidarity with them, just as in mechanics the basic disposition have much more weight than the results of the nth level. So today it is perfectly illusory to expect «revolutions» from higher research, which is nothing but an institutional luxury and a last precipitate of civilization. Also in the teaching of mathematics this is the most important thing, no matter how much the institutions try to preserve the inherited forms. But the scandal of modern mathematics is not the institutional luxury of high-level research, but the way in which the quantitative has become an instrument of control and of closure of the system on itself.

Cosmopolis

The modern Technopolis wants to conceive itself as Cosmopolis but it exists to the extent that it reduces and denies the autonomy of the powers of the Cosmos. And so, for example, it insists on linking climate with human behavior as if more important factors could not exist.

Nothing encodes the fate of our technical civilization better than the myth of the city of Tripura. The triple city has a date with itself but evades it as best it can because the alignment of its three levels also implies its destruction. Technology demands the perfection of its own sphere, however alien it may be to human nature; but perfection cannot be expected without the concurrence of that cosmic element which it has tried to evade during the symbolic thousand years that close the cycle of its development.

This open cosmic element, can be either outside or inside the great social animal; it hardly matters as is bound to remain ignored till the end. At the entrance of the temple of modern science one can read the two great precepts: «Ignore yourself», and «Everything in excess», and its countless priests and dependents try to keep them with the most abnegated jealousy. Since the worst of this science so little our own is its spurious sophistication and its exacerbated disconnection with almost anything that matters, we have always tried to re-establish some kind of link between the idea of knowledge that has been held in other ages and modern scientific knowledge, even being aware that trying to marry prudence with folly need not be of any benefit for neither sides. There is, however, an organic link between all generations, and reinforcing it when many seek to destroy it is more important than any isolated result.

A crucial feature of modern science is that it generally sustains an entitative interpretation, a combination of operational and logistic methods, and comprehensive or global principles that descend directly from the principles of Newtonian mechanics. Since these in turn have as their point of departure the closure with respect to open systems and as their tacit point of arrival the global synchronization implicit in the action-reaction principle, this circuit draws the horizon of their whole teleonomy. We have seen, however, how these same principles can be reformulated in terms of equilibrium for open systems without the need for imposed closures or metaphysical instances.

In the XVII century, at the start of the scientific revolution, people like Stevin or Newton still believed that Europe was just beginning to recover a lost original wisdom; but the progressive formalization and specialization, together with the amount of accumulated knowledge, made that idea definitely implausible. Today the most that can be admitted is that in other times man relied much more on the power of analogy, which strictly speaking is always an inverse relation; but we already see that scientific knowledge cannot do without its ascendancy either, neither in hypotheses, nor in mathematical procedures nor in interpretations, and in fact it exploits it systematically, although only in a definite direction. Newton himself came to his greatest discoveries in optics, calculus, gravity or the principles of mechanics reflecting about inverse problems.

Thus, inverse problems were at the fulcrum of his natural philosophy, and it is the direction given in this turn, more than anything else, what created the drift of modern science. That is why it is necessary to reverse the order of the sequence knowing full well that a mere return is neither possible nor desirable. Analogy has always been and will always be the most powerful tool of theory; the important thing is to know how to consistently give it another direction when it has been sufficiently seen where it leads us with its present orientation. There are still those who believe that science can be partial in its application but never in its principles; but this is to ignore completely the symbolic range of the framework of knowledge itself. And domination consists precisely in the framework that one manages to impose. Much more than the results, what matters is to control the direction in which the spirit of others works.

To give another orientation to science does not pass through the latest theoretical by-products but through the conscious restitution of what is fundamental. Today «theory» is synonymous with the speculative operation of knowledge, but we know that for the Greeks it basically meant contemplation. The very idea of number can only be linked to the good, beauty and contemplation by recovering its nature and value as proportion, which cannot be limited to the sphere of geometry. Today, restoring the meaning of proportion to number means appreciating its natural presence in calculus, which is no longer impossible after more than three centuries of endless analysis.

This unexpected emergence of the theory of proportion in analysis should be closely connected with a general theory of equilibrium in open systems, which will return us by the straightest way to the ancient Science of the Balance. Arabic algebra was literally born as «the Science of Restoring and Balancing", but separated already in al-Khwārizmī from natural philosophy, it soon rolled down the slope towards the science of substitutions and the indiscriminate logic of universal equivalence. Duly linked to the theory of equilibrium in open irreversible systems, algebra can leave the realm of universal exchange to become again the science of the irreplaceable.

The geometric phase shows that quantum mechanics is incomplete as is any conservative mechanics, but most prefer to believe that this is a secondary detail. Nor has it ever been seen that the same ray of light returns to the flame or to the light bulb, and yet everyone speaks calmly of the temporal reversibility of the fundamental laws. Thus physics would like to maintain for its object the same timeless status of mathematical truth, but not without an irreparable loss in our perception of reality. This inverted Platonism is at the origin of modern technologic optimism and its selective disconnection with many aspects that are not only essential but also vital.

The geometric phase allows to shape a wavefront modulating its potential instead of using controllable forces as is prescribed in physics. This consolidates the existence of a third domain of experience in the history of the discipline, since, alongside perturbative physics, such as that of cannons, machines or accelerators, and non-perturbative physics, such as that of astronomy, there is now physics that operates by tuning or modulation. However, everything is still interpreted in the old framework, even if the force is no longer fundamental to both quantum mechanics and relativity. But the connection between the retarded potential and the geometric phase, or between the potential in the dynamical equations and the one beyond them, directly affects the theory of equilibrium as well as the morphology of observable processes; it even has to find an explicit correlate in biological feedback processes connected to our consciousness.

Consider the relationship between respiration, its bilateral cycle and the phase shift of the retarded potential. Since, as we have seen, the retarded potential is a ubiquitous phenomenon and is equally present in biological functions, the study of its equilibrium in a case like this, which is also linked with consciousness, and its analysis in terms of finite intervals offers us a limited but precious example of what the internal side of the mathematical law entails.

It is in open systems that equilibrium can acquire something of its true dimension, and it is in a finite calculus such as constant differential calculus that questions of proportion reappear in modern mathematics and physics. The ancient Science of the Balance deals in addition with the proportion between the manifest and the unmanifest, which in physics finds some partial correspondence with extensive and intensive properties. There is, of course, a whole gulf between the intention and quantification of the one and the other, but nevertheless a certain continuity still persists which is based on the generality and multiple planes of the very concept of equilibrium. By contrast, with the tacit principle of global synchronization in conservative mechanics it would seem that everything is on the same plane of causality, which incidentally cannot be specified because it is already established beforehand.

The Science of the Balance would also be naturally related to projective geometry and morphology, and, assisted by these, would allow us to grasp a whole scale of equilibria. Since the scope of equilibrium theory is completely different when we deal with open and closed systems, like any other issue the very idea of balance can be analyzed by contrasting the four questions and four modes of investigation, as they consider open and closed systems, in order to have a broader perspective of the matter, and to better grasp how access to certain approaches has been systematically blocked. The four modes have an archetypal value because they not only trace the watersheds of thought but of will also.

Everything has an equilibrium in its specific interval of change, an equilibrium over time and an equilibrium beyond time. A theory of equilibrium in the sense indicated here must necessarily establish other connections between physics, mathematics and natural philosophy. On the other hand, modern analysis and the theory of point particles, perfectly dispensable in other frameworks, prevent the specific study of the processes of formation or individuation of an entity. Mathematically, there is more incommensurability between a point and the radius of a particle than between it and the dimensions of the known universe. We have spoken of a «symplectic morphology» not in the sense now given to the word "symplectic" in differential geometry, but in terms of the *symploké* of the individual, its complexion, which implies both a connection with the medium and a differentiation and independence with respect to it.

It is not that the individual has an experience, but that the experience has made the individual, who in turn is usually engaged in a second-order experience. This uncontrolled experience, which Nishida called «pure experience,» is the original consciousness prior to thought. There is a natural analogy between this original consciousness and a homogeneous medium prior to the conventions that determine metric, affine or projective spaces, although thoughts themselves can be conceived as projections, not on an extended space but on a line of duration that need not be an external timeline.

Nor is the analogy between consciousness and a physical field entirely unfounded, but the word «field» only indicates a portion of space with certain associated quantities, and mechanics is after all more fundamental; especially if we recognize that there are basic mechanical facts that are not captured by field theories. Nor has it been possible to demonstrate the stability of matter, but like so many other things it has been rationalized by relying on phenomenological arguments such as the exclusion principle. The idea of equilibrium is more general and has more scope than all modern theories, as long as we know how to remove the obstructions and blockages they have created.

The idea of balance connects as directly as possible the conditioned consciousness with the unconditioned, our direct physical experience, our mediated intuition and our immediate non-intuitive knowledge. In symplectic morphology, balance regulates the whole process of transformation of matter and form in the beginning, in the end and in the middle. For Yabir the highest part of the Science of the Balance was the Balance of Letters, synthesized by three of them, corresponding to these three moments. Impossible not to think of the sacred monosyllable of the dharmic traditions proposed as the name of the absolute, where being and consciousness are always in perfect equilibrium without affecting in the least the unknowable infinitude in the background.

The infinitude of language and its transformations, as well as the infinitude of everything manifested, is only a minute part of the infinitude named by the Name, and yet it is more inherent to us than any object of knowledge. The interesting thing for knowledge is to what extent it can come to terms with this metacognition; a metacognition that is not knowledge of knowledge but intimate transmutation of intellect and will.

Notes

(www.hurqualya.net)

1. Miguel Iradier, *Espíritu del cuaternario: semiosis y cuaternidad*, en https://www.hurqualya.net/espiritu-del-cuaternario-semiosis-y-cuaternidad/

2. http://milesmathis.com/are.html

3. https://www.researchgate.net/profile/Michael-Mcbeath/publication/15474292_How_Baseball_Outfielders_Determine_Where_to_Run_to_Catch_Fly_Balls/links/55a567d108ae5e82ab1fa030/How-Baseball-Outfielders-Determine-Where-to-Run-to-Catch-Fly-Balls.pdf?origin=publication_detail

4. http://infinity-theory.com/en/homepage

5. https://richardmckeon.org/content/e-Publications/e-OnPhilosophy/McK-PhilosophicSemantics&Inquiry.pdf

6. https://arxiv.org/ftp/gr-qc/papers/0512/0512125.pdf

7. Miguel Iradier, *Ironía y tragedia en la hipótesis de Riemann*, en https://www.hurqualya.net/ironia-y-tragedia-en-la-hipotesis-de-riemann/

SUNDAY EVENING

18 July, 2022

We always have intimations to the measure of our capacity, even if we neglect them for the benefit of other more exchangeable and mundane currencies. When I was a child I dreaded Sunday evenings; something in me sensed the unfathomable mediocrity of the world of the elders and their dull flight from boredom. However I was never at odds with the dreamy grayness of the asphalt or with the implausible trees that cross it, and now I understand that discomfort came more than anything from not disposing of my own time, because those who do it always find ways to deceive themselves. Over the years, the Sunday evening has become the time of the week that I appreciate most, and although I rarely honor it as I would like, it still opens a window in my mind out of the wheel of repetition.

The first time I was able to realize that this suspended interval can say something different had to be the Sunday afternoon of June 23 of 1985, under grave but not serious, intimately familiar circumstances. I remember the heat, the rays of sunlight filtering through the corded blind at the guesthouse, the complicit solitary waiting and that old television tune coming from some neighboring room that broke the silence at a given moment. The impression that afternoon left on me was so imperceptible that it took me twenty-five years to identify it as something of its own, to understand from what regions it flowed.

Unlike other calendar arbitrations, and with very rare exceptions, the sequence of the days of the week has known no interruptions in more than two thousand years; something that has had to leave its groove in the collective feeling. It is hard to believe that the social engineers, whose perversity never exceeds their foolishness —although that is not reassuring at all— have not launched a campaign to destroy this decayed relic once and for all, but in case we are suggesting improper ideas, as compensation we need to propose something in the opposite direction.

In days of yore, people who could be alone without boredom considered themselves intelligent; now that we are all entertained and alone without realizing it, there is practically no term of comparison. And in spite of everything, boredom continues to leave on people's faces the unmistakable mark of its devastation, aggravated in the long hours of the afternoons in which one does not know what to do. We complain about the brevity of life, but when in the whirlpool of consciousness doldrums arise as a modest transcript of eternity in time, we can't stand it. It is not the same to imagine exploring the infinite as letting the infinite pass through you.

Anyone can circumvent boredom by doing anything, but few welcome it and wait to see what it has to tell us. Boredom comes from the predictability of repetition, from anticipating something that has not yet happened, and from trying to avoid something already present; what desolates the vulgar mind is not the presence, never sufficiently valued, but what we may call its imaginary component. Evasion and avoidance are almost the same word, although this evidence also evades us.

Someone imagined that Abraham, father of the multitude, asked his close friend, «And how is it that you are God?» Or put another way, «How do you do to be God?» It can be assumed that the patriarch, who had bargained with God over the number of righteous men in a city, was not asking him for his credentials. And the only answer such a God can give, in full agreement with that other answer that anyone, religious or atheist, could imagine hearing —"I do nothing at all"- is silence. Silence and an imaginary answer that should please us were we able to listen without words.

Gods and Titans may be eternally striving to overthrow their rivals and seize power, like any wannabe tyrant or any political party; but a unique God could only be unique if he had never had to do anything to be unique or claim anything for itself. Surely this is why some have distinguished between God and the Godhead. But we see that I, the World and God —or the I, the World and the Law— coexist as evanescent moments of the same illusion, and that in vain we pretend to rid ourselves of one when we still believe in the reality of the others. And so one can see today men of science trying to convince us that the self or consciousness arises from neurons and molecules; this is of course a misunderstanding, but where is there anything that is not.

One can put aside the world for a moment, which will never deign to answer our questions directly, to ask to oneself: «And what do I do to be I, or even to know that I am I?» And the answer can only be the same as that of Abraham's intimate friend, so different from the God who promises an offspring as numerous as the dust of the earth: nothing at all, but on that everything depends. It is worth thinking, too, that no other is the arcane of sovereignty and power, especially when we know that all those who claim it the last thing they could afford is doing nothing.

"Abraham's secret question» and the inevitable answer it brings forth is infinitely more eloquent than any ontological argument about the existence of God or any attempt to finish off his ghost, a recurrent and endless task for those who believe in their own Self, in the World, or in the independent existence of the Law. It has been said that the pretension of demonstrating the existence of God, the key of modern theology since Anselm, contains in itself the rest of the pretensions that later science has been deploying in rigorous order of delirium. In any case, they have arisen from the same pathos, and now that science has been entirely engulfed by power, for which it was always predisposed, it is time to find a less harmful key.

That consciousness is voidness and that there can be nothing below it is, like the answer to Abraham's question, something that leaves no room for doubt and yet where uncertainty reigns supreme. One can thus understand the preference for the scientific method, where doubt and certainty are permanently negotiated and always leave loose ends to be tied up. One might think that a science that contemplates movement from the immobile and action from non-action would be really unprecedented; but that is already what happens, and rather we should see how this can happen.

Nothing comes from nothing, but it may well be that what does something ceases to do it, or that what seems to do something is only compensating some kind of reciprocal motion. Our science, as the heir of the multitude and dust, insists on ignoring the possibility of the latter and tries to convince us of the first and most basic impossibility. So much talk about the inviolability of energy conservation to conclude without blinking that everything came out from nowhere. It is like those guys of the Federal Reserve who assure us there is no free lunch, but for themselves; or like the bankers who say that they cannot give without guarantees when the one who creates the money is the one who asks for the credit. Thus cosmology is in good company, and besides, who does not prefer to think that we are something rather than nothing? Positive science can be the best ally of credulity.

Half a century ago even mathematicians protested against the Vietnam War; today it is not easy to explain how one can be a scientist and have a conscience without leaving the field. But of course it is understandable, as the disposition of science did not start yesterday and has acquired an allegedly unstoppable inertia. For the scientist to stop being the complete nullity he secretly is today, he would have to want other things. Leaving the field or an indefinite general strike would not change anything even if that were possible. Do they predict that artificial intelligence will soon replace the mathematician at proving theorems? Why not, maybe then he would rethink what he wants to devote his intelligence to. Meanwhile, the mathematician himself is the best available imitation of that long-awaited machine.

One is incurably optimistic, although for reasons different from those of the average scientist. He can feed the illusion that he is both serving society and shaping it with little concern for the contradiction that this implies in general and for his own formation in particular, but here we would like to imagine for a moment that another kind of purpose is beginning to emerge. As long as he remains so busy, the man of science will never cease to be the instrument and façade of another far more concentrated power. Of course, there is always great room for autonomy, but what good will it do even to a mathematician, the most immaterial of researchers, when in his own field he ends up bowing to the «powerful methods». We never stop to think about what this means. Can truth or reality depend on how powerful our methods are?

But the most powerful means of a theory is the analogy, and the analogy proceeds by assimilation. Some say that science, always so modest, does not raise ultimate questions as philosophy does, but is content to solve problems that have solutions. Now, not only we made too much of ultimate problems in philosophy, but, under the pretext that its speculations are clothed in calculable forms, science too has stretched its own problems to infinity, to the point of absurdity and even beyond the ridiculous. And furthermore, as we have ceased to give credit to thought for the ultimate questions, now the sciences themselves want to answer in their characteristic proactive style why the world, life or consciousness exist.

From the point of view of appearances, it is unquestionable that things have their origin and becoming. But it is science itself that evacuates the question of how phenomena come to be what they are, and replaces it with a subsidiary why that should fill the void left by the when and how much to which prediction responds. Only by attending to the how could we see how action connects with non action and how the mundane exists within that. Our well-oiled science neither cares about the how of Nature nor wants to show the how of its own procedures; it cares about both as much as the banksters care about explaining how money is actually created and how it could be created without their schemes and machinations.

If current science, which only understands motion and action, pretends to assimilate even that which does absolutely nothing, such as consciousness, a reciprocal movement is also possible in which consciousness *ASSUMES* this same science and tries to bring it to its own plane and reality. Here we have seen some of the wider avenues to make this assumption viable in the theory itself by starting from a homogeneous medium, by modifying the principles of mechanics or calculus, by considering other aspects of the idea of equilibrium, or by properly addressing morphology and individuation. That there is always room for this theoretical transmutation of principles, means and interpretations should be beyond doubt; but of this theory one also should ask what is its sphere of «technical application», and even more so what kind of practice entails.

There is a virtue and an efficacy in trying to see Nature from the side of non-action, a certain providential pertinence, which we have not yet begun to contemplate. The fact that all that is valued today in science pales before the calamity it has already become ensures that it has another value that we are not in a position today to reintegrate into our life; but the indispensable reciprocal movement tends once again to make these conditions possible if we succeed in granting it a space. Physics and mathematics have wanted to make an object out of emptiness and nothingness in a thousand different ways, but non-action, not to be confused with inaction, has little to do with this type of objectifications.

The laws by which man circumscribes the processes of Nature enact limits that must not be exceeded; non-action, on the other hand, adheres obscurely to «that which cannot be exceeded» at all, or cannot be exceeded without a corresponding reaction. If in practice it reveals itself as that in which there is neither an active nor a passive subject, in the principle itself it translates into pure indistinction between action and reaction. But even if this seems extremely vague it still has a specific translation in the domain of mechanics as a proper time of action and as another intelligence of cause and its absence.

The way of action to arrive at non-action is self-deceptive and virtually endless; techno-science, on the other hand, tries to exploit natural inaction to the maximum and convert it into action, but non-action has no opposites. Human law is teleology in disguise; it is enough to think of the whole modern theory of potential to see that man has hidden his own ends in it and has pretended that this is the very purposelessness of Nature. How to undo such a tenacious misunderstanding?

Think for a moment of the dreams of quantum computation and all its accompanying hype. Is it not a matter of exploiting the non-separability of potentials to obtain explicit calculations? Of exploiting inaction for action? All it takes is a timely technological war not to even question to what extent this may make sense. And yet there exists in calculus a certain special function, studied to exhaustion but surely even more ignored, that could be telling us in detail how action is subsumed in non-action. Even the dubious «quantum computation» could much more easily be put at the service of reading this magna encyclopedia of non-action than at the service of increasingly arbitrary ends. But, here as in everything else, would it not be infinitely preferable to seek the essence of this non-action rather than to entertain oneself with an endless study? It only remains to ask how it is that non-action matters; even if what it holds is not something that needs to be revealed or made manifest.

Non-action and true activity coincide with a greater certainty than that of any demonstrable truth; but if we are unable to conceive this, neither can we conceive of the nucleus of activity either in external Nature or in our own nature; and because we are so far from knowing their common background, to remove the capital letter from the external side of the equation is as pretentious as to believe that physics has unraveled their mysteries. With respect to this activity, physical action would be just a transfer or transaction. But it is not so much a matter of asphyxiating this other certainty with new concepts as of making it less inconceivable to our web of unreality. The worldly powers are interested in the separation of these two natures and does all in their hands to maintain it; but in no way can prevent us from reestablishing the links.

Non-action is absence of intention, but without intention a closed mechanical system is untenable. The difference between one thing and the other may seem too subtle but on the other it has an extreme potential for controversy, because what should be something diaphanous and lucid remains suffocated like a subterranean fire. A polemic that could only arise from the friction involved in ignoring that both fires are one, and although it is always preferable to avoid it, the degree of dullification of the intelligence in the mechanical makes it necessary to bring its apparent contrast to the foreground.

It seems as if both knowing and not knowing and not wanting to know have all taken the wrong turn. Enlightenment was promoted as the end of childhood, but sticking to what is indubitable, however uncertain it may be, takes us on the path of life and on the path of return that is one with it; while negotiating certainties and doubts at convenience takes us away from that path and leads us, as now we can see, to an increasing infantilization. Long ago, a poet philosopher painted a predictable retired pope, but although it was no less predictable, the pre-retired scientist managed by the administrator has taken us by surprise. In order not to be retired or exhibited like a trophy, the genuine researcher has no choice but to find his own place of retirement from all that competitive and entertaining hustle and bustle. And that place does not pass through the intrigues in the management of the great apparatus, but through the most intimate reconsideration of his own object, because that is the space that the theoretician has to open if he wants to be something more than a managed asset.

In any realm whatever, non-action and the indivisible illuminate each other, since they are only different aspects of unity.The sciences always have room to free themselves from the tutelage of power, if they are capable of conceiving another power within their theory, their application and their practice. If this is achieved, the relationship between knowledge and power can still be reversed in a spontaneous and indeliberate manner, since power, how not, also aspires to the advantages of not doing without having to apply the remedy to itself, much less to practice it.

ON THE INNER CONTROLER

18 September, 2022

«O Antaryamin! Indweller of our hearts, friend of the poor, protector of the helpless, purifier of the fallen. Forgive us our sins. Have mercy on us. Show us the simple path, the real way to the supreme abode of peace».

After the war, Norbert Wiener refused to work with military or government funds and that is to his credit, but today any scientist justifies himself by thinking that if he does not commit someone else will do it in his place; the excuse is too easy although it has a lot of truth in it. In any case, to compensate in some way for the general degradation the sciences have reached, a degradation whose extremes the father of cybernetics could hardly foresee, and the deleterious effect they have on the most varied orders of existence, something more than depriving this or another sector of our increasingly dubious and interchangeable contributions is needed.

Since 2020 it is no longer possible to ignore that fraud in the sciences affects the whole system and is consciously directed from above; because we are no longer talking about the simple control and selection of discourse, but about the omnipresent falsification and concealment of data, adulteration of explanations, and criminal practices that directly undermine the integrity of life and human dignity. Whoever tries to survive in this environment may delude himself into thinking that such misconduct does not affect his specific area of work, but even if that were true, his mere complicit silence demeans him. Corruption is not just a means, but an end in itself, the triumph for those who seek the destruction of any standard. Most are incapable of understanding the consequences of this destructive logic even if they are right in front of their eyes.

And yet it is sometimes openly admitted that the present technoscience no longer aims to explain the world or even predict it, but simply to modify it, even deliberately ignoring what the effects might be: «creating disruptions», to use their nefarious jargon. So many thousands of years of accumulated human experience to end up worse off than the child with the stick in the anthill; much worse, no doubt, because the child is still curious to see what effects it triggers, and does not deceive itself as to the nature of what he is doing. What we have then is a perverse and perverted activity, and not only in the way it treats life but also in the way it applies to its own rationality, even in mathematics itself.

We don't write more on the subject because for some of us it is too obvious and because fortunately there are authors much more gifted for denunciation, education and controversy in this detestable arena of media intoxication. In any case, we owe a great debt of gratitude to these warriors for truth[1] as the world would be much more miserable without them. Not the machines, but the machinations behind them are frightening; the machines proliferate so that the machinations of the evildoers have more impunity.

Since that distant "knowledge is power", it was written that science had to get here, or at least that is how it must seem to us now. If the question was to dare, it is clear that man dared; however, in this way science has not only been the instrument of a fall within a fall, but science itself had to experience its own fall and degradation with respect to what it was and what it can always be. We try to glimpse that with which science confronts itself, but not that from which this same science runs away, when its fall is precisely that flight. That fall extends to the past and the future, and inevitably affects how we understand those aspects of time —for it is but the fall in time itself.

In the Brihadaranyaka Upanishad, Uddalaka Aruni makes an approach to the nature of the Antaryamin, that inner controller whose very «inwardness» veils the omnipresent reality of the self. How can that which is all be hidden from us? The question itself gives us the answer. Though, on the other hand, how could we notice that which does not change? Uddalaka's famous answer is that you cannot see it, but it sees you; you cannot think it, but it can think you; you cannot understand it, but it can understand you. The consideration of this immediate absolute leaves no room for any other conclusion, nor for any kind of elaboration or science. But the Antaryamin is described in three different ways: as transcendent absolute, as objective physical reality, and as internal subjectivity.

Physics is prone to believe that, by virtue of our notions of mechanics, the world does not need any internal controller; however, the tacit principle of Newton's mechanics, a direct consequence of the third, is the global synchronization or simultaneity of action and reaction; synchronization in which all subsequent physics, including relativity and quantum mechanics, is also embedded. It may be asked then whether this global synchronizer, undoubtedly a metaphysical principle, is in itself an internal controller, or whether it rather prevents its correct appreciation. Cybernetics is the evolution of control theory, and it is anything but accidental that Wiener begins his fundamental work by contrasting Newtonian and Bergsonian time.

But the contrast presented by Wiener is already a justification for the machine to take command, because Bergsonian vitalism is a straw man that in no way provides a scientific alternative. Wiener is above all a mathematician and cannot accept a definition of time that is not mathematical; what he did not want to contemplate is that Newton's principles can be replaced by others that require neither inertia, nor constant forces, nor simultaneity of action and reaction. Long before Wiener was born, Wilhelm Weber, developing an idea of Gauss, had proposed such a dynamics that made correct predictions. Weber's dynamics is based on dynamic equilibrium, variable forces and an apparent «retarded potential», and thus what he describes is a feedback loop even in the fundamental forces. We will never know what Wiener would have thought about it; the question is whether it makes us think better at this point.

Weber dynamics has obvious advantages over the much later theory of relativity, based on simultaneity and infinitely more publicized; just as it allows a less unproblematic connection with microscopic physics. Although, as things stand, perhaps these are additional reasons to keep it buried in oblivion. But what is essential for our subject is that in Weber dynamics feedback is an immanent principle, whereas the basis of all that we understand as mechanics is an external principle added to the system itself —an entirely metaphysical principle. Therefore, to answer once and for all our previous question, Weber's misnamed law of «retarded potentials» fits perfectly with the notion of an internal controller, whereas conventional mechanics excludes it because what it demands is by definition an organization external to the system.

Moreover, if there is a simultaneity to be considered, it is precisely that of potentials, which do not require transmission time, whereas every interaction takes time necessarily. Not without bewilderment, physicists began to consider this in the fifties of the last century, and they still do not know how to judge about it; therefore they continue to affirm that quantum potentials are a special case, when it is obvious that any potential of any kind corresponds to a position and is independent of time. Presently anyone knows that the geometric phase of a potential, routinely used in control theory, operates at all scales; so if Wiener could still find excuses when he wrote his book, today there is no excuse possible. And yet the subject continues to be ignored.

As a mitigating circumstance, it may be argued that physics interrogates the world with forces and through forces, so that consideration of potentials is necessarily posterior; control theory only echoes this general situation in a restricted way. But, once it has been established that potential is not a mere auxiliary of force, but in fact it is even more fundamental, we should take the facts into account and reorder our sequence of reasoning. We have the objective evidence, it is our subjectivity that has not yet come to terms with it. It is also true that generations passed before physicists made Newton's ideas their own, and even today this is still done in the most superficial way possible.

If we speak of subjectivity it is because the three laws of mechanics also follow the ternary drift common to all signs and adhere to the conventional tripartition of temporality: inertia comes from the past, force is projected into the future and the simultaneity of action and reaction defines a fugitive and infinitesimal present without any thickness of its own. In contrast, in a relational dynamic such as Weber's, equilibrium is expressed in a threefold manner and both the temporal question and the question of causality are cautiously suspended. Only space remains unquestioned, although we know that physical reality comprises much more than mere space or extension. However, the very suspension of time and causality constitutes an «internal space» in which thought can reorder its whole conception.

Of course, under present conditions no one wants to waste time with such inopportune questions of fundamentals, however profound their implications; they could not be more opposed to the short breath of the age. But what we are talking about is that a metaphysical principle in mechanics has displaced a physical immanent principle that at least can be clearly recognized in its effects. And the consequences of this are as deep as those that follow from having turned one's back on this principle. Modern science has proceeded from the simple to the complex until it has become bogged down in all sorts of insoluble problems; and yet even the most complex phenomenon that can be observed depends critically on that indivisible principle which has been evacuated. And the non-separability in physics passes through the potential, rather than forces; there must be a way back from the complex to the simple, starting from the theory of the potential, although it has not yet been trodden.

To speak of cybernetics or control theory is literally to speak of government theory, and yet there is practically nothing in this discipline that transcends technical details. And the problem here is not esotericism or secrecy, but the sheer inability to distill principles or even general conclusions. However, Weber's dynamics is implicitly a theory of feedback in Nature which nevertheless has three principles just as simple, though less contradictory, as those of Newton's mechanics; and within their boundaries there would still be room for many further distillations.

Human practices of government, including tyranny and technocracy, assume that everyone wants to control to some extent an environment perceived as distinctly different; the idea of the internal controller, on the other hand, is based on the continuity between the agent and the environment and denies the separation between a subject and object of control. If there is a theory and practice of government that is not harmful to anyone, it must necessarily be on this path, and therefore the most exquisite attention should be paid to its principle.

But it does not take much analysis to see that the kind of control sought by technocracy is in every way diametrically opposed to natural compensatory mechanisms. Leaving aside the evidence that, to begin with, our own idea of Nature is deliberately dissociated, in Nature the adjustment between agent and environment becomes seamless, while the pseudo-elites seek by all means to maintain and increase the quantitative and qualitative differences with what they seek to control. Natural intelligence cannot separate itself from its concrete embodiment and its countless details; artificial intelligence seeks all kinds of shortcuts to make itself independent of its material context. And as others have already observed, feedback in Nature use to be negative because it tends to stability, while the capital reproduction system resolutely seeks positive feedback, amplification and destabilization, so that each output or product cycle increases the input or investment.

We could go on but one can get the picture. It is not only that the rulers do not care the least about Nature, but they do not care about government either: their so-called «governance» consists literally in creating misgovernment and sustaining it as far as the system can bear it, exploiting the growing disorder to the limit. The type of control that the modern subject tries to internalize is also based, as it could not be less, on self-exploitation via positive feedback. But to begin with, the «cybernetic hypothesis», that living beings and machines share the same principle of organization, was already patently false even for a child. Evoking the ghosts of vitalism is perfect for distracting us from the basic fact that a Nature split between inertia and force is the perfect alibi for there always to be slaves and masters. We are sold as long as the basis of this «natural philosophy», the only one we still have, is accepted undeterred.

Modern physics is based on exact analogies such as variational principles, but as these analogies build one upon the other they become opaque to the point of unintelligibility. Even just one of them becomes impenetrable; their multiple conflation results in weird creatures and if they still remain functional to some extent it is because they have been built by reverse engineering from already known data. If we dispense instead with the contradictory idea of inertia, which demands closed systems that are not closed, and consistently refer the other two principles to the idea of dynamic equilibrium, we approach by degrees the indivisible nature of non-action, of which so little is still known. The non-uniformity of force and potential allows us to satisfactorily explain such basic things as the profile of the pulse wave, and presumably also of the bilateral cycle of breath. Both cases are very concrete examples of spontaneous control, and at the same time present us seemingly innocuous but very far-reaching analogies, since they allow us to enter into an increasingly general principle. Let us remember that, as much as they claim, current theories cannot even explain the shape of the ellipses in the orbits of planets. Neither can Weber's mechanics, but at least it makes the forces conform to the observable forms and not the other way around, staying true to the principle of homogeneous proportions. It is way more interesting to properly follow the sliding knot of the feedback loop than to claim causes; for the former allows us to move with good sense along the self-regulation cosmic scale.

If we start from a homogeneous medium with unit density, which is neither full nor empty, any manifestation can be conceived as a simple change of density in the primitive medium that has been compensated by an opposite change in another region. The change of density may be equivalent to a change of potential, as well as of scale or dimension; as we have already seen, these transitions can also be studied, quite intuitively, through a certain symplectic morphology. But what do we mean by interiority when we speak of an inner controller? Certainly we are not talking about something inside material bodies, nor in space either. In a primitive homogeneous medium there must be a fundamental ambiguity between matter and space, space and consciousness. This ambiguity is of no use to the intellect seeking determinations, and yet the same principle of homogeneity underlies field theories and is more basic than the notion of force. It is a principle with intrinsic neutrality, unlike the extrinsic or metaphysical neutrality of global synchronization so convenient to generalize the dominion of mechanical manipulation.

Physicists have often claimed that time is merely subjective, but basing their science on motion measurements their arguments will never convince anyone, not even themselves. Instead we can distinguish an empty, reversible, flat time —that of the fictitious global synchronizer— which would be the outermost limit that our conception of time can reach. In truth, the global synchronizer is not at the outer limit of measurable time, but beyond it, which makes it inaccessible even to infinite approximations. At the bottom, on the other end, it would be an homogeneous medium of unit density where no motion or time are possible. And in the middle we would have the internal proper time of the systems of which the misnamed «retarded potential» would be an index. There can be no subjective sensation of time without motion being superimposed or connected with the homogeneous and timeless background from which it necessarily arises. Perhaps within this homogeneous medium, any movement and sensation of time is somewhat ephemeral or even fictitious; but there is certainly no greater fiction than that of a synchronization beyond motion when we start right from this motion and not from potentials. On the other hand, concepts such as time, mass, potential, energy or momentum, among others, can exhibit the most varied symmetries, but their meaning can hardly be revealed in the framework of superficially elegant equations that hide all kinds of false constants and heterogeneous quantities. In addition to these three levels, we have the imaginary or mental time of the interaction and drift of signs, which to a great extent seems independent of the physical laws and in another obvious sense cannot be independent of them. This time of thought is the foreground of subjectivity that use to attract our attention, not its background. Physicists and mathematicians have developed the field of symbolic dynamics, but we forget the symbolic range of dynamics whose apex is in the three principles and their association with the triple time. We can align these levels in very different ways from those that are now prevalent.

Our science wants to manipulate without hindrance, and any suggestion that there is an active intelligence in Nature, however impersonal, must be seen as a prejudice to overcome. The laws of physics, to be laws proper, cannot be immanent but above Nature, and must be imposed upon it in such a way that no appeal against their status is possible. In one fell swoop, the physicist puts himself above both Nature and the prejudices of the vulgar who still maintain a justified reverence for her. And even if this takes place at the most purely intellectual level, it gives him an enormous sense of power that twins him without any need of covenants with those who hold worldly power from unassailable positions of privilege. To its misfortune, moreover, science has ever more sophisticated methods at the service of ever grosser objectives, and this abysmal disconnection exponentially increases its harmfulness. Bacon's intentions were transparent, but since Newton few, if any, can see through the mask of neutrality and mathematical elegance.

With just four basic distinctions, a philosopher as remote as Scotus Erigena, could sustain a vision of Nature vaster than ours with all our deluges of information. The Irishman spoke of an uncreated and creative nature, of which nothing certain can be known; of the created and creative nature of forms or ideas; of the created and non-creative nature of matter, and of the uncreated and non-creative nature with its back to everything but towards which everything tends. A fourfold partition that might seem coarse at this point, but which is in perfect accord with the immutable nature of the Name and the inarticulate character of its Alpha and Omega. The fourth aspect is non-action, where the visible vanishes in the indivisible, core and ultimate destiny of everything.

If the final cause is not a force but a perspective, the idea of potential allows us to understand it much better than that of interaction. Raymond Ruyer used the concept of equipotentiality to explain the permanent formation of an organism and its stability, stressing the unbridgeable difference between unitary domains and interactions within statistical aggregates. A relational or environmental mechanics such as Weber's bridges this gap because, unlike conventional mechanics, it assumes that there are no closed systems, is valid for both point and extended particles and works at all scales. Conventional mechanics is based on laws of extremes, maxima and minima, and the absolute character of global synchronization cannot avoid pathological singularities like black holes; with «retarded potentials» such processes are forbidden and physics itself seems to be governed by optimal equilibria rather than extremes.

Ruyer exaggerated his finalism as he could not consider the spontaneous regulation that operates in relational dynamics, in the principle of maximum entropy —there is more order because order produces entropy faster— or in the ergoentropic equilibrium between maximum entropy production and minimum energy variation. However, his conception of consciousness as absolute sensation and absolute survey is far more basic than all the notions of the I that since Descartes dominate the derived realm of perception, thought, and ordered series or bodies. His glimpse of ideas as virtually eternal equipotential regions, which would nevertheless manifest themselves ephemerally in highly conditional equilibria with the environment, still deserves attention. Of course, a homogeneous medium can only be absolute sensation or pure activity with respect to any supposed alteration in it, but this is its way of remaining unobstructed in itself and for itself.

The question is whether these new avatars of the concept of equilibrium and others suggested in previous posts allow a better understanding of the innumerable possible domains between an undifferentiated medium and its manifestations. But if absolute survey is the contemplation of imbalance or dynamic equilibrium from homogeneity, one should start looking for homogeneity or proportionality in the quantities that attempt to describe it.

There should be harmony between principles, means and ends; but, moreover, for us, the homogeneous medium itself is the ultimate perspective and interpreter.

Many may think that, at most, the issue raised here is just a matter of interpretation —but it is way more than that. To begin with, it is the very principles of mechanics that we reject, knowing full well that they are far from being irreplaceable, as most still think. And not only are they not irreplaceable, but they are unacceptable as a matter of principle and from the first principle, the so-called law of inertia. What the third principle with its global synchronization does is to seal the first principle and create the mirage of a single level of causality. In any case the interpretation should turn to the principle of equilibrium which has a triple aspect; but whoever changes the interpretation and the principles, also changes the methods and selections, and in short, the whole orientation of thought and its intention. What the present science intends can no longer be clearer, the challenge is to unveil the implications of the principle of immanence. And it is only normal for the average scientist to stand here like a cow looking at a new gate.

To make of the Principle the goal, beyond the principles with which thought is endowed, is already a fundamental change. To find something that is not dead at the very core of this science of death and to adhere to it, is a triumph in every respect, and every significant step in the same direction will be one more triumph ahead. The opposite of this science of death is not vitalism; for dynamic equilibrium is beyond life and death —or rather, beyond our ideas of the living and the dead, if Life, in the broadest sense, has no opposites or limits. Birth and death are indeed opposites in nature, and in the equilibrium that lies between them a regeneration of science is also possible, and a science of regeneration whose origin is timeless but will yet know new clothes.

Notes

(www.hurqualya.net)

1. http://mileswmathis.com/focus.pdf

A RETROPROGRESSIVE LOOPBACK

18 October, 2022

Let us imagine a biofeedback machine where the three principles of mechanics are expressed in an alternative but quantitatively equivalent form, as in Weber's force law (1846): instead of inertia, there is a zero sum of forces; instead of a constant force, the intensity of force depends on the environment; instead of simultaneity of action and reaction, there is a «retarded potential» and thus an internal, proper time of the system. The first remarkable difference between these principles and the more familiar ones is that they allow, to a certain extent, the interpolation of consciousness within its framework. The second is that it applies equally to point and extended bodies or events, rendering unnecessary crucial artifices supporting special relativity, general relativity, or quantum mechanics and making them incompatible with each other. The third is that it makes possible a conception of time different from those of physics and psychology.

(Let us recall, if necessary, that as a framework the very principles of mechanics are indemonstrable within the system they frame and are a matter of choice —something Poincaré already made evident when discussing precisely Hertz's and Weber's systems). The three principles of mechanics involve, through the quantities that express them, a schematic concept of space, time and causality. Now, what kind of relation exists between the principles of mechanics and time? And here, before speaking of the subjective aspect of time —temporality— one should start considering the time of mechanics itself. Most physicists would now favor a criterion which we may call «operational» and which states that physical time is nothing but the measure of motion; and this criterion is used to argue that without assuming the constancy of forces or the equality of background and coordinate systems, physics would not be possible. This is typical of Newton's system, which, as later relativity theory would do, conflates absolute and relational concepts; however when we stick to a purely relational idea of dynamics as in Weber's law, we can consistently dispense with both universal constants and global synchronization in addition to inertia.

If one can establish unequivocal temporal sequences with Newton's three principles, the same can be done with the three relational principles that Weber does not enunciate but Andre K. T. Assis makes explicit —even if the later is still far from drawing the consequences of these principles. Of course, the principles of mechanics alone, whatever the choice, don't suffice to account for the inexhaustible complexity of natural flows with a possible temporal interpretation. They are a general frame that often will require a strong statistical component, replacing simple ideal bodies for assemblies — something already contemplated in the distinction between material particle and material point of Hertz's mechanics. As far as I know, Vladimir V. Aristov is the most prominent proponent of a relational statistical approach to time.

Aristov has done much to enrich the hitherto extremely schematic conventional description of time in physics. This one deals with time intervals — when it does not think of instants as points on a numerical axis— but lacks a concept of time as a state. Of course the big question remains of how time irreversibility arises, something that is not even considered in the so-called fundamental equations. Aristov proposes a model with three reference points instead of two, as well as a number of statistical and axiomatic contributions that also cover biological issues and are needed to make the quantitative treatment of time less simplistic.

However necessary such precisions may be, it is very likely that the most intangible component of time, namely its subjective aspect, does not depend on complexity either. In previous posts we have suggested that the «proper time» or retardation associated with Weber's mechanics might lie halfway between the purely external limit of global synchronization and the undifferentiated homogeneous medium —and that the subjective sense of the passage of time requires both a constant and a variable relation between a physical proper time and the unit density of background homogeneity, even if for particles the homogeneous background may be considered as a statistical average. These would be three different aspects or planes of time, similar, to some extent, to C. S. Peirce's categories of thirdness, secondness and firstness in the moments of semiotics or drift of signs in the mind, which also entails a temporal process, though certainly irreversible in contrast to the three laws of mechanics, which in addition can also be put in correspondence with these categories.

Relational mechanics can take various forms which need not be equivalent. If one asks various physicists engaged in this approach what its essential feature is compared to more promoted approaches such as field theories, the most common answer is that it dispenses with space-time as an independent category. Others might say that, at least in Weber's mechanics, the use of homogeneous quantities should allow for a better understanding of the abundant unexplained numbers in physics. But if we dwell here on the latter as the model par excellence of relational physics, it is above all because it frees us from the idea of inertia, as well as from the global synchronization and universal constants that serve as its safeguard; and with it from many other artifices, such as the cumbersome, scholastic arbitrations with the various reference systems.

But the importance of getting rid of inertia goes far beyond questions of convenience or simplicity. What is at stake is the foundation of our idea of Nature, hijacked by a certain idea of physics, as well as the idea of the living and nonliving, and even of the general drift of our logic and our mind. It took two thousand years to mature the concept of inertia, but less than a century to sit quietly on it — because it was only wanted as a starting point. One can always argue that Weber's mechanics deviates from relativity at high speeds and energies, just as one can argue that a theory like relativity always has plenty of artifices to rearrange the data in its favor; but at this point we will not enter into such controversies. Of course, there is no room for black holes and many other paradoxes and pathologies without which the fantastic genre is orphaned, but so much the better. In fact, a black hole is not a consequence of relativity, but of the absolute character of a gravitational force regardless of the conditions of the system, which can only be a fiction. On the other hand, there are many other things to keep track of that are still awaiting detection. But first and foremost it is a matter of principle, and since principles already affect the final interpretation, they also affect everything else.

Husserl undertook penetrating analyses of the three aspects of time, both in music and in relation to transcendental subjectivity —with the self-referential moment of a consciousness not yet driven by intention, a kind of 'time zero' from which the moments of the present, past and future would escape. Husserl's exposition of temporality is among the most profound reflections on the issue, yet the presentation of the self-referential experience of consciousness as a painful short-circuit or trauma is too reminiscent of the contemporary lucubrations of the father of psychoanalysis. Why should self-consciousness be painful? What prevents it from resting in itself? Here one could argue that this inability is due to the mere inertia of habit or conditioning.

Which would bring us again to biofeedback, that, rather than a way of controlling biological functions, is a means of attenuating their conditioning — it is a negative feedback, as is often the case in natural regulatory mechanisms. A signal of a potential, associated with a system state, seems much more suitable for inducing negative feedback than a signal of force, since the type of influence here is involuntary, not efferent or motor. And what does this have to do with a relational mechanics such as Weber's? Weber's or Assis' mechanics substitutes inertia for dynamic equilibrium, which must exist always by definition. On the other hand, the equilibrium between action and reaction takes place through a time interval, coexisting with the adjustment of the force to its medium, be it external or internal.

In fact there is no distinction between both points of view. In Newtonian or any other mechanics with inertia and the global synchronization that enforces it, everything is externally arranged even if it involves a flagrant contradiction, since after all inertia begs us to assume a closed system that at the same time is not closed. This creates the possibility of temporally reversible laws, which must surely be another fiction, and deepens the divide between physical laws and what Peirce called «the law of mind». Which brings us to another question that is never asked but is nonetheless far-reaching: is Weber's relational mechanics, taken in its most generic sense, reversible or irreversible? The fact that Weber's force law and its associated potential emerged in the context of electrodynamics might lead one to think that it is exclusive of reversible systems, but this is a prejudice without further justification, and we have already seen that even Maxwell equations belong to two thermodynamically different categories. In the absence of proper studies on the subject, everything suggests that, assuming that this mechanics does not depend on the chimerical condition imposed by the inertia principle, it allows both reversible and irreversible behaviors. However, contrary to the most widespread opinion, we believe that the latter is the general case, and therefore also the most fundamental.

If relational dynamics admits the irreversible, it allows to transfer to the internal, and also to the mental sphere, a rectification over time between the three moments of its equilibrium, something that in classical mechanics can never be made explicit. In contrast to the relational approach, we have seen other anholonomic formulations of mechanics, of the Cartesian type, which do not have inertia either, but instead make all movement dependent on acceleration or imbalance. Opposed in approach, both dynamics can coincide on the common ground of the shift of the potential, which in modern physics has been introduced through the back door in what is known as the «geometric phase». And we have also seen the ergoentropic equilibrium between minimum energy variation and maximum entropy proposed by Mario Pinheiro for an irreversible reformulation of mechanics. We cannot go into the relationship between these three formulations here, although the subject deserves an in-depth study.

Within Peirce's semiotic scheme, the retroprogressive arises by tracing back from the tertiary to the secondary categories, and from these to the primary one. The third law of action-reaction defines the conditions of interaction, mediation and measurement; the second law, which defines the action or force, can only have for time an inditial, incomplete character; the first law defines or cuts out the contour of the immediate from which we start. But the point that has not been made is that the same laws of mechanics, being a particular case for the internal use of physics, end up dragging down the drift of the world and the mind into which we have plunged ourselves —since, as a general disposition, they have also determined the entire flight of technology. If technoscience went all out in the external control of the world, it is only natural that the external will now drag us along no matter what our will may want.

Biofeedback can be used as a means of testing the internal, subjective side of certain principles of dynamics, while the relational principles of a dynamics such as Weber's allow us to see that this internal side also admits an external correlate. Looking for a general context for the man/machine interface, approaches such as «endophysics», a physics «from within» that would combine elements of quantum mechanics, relativity and chaos theory, have been proposed in recent decades; but perspectives of this kind remain tied to the dominant conception in physics of the observer, representation and controllable forces. The interface is presented as a «cut» at various levels, but the great and definitive cut that physics has introduced is inertia and the inertial system, and it is here that we must begin. The «endophysical approach» has not been developed further, but the work to find interfaces between the living and machines grows steadily, and today we find that the geometric phase of potentials is routinely used as a tuning factor in robotics and control theory.

The real «interface» would be the shift of the potential, but this does not require any «cut», for in this relational mechanics there is no conceptual separation between the internal and the external, the «living» and "nonliving», the inert and the driving forces. And the "shift of the potential» is not unique to quantum mechanics as is still assumed, but necessarily occurs at all scales since the retardation can only exist with respect to the fictitious global synchronizer of Newtonian mechanics.

As Ruyer would have put it, there is the observable and the participable; the former is the object of intentional consciousness, the latter is not, but it is no less important for that. It can even be said that civilization is the exploitation of the surplus of the participable for the benefit of the observable, which always ends up having an owner. The race for the fusion of man and machine is a compulsive escape that tries to compensate for the self-inflicted cut accomplished with mechanics, but it is clear that our use of machines and tools goes back much earlier than the generalization of the three principles —as is clear that the «instrumentation principle» we sometimes talk about is not the late «instrumental reason» of modern thought but a way of externalizing force with or without tools.

Thus, the internal coordination with what we may, not without malice, call «Weber's mechanism", goes beyond physics or technology. To what extent have the specific form of the principles of mechanics been able to influence our subjective temporality? It seems an impossible question to answer, because temporality never moves in vacuum, and yet Newton's principles are conceived as motion in vacuum. The three classical principles are purely external, in no way can they be internalized. Their influence extends first and foremost as heteronomous drift. In Newtonian mechanics, nothing moves without being moved by something else; in relational mechanics a body can thrust itself in perfect consistence, without any contradiction.

Temporality does not move in vacuum, otherwise it would never have taken hold as an internal form of a culture —as when we speak of vertical, linear, or circular conceptions of time. The three classical principles may represent a causal sequence, but they do not delve into the subjective sensation of time. The three principles of relational mechanics can do so, and yet this «mechanics» does not need, or rather excludes, causal representations —the perspective it provides is really acausal, and thus amechanical. But the question of causality would now take us too far; on the crucial arguments about Newton's bucket experiment, we can explain the curvature of water by both

Pinheiro's ergoentropic equilibrium and potential energy, but hardly by absolute space as the proponent of the experiment did.

In the face of so much misunderstanding, it is worth remembering that our present civilization has had only one natural philosophy, which is none other than Newton's. All subsequent «revolutions» do not change the essential question, nor will they change it, because what it takes to do so is not speculative physics, but its antipode, fundamental physics, as well as, of course, natural philosophy itself. To contemplate the world without inertia is equivalent to continually renewing it, but consciousness can barely maintain itself for an instant in such a state of suspension. How can this be, if we said that the laws of classical mechanics are not susceptible to internalization? We have already seen that even the three principles of classical mechanics can be understood both on the same level and on different degrees or levels, according to Peirce's semiotic implications. It can be argued that intentional consciousness is necessarily external, albeit to very different degrees, and that contemplating non-inertiality drives us back in the direction of that other consciousness prior to intentionality which Husserl called transcendental subjectivity. And within this consciousness we can perhaps distinguish between a stable equilibrium and an unstable one, which can also be associated with certain principles and evolution of dynamics.

Breathing, for example, is both a voluntary and involuntary process, just as it is mechanical and amechanical simultaneously. The bilateral nasal cycle is most likely associated with a retarded potential, and following its axis we can access the amechanical aspect of the breathing phenomenon. This in turn is linked to the perception of time, and to the transformation of intention into attention, and of attention into self-awareness. It is not difficult to glimpse in all this the possibility of both a gradual and a direct path.

The life of the spirit not only has no need of technologies, but rather opposes them in general. But here we glimpse a passage beyond the principle of instrumentation which allows us to see science and technology in a different light; and the fact that all this is taking place at the very heart of mechanics has a special significance which we want to endure.

References

Henri Poincaré, Nicolae Mazilu, *Hertz's Ideas on Mechanics* (1897)

V. V. Aristov, *Relative Statistical Model of Clocks and Physical Properties of Time* (1995)

Nikolay Noskov, *The phenomenon of retarded potentials*

K. T. Assis, *Relational Mechanics and Implementation of Mach's Principle with Weber's Gravitational Force (2014)*

Alejandro Torassa, *On classical mechanics* (1996)

Koichiro Matsuno, *Information: Resurrection of the Cartesian physics* (1996)

Mario J. Pinheiro, *A reformulation of mechanics and electrodynamics* (2017)

www.ingramcontent.com/pod-product-compliance
Lightning Source LLC
LaVergne TN
LVHW100252210926
848480LV00003B/907

* 9 7 9 8 2 2 7 9 9 0 5 8 7 *